AF356224

Horticultural Crops under Stresses

Horticultural Crops under Stresses

Editors

Alberto Soares De Melo
Hans Raj Gheyi

Basel • Beijing • Wuhan • Barcelona • Belgrade • Novi Sad • Cluj • Manchester

Editors

Alberto Soares De Melo	Hans Raj Gheyi
Department of Biology	Department of Agricultural
State University of Paraíba	Engineering
Campina Grande	Federal University of
Brazil	Campina Grande
	Campina Grande
	Brazil

Editorial Office
MDPI
St. Alban-Anlage 66
4052 Basel, Switzerland

This is a reprint of articles from the Special Issue published online in the open access journal *Plants* (ISSN 2223-7747) (available at: https://www.mdpi.com/journal/plants/special_issues/horticultural_crops_under_stresses).

For citation purposes, cite each article independently as indicated on the article page online and as indicated below:

Lastname, A.A.; Lastname, B.B. Article Title. *Journal Name* **Year**, *Volume Number*, Page Range.

ISBN 978-3-7258-0899-1 (Hbk)
ISBN 978-3-7258-0900-4 (PDF)
doi.org/10.3390/books978-3-7258-0900-4

Contents

About the Editors

Alberto Soares De Melo

Alberto Soares De Melo is an associate professor at the State University of Paraíba—UEPB, Brazil, Department of Biology, and the Postgraduate Program in Agricultural Sciences. He graduated as an Agricultural Engineer from the University of Viçosa, Brazil, in 1994. He received his doctorate in Natural Resources from the Federal University of Campina Grande, Brazil, in 2007. He is also a Research Productivity Fellow of the National Council of Scientific and Technological Development—CNPq. His research interests include: the ecophysiology of plants grown in semi-arid ecosystems, with an emphasis on the efficient use of water, and the use of elicitors in the mechanism of water stress in plants. He has published more than 120 peer-reviewed articles in the field of plant science. He is a reviewer of projects for funding agencies such as CNPq, the Research Support Foundation of the State of Pernambuco—FACEPE, and the Research Support Foundation of the State of Paraíba—FAPESQ-PB and a member of the UEPB Publications Editorial Board.

Hans Raj Gheyi

Hans Raj Gheyi is currently an emeritus professor at the Federal University of Campina Grande, Brazil. He served as a visiting professor at the Federal University of Recôncavo da Bahia—UFRB, Brazil, from 2010 to 2020 and is the chief editor of the *Revista Brasileira de Engenharia Agrícola e Ambiental (Brazilian Journal of Agricultural and Environmental Engineering)* published by the Federal University of Campina Grande. He graduated with a degree in Agriculture from the University of Udai Pur, India, in 1963 and obtained his master's degree in Soil Sciences from the Punjab Agricultural University, India, in 1965. He received his doctorate in Agronomy Sciences from the Université Catholique de Louvain, Belgium, in 1974. He is a research fellow of the National Council for Scientific and Technological Development—CNPq. He has experience in agricultural engineering, with an emphasis on soil and water engineering, working mainly on the following topics: salinity, irrigation, soil–water–plant relationships, salt stress, plant tolerance to water deficit, water quality, and wastewater reuse. He has published more than 600 peer-reviewed articles on these topics.

Preface

It is with great pleasure that we present this reprint of the Special Issue "Horticultural Crops under Stresses" from the journal *Plants*. This collection of 13 scientific articles addresses a topic of increasing importance in the context of climate change and contemporary agricultural challenges. Horticultural crops, essential for nutrition and food security, face a variety of abiotic stresses that can compromise their productivity and quality.

The scope of this reprint is broad, covering topics from phenotyping and the modeling of stress responses to innovative management strategies, such as the use of biostimulants and micromolecules. Our goal is to provide insights into recent advances in research and agricultural practices that contribute to the resilience and sustainability of horticultural crops.

This reprint is intended for a diverse audience, including researchers, agronomists, students, and professionals in the field of agricultural and environmental sciences. We extend our sincere thanks to all of the authors who contributed their valuable work and the reviewers who ensured the scientific quality of this special edition. We also express our gratitude to the editorial team of the journal *Plants* for their continuous support.

We hope that this reprint will be a source of inspiration and knowledge for all those interested in addressing the challenges of horticultural crops under stress and promoting more sustainable agricultural practices.

Alberto Soares De Melo and Hans Raj Gheyi
Editors

Editorial

Horticultural Crops under Stresses

Alberto Soares de Melo [1,*] and Hans Raj Gheyi [2]

[1] Center of Biological and Health Sciences, Universidade Estadual da Paraíba, Campus I, Campina Grande 58429-500, PB, Brazil
[2] Center of Tecnologia and Natural Resources, Universidade Federal de Campina Grande, Campina Grande 58429-900, PB, Brazil; hgheyi@gmail.com
* Correspondence: alberto.melo@servidor.uepb.edu.br

Citation: de Melo, A.S.; Gheyi, H.R. Horticultural Crops under Stresses. *Plants* **2023**, *12*, 3400. https://doi.org/10.3390/plants12193400

Received: 18 September 2023
Revised: 23 September 2023
Accepted: 25 September 2023
Published: 26 September 2023

Climate change causes alterations in the spatio-temporal temperature and rainfall distribution. These changes have led to soil water restriction and decreased food production. Understanding plant plasticity will result in new genotypes with attributes of interest to farmers and plants adapted to environments under abiotic stresses.

In arid and semi-arid regions, abrupt changes in the soil–plant–atmosphere continuum cause oxidative stress, germination failures, reduced growth, and changes in gas exchange and plant stands, decreasing grains and fruit production. Seedlings grown under water stress conditions associated with high temperatures, a common situation in these regions, have nutrient-uptake restrictions. Another striking phenomenon is increased reactive oxygen species (ROS) production. ROS production changes cell membrane properties and decreases plants' photosynthetic efficiency. Oxidative stress decreases dry mass production and leaf water status, compromising crop production. Biochemical, physiological, and agronomic responses depend on plant tolerance or sensitivity to abiotic stresses. Thus, studies that explore fitness due to genotypic diversity in hostile environments are needed. Using technologies that enable sustainable agricultural production, even in adverse conditions, will provide farmers with food and economic security.

The literature shows mechanisms for inducing tolerance to abiotic stresses using elicitor application technologies, promoting resilience and stress memory, with biochemical and physiological benefits in plants. *Seed priming* is a technology that presents positive results. *Seed priming* uses several mitigating agents such as silicon, light radiation, gibberellic acid, hydrogen peroxide, and polyethylene glycol 6000. Other technologies had positive results, such as balanced mineral fertilizer for stressful conditions, glycine betaine, salicylic acid, methionine, and bio-input applications aiming to mitigate the harmful effects of salinity, thermal oscillation, and water deficit expected in arid and semi-arid regions.

This Special Edition of *Plants* comprises 12 articles, highlighting the promising results of agrotechnologies in inducing tolerance and diagnosing the nutritional status of plants under abiotic stress. They provide knowledge of plants' physiological and biochemical processes in challenging environments and elucidate the response mechanisms of elicitors for agriculture. However, many knowledge gaps in knowledge of regarding horticultural cropping systems, especially under adverse conditions, require further investigation due to the environmental dynamics imposed on plants.

In summary, this collection reflects the efforts of multiple researchers in the field of plant sciences, who collaborated with different insights to investigate horticultural plants' responses under abiotic stress conditions. Therefore, this Special Issue will contribute to instigating new studies that enable the incorporation of scientific and technological knowledge into production processes in the face of climate change.

Conflicts of Interest: The authors declare no conflict of interest.

Disclaimer/Publisher's Note: The statements, opinions and data contained in all publications are solely those of the individual author(s) and contributor(s) and not of MDPI and/or the editor(s). MDPI and/or the editor(s) disclaim responsibility for any injury to people or property resulting from any ideas, methods, instructions or products referred to in the content.

Article

Foliar Spraying of Glycine Betaine Alleviated Growth Inhibition, Photoinhibition, and Oxidative Stress in Pepper (*Capsicum annuum* L.) Seedlings under Low Temperatures Combined with Low Light

Nenghui Li, Kaiguo Pu, Dongxia Ding, Yan Yang, Tianhang Niu, Jing Li * and Jianming Xie *

College of Horticulture, Gansu Agricultural University, Yingmen Village, Anning District, Lanzhou 730070, China; linh@st.gsau.edu.cn (N.L.); pukaiguo1998@163.com (K.P.); dingdongxia100741@outlook.com (D.D.); yy@st.gsau.edu.cn (Y.Y.); niutianhang@outlook.com (T.N.)
* Correspondence: lj@gsau.edu.cn (J.L.); xiejianming@126.com (J.X.); Tel.: +86-15193134766 (J.L.); +86-13893335780 (J.X.)

Citation: Li, N.; Pu, K.; Ding, D.; Yang, Y.; Niu, T.; Li, J.; Xie, J. Foliar Spraying of Glycine Betaine Alleviated Growth Inhibition, Photoinhibition, and Oxidative Stress in Pepper (*Capsicum annuum* L.) Seedlings under Low Temperatures Combined with Low Light. *Plants* **2023**, *12*, 2563. https://doi.org/10.3390/plants12132563

Academic Editors: Alberto Soares De Melo and Hans Raj Gheyi

Received: 26 May 2023
Revised: 30 June 2023
Accepted: 3 July 2023
Published: 6 July 2023

Abstract: Low temperature combined with low light (LL stress) is a typical environmental stress that limits peppers' productivity, yield, and quality in northwestern China. Glycine betaine (GB), an osmoregulatory substance, has increasingly valuable effects on plant stress resistance. In this study, pepper seedlings were treated with different concentrations of GB under LL stress, and 20 mM of GB was the best treatment. To further explore the mechanism of GB in response to LL stress, four treatments, including CK (normal temperature and light, 28/18 °C, 300 µmol m^{-2} s^{-1}), CB (normal temperature and light + 20 mM GB), LL (10/5 °C, 100 µmol m^{-2} s^{-1}), and LB (10/5 °C, 100 µmol m^{-2} s^{-1} + 20 mM GB), were investigated in terms of pepper growth, biomass accumulation, photosynthetic capacity, expression levels of encoded proteins *Capsb*, cell membrane permeability, antioxidant enzyme gene expression and activity, and subcellular localization. The results showed that the pre-spraying of GB under LL stress significantly alleviated the growth inhibition of pepper seedlings; increased plant height by 4.64%; increased root activity by 63.53%; and decreased photoinhibition by increasing the chlorophyll content; upregulating the expression levels of encoded proteins *Capsb A*, *Capsb B*, *Capsb C*, *Capsb D*, *Capsb S*, *Capsb P1*, and *Capsb P2* by 30.29%, 36.69%, 18.81%, 30.05%, 9.01%, 6.21%, and 16.45%, respectively; enhancing the fluorescence intensity (OJIP curves), the photochemical efficiency (Fv/Fm, Fv$'$/Fm$'$), qP, and NPQ; improving the light energy distribution of PSII (Y(II), Y(NPQ), and Y(NO)); and increasing the photochemical reaction fraction and reduced heat dissipation, thereby increasing plant height by 4.64% and shoot bioaccumulation by 13.55%. The pre-spraying of GB under LL stress also upregulated the gene expression of *CaSOD*, *CaPOD*, and *CaCAT*; increased the activity of the ROS-scavenging ability in the pepper leaves; and coordinately increased the SOD activity in the mitochondria, the POD activity in the mitochondria, chloroplasts, and cytosol, and the CAT activity in the cytosol, which improved the LL resistance of the pepper plants by reducing excess H_2O_2, O_2^-, MDA, and soluble protein levels in the leaf cells, leading to reduced biological membrane damage. Overall, pre-spraying with GB effectively alleviated the negative effects of LL stress in pepper seedlings.

Keywords: pepper; low temperature combined with low light; glycine betaine; antioxidant; photoinhibition; biological membrane; subcellular localization

1. Introduction

Plants live in an ever-changing environment that is often unfavorable for growth and development or stressful, which can include biotic stress (pathogen infection) and abiotic stresses such as drought, high and low temperature combined with low light, nutrient deficiency, and excessive salt or toxic metals in the soil [1]. Among these, low temperature combined with low-light stress is a major environmental factor that limits crop productivity in agriculture, resulting in threatened economic benefits [2]. In addition, low temperature combined with low light inhibits plant biological processes from seed germination to seedling growth, flowering, and fruiting [3–5]. It is worth noting that overwintering vegetables are often damaged by low temperatures combined with low light in facilities in northwestern China.

Pepper (*Capsicum annuum* L.), the second largest edible crop of Solanaceae after tomatoes, has an annual yield of more than 37 million tons and great economic benefits [6]. As the most important off-season vegetable cultivated in solar greenhouses in northwestern China, it is often exposed to low temperatures combined with low-light conditions in winter or early spring months. Studies have shown that the low temperatures combined with low-light stress damage the cell membrane, then increase osmotic substances, thereby weakening photosynthesis and destroying leaf tissue [5]. Moreover, antioxidants accumulate [7], key gene expression is downregulated, and antioxidant enzyme activity reduces in pepper leaves [8].

When plants are exposed to stresses, morphology (height, stem diameter, leaf count, and dry biomass) change is the most intuitive outcome of their response to an external stimulus. Similarly, the increase in MDA level and electrolyte leakage rate (EC) typically serves as the main reason for damage-induced biological membrane rigidification [9–11]. As a typical indication of photosynthesis, chlorophyll fluorescence parameters such as Fv/Fm, qP, NPQ, and Y(II) provide a detailed insight into the electron transfer process in photosystem II and beyond [12]. Likewise, the OJIP phase of the Chl fluorescence induction curve can be the most useful surrogate value for whole-plant seedling vigor [13]. In addition, the leaf relative water content (RWC), which includes free and bound water content (FWC, BWC), is an indicator of stress and tolerance in spring wheat [14] and cassava [15], and a reduction in RWC is a common consequence of stress [16].

LL stress can cause oxidative damage to plants and produce excessive reactive oxygen species (ROS), such as superoxide anion (O_2^-), H_2O_2, and hydroxyl (-OH); therefore, plants have evolved enzymatic antioxidant systems, such as superoxide dismutase (SOD), peroxidase (POD), catalase (CAT), and ascorbate peroxidase (APX), to cope with stress, either by stress avoidance or stress tolerance [17]. Under abiotic stress (heavy metals, drugs, drought, temperature extremes, etc.), plants are generally exposed to oxidative stress and produce excess free radicals, therefore, they respond to stress by regulating the activity of enzymes such as SOD, POD, and CAT [5,18–20]. In the ascorbate–glutathione cycle, APX reduces H_2O_2 by using ascorbate as an electron donor. Hamid et al. found that hormones regulate leaf senescence mainly by modulating H_2O_2 through different mechanisms, such as enzymatic antioxidants (SOD, POD, and CAT) and nonenzymatic antioxidant defense systems, under water stress in wheat [21]. Furthermore, the subcellular localization of antioxidant enzymes remains an important indicator of abiotic stress, and studies have shown that organelles (mitochondria, chloroplasts, and cytosol) mitigate stress at extreme temperatures by coordinating the intracellular activity of antioxidant enzymes [22,23].

However, when pepper seedlings experience low temperatures combined with low-light stress, the traditional solution is to install a passive climate control system in the greenhouse; however, the installation costs are high [24]. Therefore, it is imperative to mitigate low temperatures combined with low light by spraying with exogenous osmolytes.

Glycine betaine (GB) has been widely studied as an osmolyte in plants, bacteria, animals, and humans [25]. GB is also a protein stabilizer known as an osmoprotector that plays an important role in osmoregulation and is one of the main nitrogenous compatible osmolytes in Poaceae. When plants encounter abiotic stresses, exogenous GB applications have been shown to induce the expression of genes that are involved in oxidative stress responses that limit the accumulation of ROS and lipid peroxidation in cultured plant cells under drought, salinity, heavy-metal, chilling, and waterlogging conditions, and even stabilize photosynthetic structures under stress.

Although research on GB under abiotic stresses is sufficient, the study of low temperatures combined with low light with exogenous GB in pepper seedlings is still lacking. Therefore, the aim of the present study was to explore some of the physiological functions that regulate plant tolerance, especially in terms of photosynthesis, osmoregulatory substances, and antioxidant enzyme activity, and their ability to coordinate subcellular distribution.

2. Results

2.1. Changes in Growth, Water Content, and Antioxidant Enzyme Activity under LL Stress by GB at Different Concentrations

In Figure 1A, the pepper seedlings wilted severely (T0) after 24 h of LL (low temperature combined with low light); however, spraying with different concentrations of GB obviously alleviated the wilting. The cotyledons of the pepper in the T1 treatment were pendulous, the tips of the second true leaves were dehydrated and wilted in the T3 treatment, all of the true leaves curled in the T4 treatment, and, interestingly, the leaves in the T2 treatment were under normal conditions. The results of the LL treatment for 7 days regarding the accumulation of dry material in the aboveground part of the seedlings are shown in Figure 1B. At different concentrations of GB, the leaf count, plant height, and stem diameter of the pepper seedlings showed the same trend, which gradually increased from T0 to T2 and decreased from T2 to T4. Among these, the value of T2 was significantly higher than that of the other treatments in terms of the leaf count and stem diameter; however, the plant height was not significantly different from that of the others. The relative intracellular water content of the leaves of the pepper seedlings was also used to measure the degree of damage caused by LL stress. The BWC (bound water content), FWC (free water content), RWC (relative water content), and WSD (water saturation deficit) in Figure 1C show that low temperature combined with low light increased BWC and decreased FWC; however, in T2, with the sparse nature of GB, the BWC was significantly higher than that of T0, and the FWC was significantly lower than that of T0. The RWC was the highest and the WSD was the lowest in T0 compared with the other four treatments; however, T2 was significantly different from T0, T1, T3, and T4. It is also noteworthy that the RWC of T0, T1, T3, and T4 increased by 20.75%, 9.14%, 16.18%, and 11.63%, compared to T2, respectively, while the WSD of T0, T1, T3, and T4 was lower that of T2 by 83.07%, 7.32%, 66.09%, and 47.48%, respectively.

As shown in Figure 1D, the antioxidant enzyme activities of SOD, CAT, and APX were clearly observed in the pepper seedlings in response to LL stress at different GB concentrations. Compared with the other treatments (T0, T1, T3, and T4), the activities of SOD, CAT, and APX in T2 were significantly different and higher. The SOD activity of T2 increased by 368% in T0, 83.86% in T1, 64.51% in T3, and 25.86% in T4; the CAT activity of T2 increased by 20.96% in T0, 40.27% in T1, 22.92% in T3, and 27.43% in T4; and the APX activity of T2 increased by 9.57% in T0, 69.43% in T1, 33.31% in T3, and 60.98% in T4.

In conclusion, exogenous spraying with 20 mM of GB (T2) was the optimal concentration to respond to LL stress in subsequent experiments.

2.2. GB Affected the Growth, Dry Biomass, and Root Activity under LL Stress

Figure 2 shows that exogenous GB had no effect on the growth of the pepper seedlings at a normal temperature (CK) after 7 days of treatment; however, compared with LL treatment, exogenous GB treatment (LB) significantly increased the plant height by 4.64%, increased the dry and fresh weight of the pepper shoot by 13.55% and 7.53%, respectivley, and enhanced the root activity by 63.53%. The dry and fresh weight of the pepper root was not significantly different from the treatments with LL.

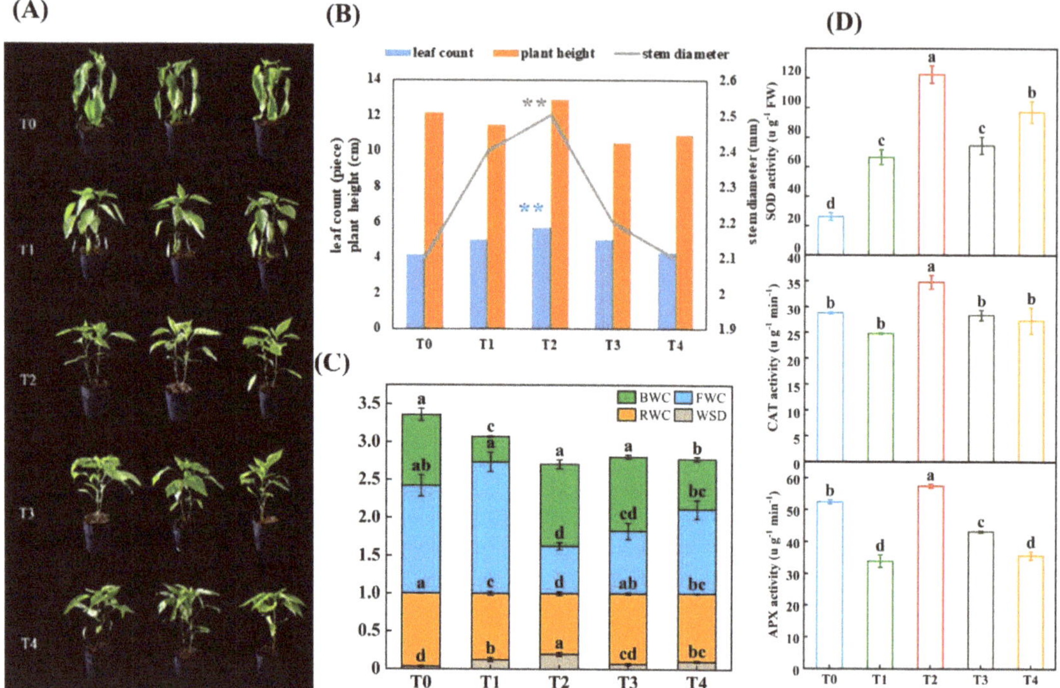

Figure 1. The growth and physiological indices affected by LL stress in pepper seedlings. Photographs and data of pepper plants under low-temperature conditions were obtained after 7 days. (**A**) Morphological changes. (**B**) Accumulation of dry material. (**C**) Related water content, water saturation deficit, free water content, and bound water content. (**D**) Enzyme activity contained SOD, CAT and APX. The results show the mean ± SE of three replicates, and the different letters denote the significant difference among the treatments ($p < 0.05$), according to Duncan's multiple tests. T0, control; T1, 10 mM GB; T2, 20 mM GB; T3, 40 mM GB; and T4, 80 mM GB. Data are means ± SDs (n = 3). Different lowercase letters represent significant differences ($p \leq 0.05$) in the same period among treatments, according to Duncan's test. The "**" in blue and gray color represented leaf number and stem diameter under T2 treatment were significantly different with T0, T1, T3 and T4.

2.3. GB Affected the Photosynthetic Pigment Content and Analysis of the OJIP Curve under LL Stress

In order to understand how GB alleviates the PSII photoinhibition induced by LL stress, the photosynthetic pigment content and fluorescence intensity (OJIP curve) were determined for each treatment (Figure 3). At normal temperatures, exogenous GB significantly increased the chla and chlT contents but had no effect on the chlb and car contents. In contrast, under LL stress, the addition of exogenous GB significantly increased the chla, chlb, chlT, and car contents of the pepper leaves by 21.32%, 20.92%, 21.23%, and 20.37%, respectively, compared to the LL treatment. The OJIP curves showed that GB enhanced the fluorescence intensity at the normal temperatures, and LL stress significantly decreased the fluorescence intensity at the J-I-P stage. However, GB significantly increased the fluorescence intensity at LL and was greater than that under LL stress.

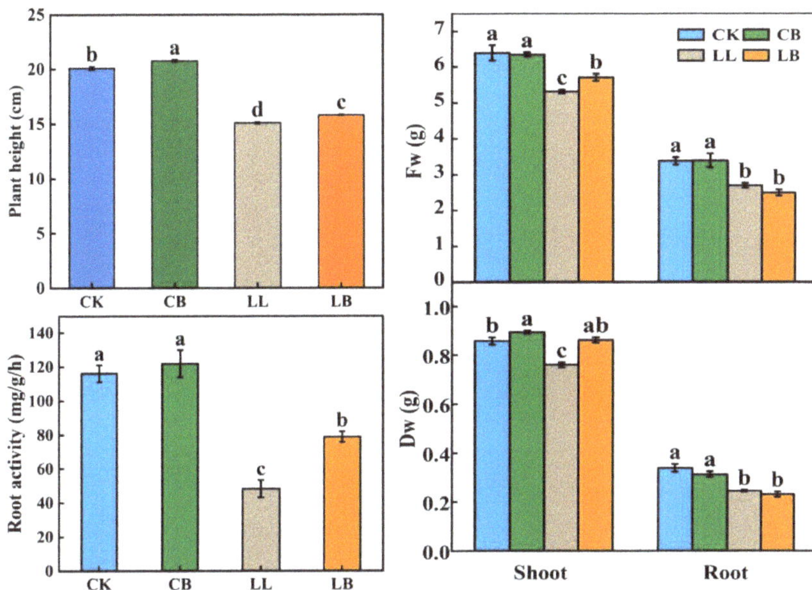

Figure 2. Impact of exogenously pre-spraying with GB on plant height, fresh and dry weight, and root activity of pepper leaves under LL stress. Photographs and data of the pepper plants under LL conditions were obtained after 7 days, which contained plant morphology, dry and fresh weight of the shoot and root, and root activity. CK: Normal conditions, spring RO water. CB: Normal conditions, spring 20 mM GB. LL: Low temperature combined with low light condition, spring RO water. LB: Low temperature condition, spring 20 mM GB. Different lowercase letters represent significant differences ($p \leq 0.05$) in the same period among treatments, according to Duncan's test.

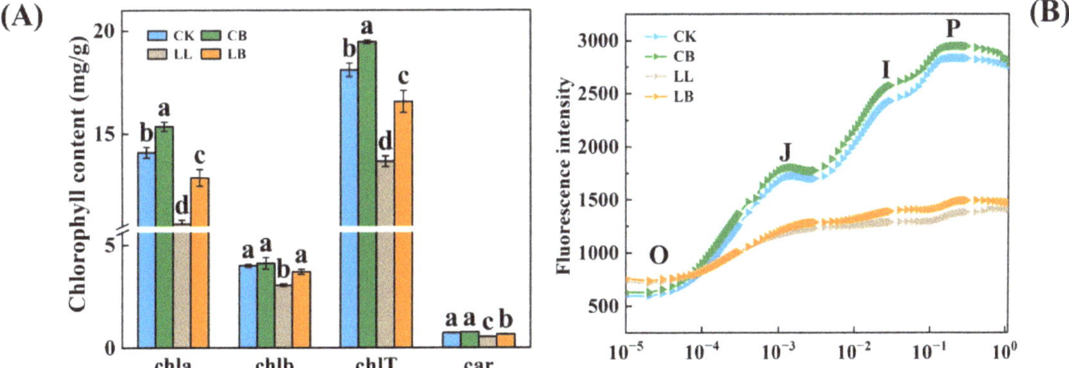

Figure 3. GB affected the photosynthetic pigments content and fluorescence intensity (OJIP curve) by LL stress in pepper seedlings. Data of pepper plants under LL conditions were obtained after 7 days. (**A**) chlorophyll content, contained chla, chlb, chlT (chla + b), and car (total carotenoids) content. (**B**) changes in OJIP curve. Data are means ± SDs (n = 3). Different lowercase letters represent significant differences ($p \leq 0.05$) in the same period among treatments, according to Duncan's test.

2.4. GB Affected Chlorophyll Fluorescence Parameters and Energy Distribution under LL Stress

In Figure 4A, Fv/Fm (the maximum photochemical efficiency of PSII) and Fv′/Fm′ (the efficiency of excitation energy capture by open PSII reaction centers) levels showed the same trend. The pre-spraying of GB under normal conditions effectively increased the

Fv/Fm levels, with no difference in Fv′/Fm′ levels, and both levels decreased significantly under LL stress; however, the addition of GB significantly increased the Fv/Fm and Fv′/Fm′ levels, even when there was no difference compared to normal temperature. As seen in Figure 4B, the values of the Y(NO) and Y(NPQ) levels were significantly lower under LL stress, and the addition of GB under LL stress significantly increased the values of Y(NO) and Y(NPQ) and was not different from CK and CB treatments at the normal temperature. The order of the Y(II) levels was CK > CB > LB > LL. As shown in Figure 4C, the NPQ level was significantly reduced by the LL treatment, which was significantly increased the LB treatment and was not different from the CK and CB treatments. Compared with the CK treatment, the qP level was significantly reduced in the CB treatment; however, compared with the LL treatment, its level was significantly increased in the LB treatment and was not different from the CB treatment. A portion of the energy reaction in PSII is shown in Figure 4D. The LL stress resulted in a significant decrease in P and Ex and a significant increase in D. However, the application of GB under LL stress reversed this phenomenon, with P and Ex significantly increasing and D significantly decreasing in the CB treatment.

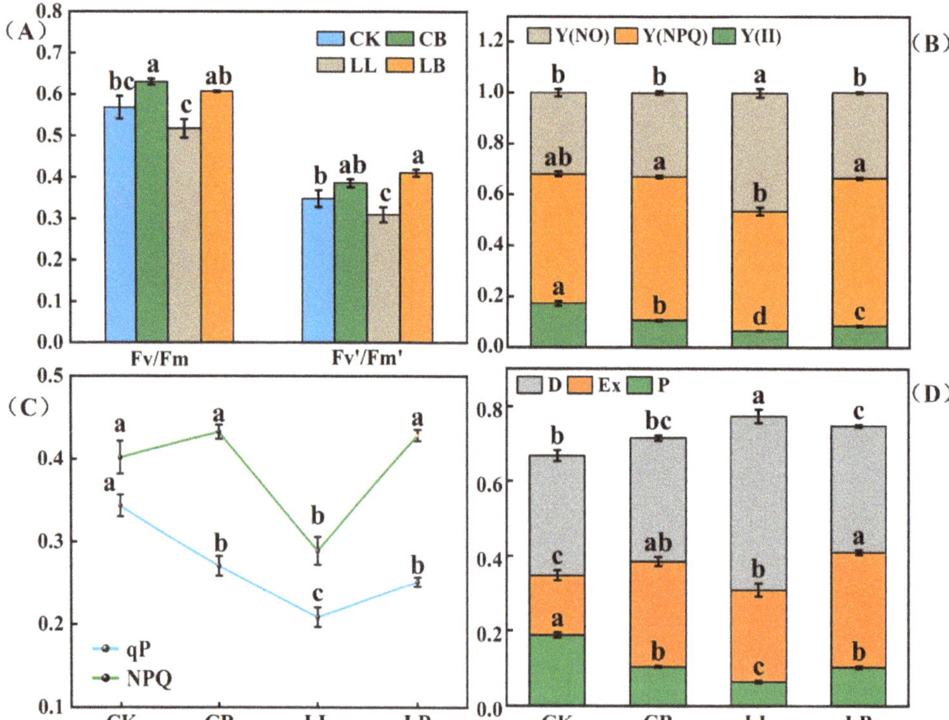

Figure 4. GB affected the chlorophyll fluorescence parameters and light energy distribution of PSII. (**A**) Fv/Fm, the maximum photochemical efficiency of PSII and the efficiency of excitation energy capture by open PSII reaction centers, Fv′/Fm′. (**B**) Light energy distribution of PSII. [Y(II)] was the actual quantum yield and Y(NPQ) and Y(NO) were the quantum yield of nonregulated and regulated energy dissipation, respectivley. (**C**) Non-photochemical quenching, NPQ; Photochemical quenching, qP. (**D**) The portion of energy reaction in PSII, where P is the portion of photochemical reaction, Ex is the portion of non-photochemical reaction, and D is the portion of antenna heat dissipation. Data are means ± SDs (n = 3). Different lowercase letters represent significant differences ($p \leq 0.05$) in the same period among treatments, according to Duncan's test.

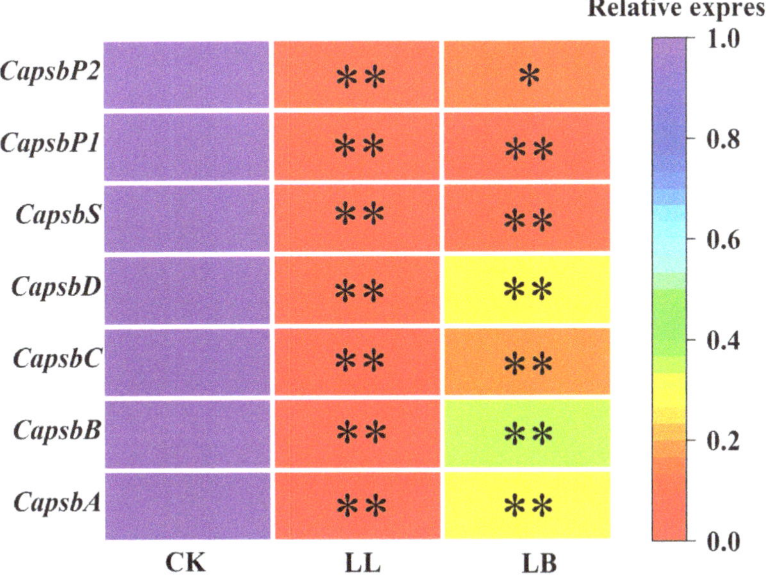

Figure 5. Expression patterns of differentially expressed genes related to the PSII reaction center proteins in pepper seedling leaves under LL stress for 7 days. The expression levels of the control group in the three treatment groups were used as the reference for the corresponding gene expression levels, and actin served as the internal standard. The figures show the expression levels of Capsb A, Capsb B, Capsb C, Capsb D, Capsb S, Capsb P1, and Capsb P2, respectively. The color scale corresponds to relative expression values. The more purple the block is, the higher the expression, and the more red, the lower the expression. Each row represents a unigene. CK, LL, and LB represent different treatments. Data are means $\pm$ SDs (n = 3). Different lowercase letters represent significant differences in the same period among treatments, according to Duncan's test. the "*" represent significant differences ($p \leq 0.01$), "**" represent significant differences ($p \leq 0.05$).

2.5. GB Affected the Expression of Genes Encoding the PSII Reaction Center Proteins under LL Stress

Figure 5 shows that the expression levels of Capsb A, Capsb B, Capsb C, Capsb D, Capsb S, Capsb P1, and Capsb P2 genes were significantly downregulated after LL treatment, especially the relative expression of *Capsb S* and *Capsb P1*, with values of 0.8% and 0.29%, respectivley; however, this result was reversed by pre-spraying with GB. Compared with LL treatment, the expression of Capsb A, Capsb B, Capsb C, Capsb D, Capsb S, Capsb P1, and Capsb P2 genes were upregulated at LB, with values of 30.29%, 36.69%, 18.81%, 30.05%, 9.01%, 6.21%, and 16.45%, respectively.

2.6. GB Affected the Osmotic Substances under LL Stress

To investigate how GB regulates leaf membrane permeability and osmotic substances in pepper seedlings under LL stress, EC, proline, soluble protein, and endogenous GB contents were measured (Figure 6A–D). There was no difference in the content of EC, proline, or soluble protein in the CK and CB treatments; however, they were significantly increased in the LL treatment, and their levels were significantly lower than those of LL with the addition of exogenous GB.

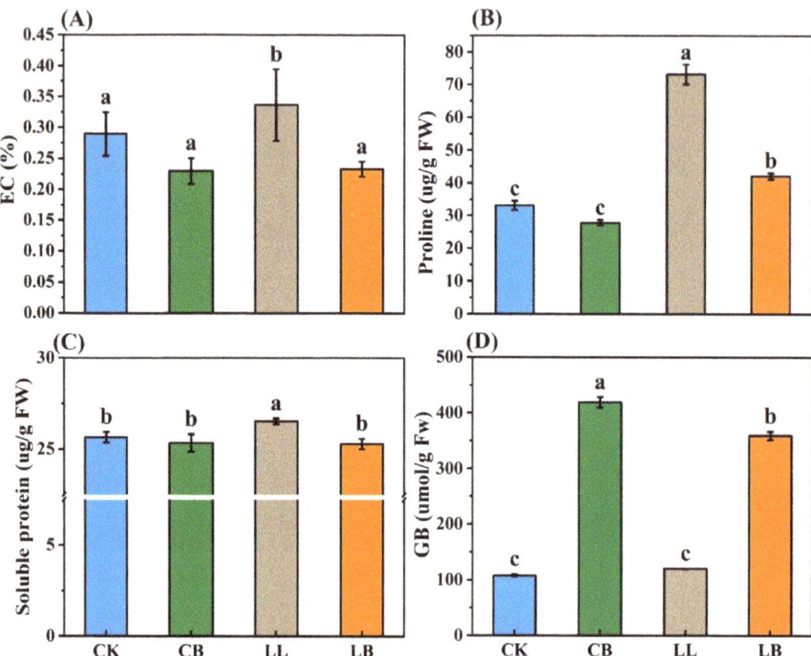

Figure 6. GB affected the osmotic substances by LL stress in pepper seedlings. Data of pepper plants under low-temperature conditions were obtained after 7 days. (**A**) EC: electrical conductivity. (**B**) Proline. (**C**) Soluble protein. (**D**) GB (glycine betaine) content. Different lowercase letters represent significant differences ($p \leq 0.05$) in the same period among treatments, according to Duncan's test.

In addition, there was no difference in endogenous GB content under CK and LL treatments. However, when exogenous GB was sprayed on the leaves, the GB content was significantly higher in the CB and LB treatments, and the GB content in the LB treatment was significantly lower than that of CB.

2.7. GB Affected the ROS Scavenging-Ability under LL Stress

As shown in Figure 7A, the ROS-scavenging ability of peppers also reflected the vital role of exogenous GB under LL stress. At the normal temperature, the CK and CB treatments showed no difference in the activity of SOD or POD. Compared with the normal temperature, the activities of SOD and POD were significantly decreased under LL stress; however, their activities significantly increased under LB treatment, and the SOD activity of LB was similar to that of CK. For CAT activity, the results showed that the treatment with exogenous GB addition was significantly higher than that of the treatment without GB addition at the normal temperature and LL stress, and CAT activity was higher under the LB treatment than under the CB and CK treatments.

In Figure 7B, the results showed that the expression of CaSOD and CaPOD genes were significantly upregulated by the exogenous pre-spraying of GB under normal conditions, and sharply downregulated under LL stress. However, it is notable that the GB pre-spraying reversed the low level of expression caused by LL stress and significantly upregulated the expression of CaSOD, CaPOD, and CaCAT genes under LB treatment.

Furthermore, H_2O_2 and O_2^- contents significantly increased under LL treatment, but significantly decreased after the pre-spraying of GB (LB treatment). Likewise, the shades of color in NBT and DAB histochemical staining showed the levels of H_2O_2 and O_2^- (Figure 8C), indicating that the color of LL was darker than that of LB, and the colors of CK and CB were light. These results suggest that exogenous GB enhances the ROS-scavenging ability under LL stress.

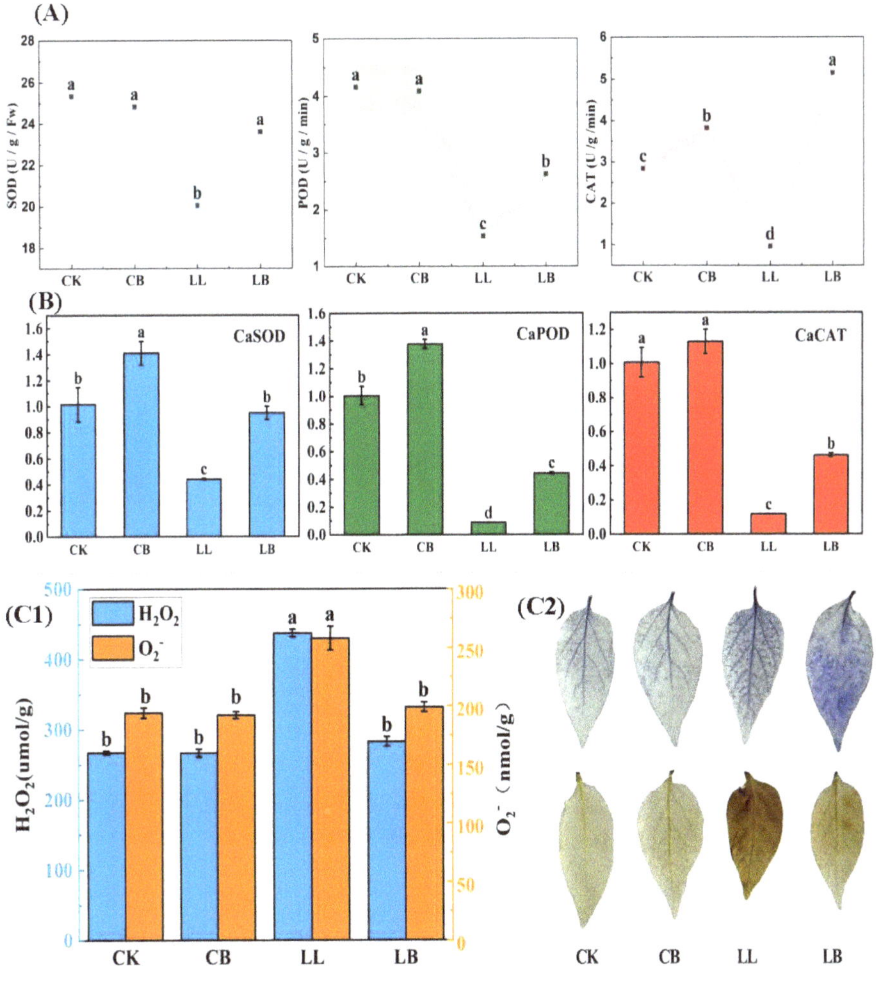

Figure 7. GB affected the antioxidant enzymes, relative expression, and histochemical staining by LL stress in pepper seedings. Data from pepper plants under low-temperature conditions were obtained after 7 days. (**A**) Contained SOD activity, POD activity, and CAT activity. (**B**) The relative expression of CaSOD, CaPOD, and CaCAT. (**C1**) H_2O_2 and O_2^- content. (**C2**) NBT and DAB histochemical staining. Data are means ± SDs (n = 3). Different lowercase letters represent significant differences ($p \leq 0.05$) in the same period among treatments, according to Duncan's test.

2.8. GB Affected the Subcellular Localization of Antioxidant Systems in Pepper Leaves under LL Stress

Each subcellular component can coordinate antioxidant protection. Therefore, we examined how the subcellular components coordinate the activities of the antioxidant enzymes SOD, POD, and CAT under the application of exogenous GB under LL stress. In Figure 8A, the data show that, under LL stress, the addition of exogenous GB significantly increased the SOD activity in the mitochondria, whereas it made no difference in the chloroplasts, but significantly decreased the SOD activity in the cytosol. Furthermore, the SOD activity in the chloroplasts and cytosol was significantly higher under the low temperature combined with low light treatment compared to that of the normal temperature treatment. Figure 8B shows that the CAT activity in the cytoplasmic solutes was significantly higher than that in the mitochondria and chloroplasts, and there was no significant difference in the pre-spraying with GB at the normal temperature; however, the addition of GB significantly increased the CAT activity in the chloroplasts, mitochondria, and cytosol under LL stress. Moreover, the addition of GB under LL stress significantly increased the activity of POD in the chloroplasts, mitochondria, and cytosol, whereas the difference in the levels of activity at normal temperatures was not extremely significant between the CK and the CB treatments (Figure 8C). In addition, the MDA and soluble protein contents in the chloroplasts, mitochondria, and cytosol were all significantly higher in the LL treatment than in the other treatments (Figure 8C,D).

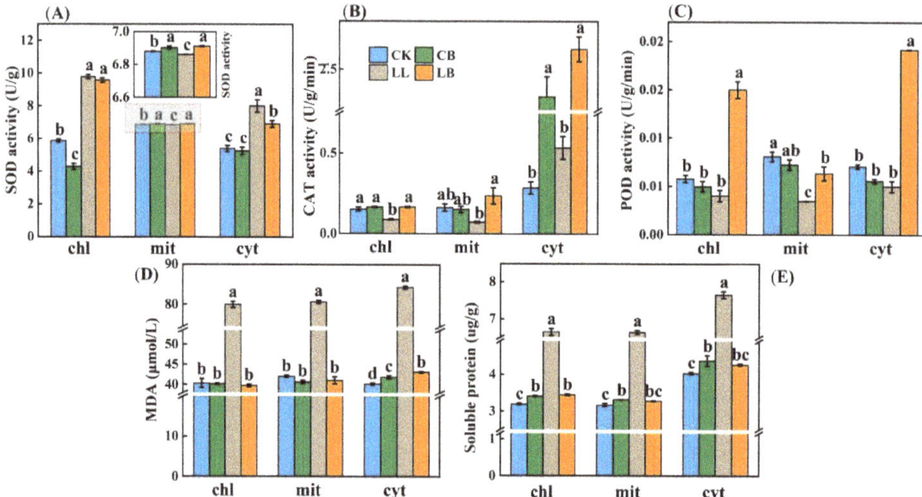

Figure 8. GB affected the subcellular localization of antioxidant systems in pepper leaves. Data from pepper plants under LL conditions were obtained after 7 days. (**A**) SOD activity in mitochondria (mit), chloroplast (chl), and cytosol (cyt). (**B**) CAT activity in mit, chl, and cyt. (**C**) POD activity in mit, chl, and cyt. (**D**) MDA content in mit, chl, and cyt. (**E**) Soluble protein content in mit, chl, and cyt. Data are means $\pm$ SDs (n = 3). Different lowercase letters represent significant differences ($p \leq 0.05$) in the same period among treatments, according to Duncan's test.

In conclusion, the addition of exogenous GB can respond to LL stress by coordinately increasing the activity of SOD in mitochondria, increasing the activity of POD and CAT in the chloroplasts, mitochondria, and cytosol, and decreasing the levels of MDA and soluble proteins in the chloroplasts, mitochondria, and cytosol.

3. Discussion

Abiotic stresses have a negative impact on crop performance. In particular, LL stress is a major constraint on crop yield and quality [26] and effects plant physiology at the both whole-plant and cellular levels at all stages of developmental, from seedling to senescence [27]. Therein, the seedling stage of plants is sensitive to adverse environmental factors. In this study, it has been shown that LL-stress-induced growth inhibition is associated with decreased photosynthesis and antioxidant enzyme activity in pepper plants. As an osmolyte, GB effectively improves plant stress resistance [28,29], increasing the root activity of pepper seedlings, decreasing photoinhibition, lowering the levels of osmoregulatory substances, increasing the activity of antioxidant enzymes, and coordinating the activity of subcellular antioxidant enzymes.

Morphological indexes are the most intuitive expression of a plant's response to an external stimulus. Dry biomass, as the morphological establishment trait, is equally sensitive when plants face stress [30]. Meanwhile, root activity affects the ability of the plant to absorb nutrients and, thus, the morphological establishment of the aboveground parts under abiotic stress [31]. Studies have shown that drought stress negatively affects root morphology and reduces photosynthetic pigments (Chl), which severely inhibits the growth and biomass production of tomato seedlings [32]. Likewise, it was found that the exogenous spraying of GB significantly enhanced the growth characteristics, biomass, proteins, and chl content of wheat under chromium (Cr) stress, which was consistent with the results of this research. Exogenous GB increased the shoots' fresh weight and root activity, thereby increasing the dry biomass of the pepper seedlings under LL stress (Figure 2). Thus, we found that a low temperature of 10/5 °C decreased the root activity of the pepper plants in this experiment, which inhibited the morphological establishment of the aboveground parts by reducing the uptake of nutrients, and that the application of exogenous GB may have increased the nutrient NPK [33], thereby increasing the dry matter accumulation of the pepper seedlings and reducing growth inhibition.

Moreover, LL stress affects photosynthesis in plants. The typical indicators are the photosynthetic pigment (chl) content and chlorophyll fluorescence parameters, which describe the mechanism of photosynthesis and the physiological status of photosynthesis in plants and are considered intrinsic probes for studying the relationship between the photosynthesis of plants and the environment [34]. When high salt concentrations enter plant cells, the membrane system and function of cyst-like bodies in the chloroplasts can be disrupted [35]. Heat stress impairs the energy flux reaching the reaction centers of PSII and the ability of the active reaction centers to photon capture [36], and a low temperature combined with low light significantly reduces Fv/Fm, Y(NO), NPQ, Fv′/Fm′, and [Y(II)] in watermelon plants [37]. Interestingly, one study has reported that cadmium (Cd) stress decreased Fo, Fm, and Fv/Fm levels; however, this was reversed after the foliar application of GB [38]. Another study investigated whether exogenous GB could preserve the photochemical activity of PSII, maintain higher Fv/Fm, and recover more rapidly from photoinhibition under drought stress [39], the results of which were similar to those of this experiment. The pepper seedlings had reduced Fv/Fm, Y(NO), NPQ, Fv′/Fm′, Y(II), and qP contents (Figure 4A–C). Finally, in our study, we examined pepper seedlings under LL stress, which reduced P and Ex and increased D; however, GB resisted the low temperature combined with low-light stress by reducing D and increasing P and Ex (Figure 4D), which was consistent with the study that reported that the energy distribution ratio of the PSII reaction center changed significantly (P and Ex showed an overall downward trend, while D was promoted) under salt stress [40]. Similarly, the OJIP curve, which represents the fluorescence intensity, can indicate the light and efficiency of the plant [41]. Studies have shown that the PSII system of shade-intolerant plants is more vulnerable, with J-P sites having a lower fluorescence intensity in low light [42]. Salt stress has also been found to reduce the intensity of all of the peaks of the plant chlorophyll fluorescence transient (OJIP) curve, indicating a significant deactivation of energy on the donor and acceptor side of the PSII [43]. Again, LL stress reduced the intensity of the J-P dot, but pre-spraying with GB increased the intensity of the J-P dot in the pepper plants under LL treatments (Figure 3). We hypothesize that GB may mitigate the energy deficiency caused by LL stress on the donor and acceptor side of the PSII and respond to LL stress by improving the sufficiency of light energy utilization and reducing the dissipation of excess energy from the pepper seedlings.

Additionally, the *psbA* and *psbD* genes encode the turnover proteins D1 and D2, and the *psbB* and *psbC* genes encode the internal light-harvesting complex proteins CP43 and CP47, which are located in the core of the reaction center, to recover and maintain the stability of the PSII in plants after some stress [44]. The *PsbP* proteins are required for a normal cystoid structure and are also required for the assembly/stability of the PSII complex and photoautotrophy in Arabidopsis [45,46]. The *PsbS* genes are rapidly activated in green organisms by light stress. In this work, pre-spraying with GB significantly upregulated the transcript expression levels of *Capsb A, Capsb B, Capsb C, Capsb D, Capsb S, Capsb P1,* and *Capsb P2* under LL stress (Figure 5). Therefore, we hypothesize that the enhanced tolerance to LL stress in pepper plants is due to the redistribution of the PSII center energy by GB using the encoded protein *CapsB*, which reduces the dissipation of excess energy and allows more light energy to be captured by the PSII center, thereby increasing the photosynthetic efficiency and accelerating the recovery of the PSII from the photoactivated state, allowing it to recover more quickly from photoinhibition [47].

Osmotic regulatory substances are also indispensable in characterizing abiotic stress. The levels of EC, soluble protein, and proline increased under drought stress in soybean leaves [48] and decreased under low temperature combined with low light in pepper leaves [5]. Remarkably, exogenous GB reversed this condition, and the levels of EC, soluble protein, and proline decreased under drought stress. Additionally, in this study, exogenous GB (LB treatment) was found to attenuate the increase in EC, soluble protein, and proline levels in the pepper leaves under the LL treatment; however, we found that the proline levels were significantly higher in the LB treatment than in the CK and CB treatments at the normal temperature (Figure 6A–C). This may be due to the fact that the exogenous GB responds to LL stress mainly by regulating proline metabolism. As an osmotic regulatory substance, GB can alleviate stress under stress conditions by self-accumulation, and studies have shown that GB acts as an accumulator inSolanaceae, Asteraceae, Convolvulaceae, and Amaranthaceae under salt stress [4,49]. Indeed, the experiment results accumulated with the increase in exogenous GB (Figure 2). Notably, the pepper leaves did not accumulate endogenous GB under LL stress (which is in agreement with previous studies) [50], and, compared to the CB treatment, the GB content significantly decreased under the LB treatment, as shown in Figure 6D. It was assumed that the pepper leaves would respond to LL stress by consuming the exogenous GB, converting it into other substances or redistributing the exogenous GB, and that exogenous GB could alleviate the LL stress by reducing the leaf membrane permeability and the osmotic substance levels (proline and soluble protein) in the pepper leaves under low temperatures combined with low-light conditions.

Plants have evolved a number of enzymatic antioxidant systems to control the production and scavenging of ROS in order to cope with stress and to avoid photooxidative damage, either by stress avoidance or stress tolerance [51]. SOD, POD, and CAT are typical ROS scavenger enzymes that can effectively balance excess H_2O_2 and O_2^- in plant cells. Studies have suggested that the inhibition of plant growth by salinity may be related to increased oxidative damage due to the accumulation of ROS, but the application of GB can scavenge excess free radicals by enhancing the ROS enzyme activity. It has also been shown that GB can activate the expression of ROS scavenger genes under abiotic stress. In addition, in this experiment, the pepper seedings accumulated a lot of ROS under LL stress; however, the application of GB upregulated the expression of *CaSOD, CaPOD,* and *CaCAT,* enhanced the activity of the antioxidant enzymes (SOD, POD, and CAT), and enhanced the ability of ROS scavenging (Figure 7). Thus, GB activates the enzymes and genes that scavenge reactive oxygen, and the highly active enzyme scavenges excess ROS under low temperature combined with low-light conditions, improving the LL tolerance of peppers.

Finally, owing to abiotic stress, high levels of reactive oxygen species (ROS) are formed in various organelles, leading to cellular damage. The main organelles are the mitochondria, chloroplasts, and cytosol. The PSI and PSII are the reaction centers in the chloroplast, and controlling the ROS levels in the chloroplast is essential for plant survival under stress. ROS is formed less in the mitochondria than in chloroplasts, which are also sites of photorespiration that are surrounded by cytosol [52,53]. At a normal temperature, plants rely on free radical scavenging systems inside organelles such as chloroplasts and mitochondria to keep intracellular free radicals at low levels and to maintain normal physiological metabolism. However, under LL stress, the plant employs the ROS scavenger enzyme activity system to scavenge free radicals and coordinate ROS homeostasis between the organelles [22]. The POD and CAT activities were the strongest in the cytosol of rhododendron leaves under high-temperature stress, consistent with the findings of the present study, and differences in subcellular localization occurred after exogenous hydrogen high temperatures [23]. Similarly, LL stress mobilizes SOD enzymes in the chloroplasts and cytosol to scavenge free radicals in pepper leaves, while exogenous GB under LL stress reduces cell damage by mobilizing SOD in the mitochondria and POD and CAT in the chloroplasts, mitochondria, and cytosol to scavenge excess free radicals (Figure 8A–C). Moreover, in response to stress, GB regulated the osmotic substances MDA and soluble proteins in various organelles (Figure 8D,E). Therefore, we hypothesized that GB coordinates the activities of the free radical scavenging enzymes in each organelle under LL stress and that each enzyme activity is distributed differently and acts differently in different species.

4. Materials and Methods

4.1. Experimental Materials and Growth Conditions

Pepper (*Capsicum annuum* L.) seeds of 'HangJiao No. 2' (provided by Tianshui Agricultural Science Research Institute) were soaked in heated water (50–55 °C), stirred for 30 min, immersed in 25 °C water for 6 h, and allowed to germinate in an artificial climate chamber at 28 °C in the dark for 4 days. When the radicle reached 1–2 mm, two seeds were placed in a nutrient tray (9 cm × 9 cm) containing seedling substrate, vermiculite, and peat (*v/v* 3:1:1) at a temperature of 28 °C/18 °C (day/night), a PPFD of 300 μmol m^{-2} s^{-1}, a relative humidity of 65%, and a photoperiod of 12/12 h (day/night) in an artificial climate chamber (Ningbo Southeast Instrument, Ningbo, China). A total of 40 pots were grown for each treatment.

4.2. Treatments

4.2.1. Different GB Concentration Treatments

When the sixth true leaf was fully expanded, the seedlings were pretreated with five different concentrations of GB solution (CAS:107-43-7; Shanghai Yuanye Biotechnology Co. Ltd., Shanghai, China), including 0 mmol L^{-1} GB (T0), 10 mmol L^{-1} GB (T1), 20 mmol L^{-1} GB (T2), 40 mmol L^{-1} GB (T3), and 80 mmol L^{-1} GB (T4), and then transferred to an artificial climate chamber for 7 days at a temperature of 10 °C/5 °C (day/night), PPFD of 100 μmol m^{-2} s^{-1}, relative humidity of 65%, and photoperiod of 12/12 h (day/night). It was a preliminary experiment, which was expanded upon in the following experiments.

4.2.2. Alleviating Effects of Exogenous GB on Pepper Plants under LL Stress

Based on the above parameters, the following four treatments were set when the sixth true leaf of the seedlings was expanded:

Normal temperature and light (CK, 28/18 °C, 300 μmol m^{-2} s^{-1});

Normal temperature and light + GB (CB, 28/18 °C, 300 μmol m^{-2} s^{-1}, 20 mM GB);

Low temperature combined with low light (LL, 10/5 °C, 100 μmol m^{-2} s^{-1});

Low temperature combined with low light + GB (LB, 10/5 °C, 100 μmol m^{-2} s^{-1}, 20 mM GB).

Before LL stress, the CB and LB treatments were sprayed with GB for five successive days (solution containing 0.01% Tween-80 as a surfactant). The CK and LL treatments were replaced with equal amounts of double-distilled water (containing 0.01% Tween-80) and then placed in an artificial climate chamber for 7 days. The third and fourth leaves were selected for subsequent physiological and biochemical analyses after 7 days at 8 am. The experiments for each treatment were repeated three times.

4.3. Determinations

4.3.1. Analysis of Morphology and Root Activity

The plant height, stem diameter, number of leaves, and fresh and dry weights of the shoots and roots were measured at 24 h and 7 days after treatment. A tape measure was used to measure the length of the pepper seedlings from the base to the growing point; the stem thickness was measured using a vernier caliper to measure the thickness of the stem below the cotyledons; and the number of fully development leaves was recorded. Whole seedlings were collected, cleaned with water, dried, and separated into roots and shoots to obtain a fresh weight, and dried to a constant weight at 80 °C to obtain a dry weight.

A total of 0.5 g of root tip was weighted into a 10 mL test tube, 10 mL of an equal mixture of 0.4% TTC solution and phosphate buffer was added to completely immerse the roots in the solution, and it was kept in the dark at 37 °C for 2 h, then, 2 mL of 1 mol/L sulfuric acid was added to stop the reaction. Then, the root tips were removed, blotted dry, ground in a mortar with 4 mL of ethyl acetate and a small amount of quartz sand, filtered, and the absorbance values (OD) of the extracts were measured at 485 nm with a UV-1780 spectrophotometer (Shimadzu, Kyoto, Japan) to calculate the root activity.

4.3.2. Related Water Content of Pepper Leaf Cells

The fresh leaves were washed with steamed water and weighed (FW), soaked in 60% sucrose solution for 6 h, kept in the dark at 4 °C, removed, washed several times, and weighed as saturated water (SW). Then, the leaves were placed in an oven at 70 °C for 48 h, and the dry leaf tissue was weighed as DW.

To determine the relative water content (RWC), the fresh leaves were washed with double-steamed water and weighed (Wf), the weighed fresh leaves were soaked in double-steamed water for 24 h, were weighted as a saturated quality (Wt), and then kept in a drying oven at 70 °C for 48 h (Wd). RWC (%) = (Wf − Wd)/(Wt − Wd) × 100. The free water content (FWC) was evaluated as FWC (%) = (FW − SW)/DW × 100%, bound water content (BWC) (%) =TWC-FWC, total water content (TWC) (%) = Wf − Wd, and water saturation deficit (WSD) (%) = 100% − RWC [54].

4.3.3. Determination of Photosynthetic Pigment Content and Chlorophyll Fluorescence Parameter Determination

Photosynthetic Pigment Contents of chla, chlb, chlT, and car

Each sample (0.1 g fresh samples) was transferred to a 20 mL tube immersed in 10 mL 80% acetone and then placed in the dark for about 48 h (shaking every 12 h) until the leaves had turned white. Finally, the absorbance values (OD) of the extracts at 663 nm, 645 nm, and 440 nm were measured using a UV-1780 spectrophotometer (Shimadzu, Japan) to calculate the content of chla, chlb, chlT, and car, respectively [55].

Chlorophyll Fluorescence Parameters

After 7 days of low temperature combined with low-light treatment, an Imaging-PAM fluorimeter (Walz, Effeltrich, Germany) was used to measure the indexes. Previously, the whole seedlings were kept in the dark for at least 30 min, and the third young leaf was selected for measurement. Fo (minimum fluorescence yield of dark-adapted leaves) and Fm (maximum fluorescence yield of dark-adapted leaves) were obtained by applying a saturation pulse of 2700 μmol m^{-2} s^{-1}. Subsequently photochemical light (81 μmol m^{-2} s^{-1}) was turned on every 0.8 s for 20 s, and photoadaptation was performed for 5 min. The actual photochemical efficiency of photosystem II [Y(II)], maximum photochemical efficiency of photosystem II (Fv/Fm), quantum yield of regulatory energy dissipation of photosystem II [Y (NPQ)], quantum yield of nonregulatory energy dissipation of photosystem II [Y(NO)], and portion of energy reaction in PSII were P = (Fv'/Fm') × qP, Ex = (1 − qP)/(Fv'/Fm'), D = 1 − (Fv'/Fm') [35].

Measurement of the "OJIP" Curve and the JIP Test

Chlorophyll fluorescence induction kinetics (OJIP curves) were captured using a Plant Efficiency Analyzer (Handy PEA, Hansatech, UK), resulting in light response curves and fitting parameters. The leaves were placed in complete darkness for 30 min and then continuously illuminated with 3000 μmol m^{-2} s^{-1} of red light to induce a fast chlorophyll fluorescence curve. Instantaneous fluorescence was recorded from 10 μs to 300 s using an OJIP curve diagram to reflect the details of the PSII linear electron transfer. In the OJIP curve, "O" represents the "origin" (minimal fluorescence), "P" represents the "peak" (maximum fluorescence), and "J" and "I" represent the "inflection points" between the "O" and "P" levels. Fo is the fluorescence intensity at the "O" level, while Fm is the intensity at the "P" level, and Fv = (Fm − Fo) is the variable fluorescence [54].

4.3.4. Analyses of Biological Membrane Damage

The membrane permeability was evaluated using relative electrical conductivity (EC) [56]. Leaf discs (0.1 g, 0.6 cm in diameter) were weighed and placed in glass test tubes with 15 mL deionized water and vacuumed for 30 min using a vacuum pump. The tubes were sealed and shaken on a shaking table for 3 h. Then, the initial electric conductivity (L1) was determined with a Ddsj-308f-type conductivity meter (Shanghai INESA Scientific Instrument Co., Ltd., Shanghai, China) after it was set to 25 °C for 10 min. Then, the final electrical conductivity (L2) was measured after the tubes were boiled in a water bath for 15 min and then cooled to 25 °C. The electrical conductivity (L0) of deionized water was used as the blank. The REC was evaluated as REC (%) = (L1 − L0)/(L2 − L0) × 100%.

The malondialdehyde (MDA) content can be used as a representative index to evaluate the degree of membrane lipid peroxidation. The leaf samples (0.3 g) were ground into homogenates on ice with 5% trichloroacetic acid (5 mL). Then, the samples were centrifuged at 3000 r/min for 10 min. The supernatant (2 mL) was mixed with 2 mL of 0.67% thiobarbituric acid and dissolved in 5% trichloroacetic acid. The samples were heated at 100 °C for 30 min and then immediately placed into ice water to cool. Finally, the absorbance of the mixture was measured at 450 nm, 532 nm, and 600 nm using a UV-1780 spectrophotometer (Shimadzu, Japan).

The proline and soluble protein contents were determined according to Huang et al., with some fine adjustments [57]. A total of 0.5 g of the leaves was weighted, 10 mL of 3% sulfosalicylic acid solution was added and quickly ground on ice to a homogenous slurry, filtered, and then 2 mL of filtrate was transferred to a test tube, 2 mL of ninhydrin solution and 2 mL of glacial acetic acid were added, the test tube was placed in boiling water for 1 h, quickly cooled on ice, 4 mL of toluene was added, and it was thoroughly mixed. The determined absorbance at 520 nm was used to calculate the content of the proline using a spectrophotometer.

The leaf sample (0.2 g) was weighed into a precooled mortar, 1.6 mL of 50 mmol/L of precooled phosphate buffer (pH 7.8) was added, and the sample was ground in an ice bath to form a homogenate. It was then transferred to a centrifuge tube at 4 °C and centrifuged at 12,000× g for 20 min. The supernatant was the crude protein extract. A total of 100 mL of enzyme solution was taken and added to 2.9 mL of Coomassie brilliant blue G-250 solution. After a reaction time of 2 min, OD595 was measured to calculate the content of soluble protein.

4.3.5. Determination of the ROS-Scavenging Capability

The H_2O_2 and O_2^- contents were determined by histochemical staining with nitroblue tetrazolium (NBT) and 3,3-diaminobenzidine (DAB), respectively [51]. The leaves were washed with distilled water to remove extraneous material. Then, the leaves were collected in conical flasks, and each treatment included 4 leaves immersed in either 0.1 mg·mL^{-1} NBT (pH 7.8; 0.2 h) or 1 mg·mL^{-1} DAB (pH 7.0; 24 h) staining solutions at 25 °C in the dark to detect H_2O_2 and O_2^-, respectively. Finally, the samples were decolorized with lactic acid:

glycerol: ethanol = 1:1:3 (*v/v*) in a boiling water bath for 5–6 min and photographed using an EPSON Expression 11,000 XL color image scanner (WinRHIZO Pro LA2400, Canada).

The leaf tissues (0.5 g) were ground in 5 mL ice-cold 0.05 mol sodium phosphate buffer (containing 5 mM EDTA-Na$_2$, 2 mM AsA, and 2% (*w/v*) PVP, pH 7.8) using a prechilled mortar and pestle. The homogenate was centrifuged at 12,000 r/min for 20 min at 4 °C. The antioxidant activity of the supernatant was determined using a UV-1780 spectrophotometer (Shimadzu, Japan). The superoxide dismutase (SOD) activity was measured according to the method of Geng et al. The SOD activity was measured using the nitrotetrazolium blue chloride (NBT)-illumination method, the peroxidase (POD) activity was measured using the o-methoxyphenol method, the catalase (CAT) activity was measured by ultraviolet absorption, and the ascorbic acid peroxidase (APX) activity was estimated by monitoring the decline in absorbance at 290 nm, due to the oxidation of AsA (ascorbic acid) oxidation by H$_2$O$_2$ [23].

4.3.6. Determination of the Subcellular Location of ROS-Scavenging Enzymes

Subcellular antioxidant enzyme activity in pepper leaves was reported by Geng et al., and the subsequently determined numerical by visible spectrophotometric determination [23]. The mitochondria, chloroplasts, and cytoplasmic stroma were separated using differential centrifugation. A total of 10 g of leaf sample was taken, 30 mL of pre-cooled extraction buffer (0.05 mol/L Tris-HCl, 0.35 mmol/L sorbitol, 2 mmol/LEDTA, and 2.5 mmol/LDTT, pH = 7.5) and a little quartz sand were rapidly ground on ice, and then the sample was filtered into 4 layers and centrifuged at 500 g/min for 5 min. The supernatant was centrifuged again (2000 rpm/min, 10 min) and the resulting precipitate was the chloroplast fraction. The supernatant was further centrifuged (12,000 rpm/min, 20 min), and the resulting precipitate was the mitochondrial fraction, and the supernatant was the cytosolic fraction. The chloroplasts and mitochondria were washed once with the extraction solution, centrifuged again, and the precipitate was suspended in 5 mL of extraction buffer. The suspensions were then used to determine the antioxidant enzyme activities in each organelle. The NBT method was used to determine the SOD enzyme activity, the guaiacol method was used to determine the POD enzyme activity, and the UV absorption method was used to determine the CAT enzyme activity.

4.3.7. Content of Glycine Betaine

The sample (0.5 g) was homogenized with 5 mL of 0.05% toluene, shaken at 25 °C for 24 h, and centrifuged at 3000× *g* for 20 min. The supernatant was mixed with 3 mL of 1 M hydrochloric acid and KI-I$_2$ solution, mixed thoroughly, and shaken at 0 °C for 90 min. Next, 10 mL of 1,2 dichloroethane was added, and, after the separation of the organic layer, the absorption was measured at 365 nm [58].

4.3.8. Quantitative Polymerase Chain Reaction (PCR) Analysis

The total RNA of the different treatments was extracted using the RNA extraction kit provided by Tengen Biotechnology and then reverse-transcribed with a cDNA synthesis kit provided by TaKaRa, Kusatsu City, Japan, to obtain cDNA. Primer 5 software was used to design all primers (Table 1), which were synthesized by Sangon Biotech. The qRT-PCR was performed on BioRad (CFX96, Hercules, California, USA) using the TransStart Top Green qPCR SuperMix kit (TransStart Green, Beijing, China). The reaction mixture contained 2 µL cDNA, 0.6 µL forward primer, 0.6 µL reverse primer, 10 µL qPCR SuperMix, and 6.8 µL RNase-free ddH$_2$O. The reaction conditions were 94 °C for 30 s, 95 °C for 5 s, and 60 °C for 30 s, with a total of 40 cycles. The technique was repeated three times for each sample. Actin was used as the reference gene, and the relative gene expression was calculated by the 2$^{-\Delta\Delta Ct}$ method.

Table 1. The sequences of primers used for the qRT-PCR.

Gene Name	Sequence (5′–3′)	GenBank Accession Number	Encoded Target	Amplicons Size (bp)
Actin	F: GTCCTTCCATCGTCCACAGG R: GAAGGGCAAAGGTTCACAACA	XM_016722297.1	*Capsicum annuum* actin	1134
CaSOD	F: GTGAGCCTCCAAAGGGTTCTCTTG R: AAACCAAGCCACACCCAACCAG	AF036936.2	Manganese superoxide dismutase	687
CaPOD	F: GCCAGGACAGCAAGCCAAGG R: TGAGCACCTGATAAGGCAACCATG	FJ596178.1	Peroxidase	975
CaCAT	F: TTAACGCTCCCAAGTGTGCTCATC R: GGCAGGACGACAAGGATCAAACC	NM_001324674.1	Catalase	1479
CaPsbA	F: GAATAGGGAGCCGCCGAATACAC R: TATTCCAGGCTGAGCACAACATCC	NC_018552.1:565–1626	Turnover proteins D1	1062
CaPsbB	F: TGGGTTTGCCTTGGTATCGTGTTC R: GCCCAACCAGCAACCAGAGC	NC_018552.1:75670–77196	Internal light-harvesting complex proteins CP43	1527
CaPsbC	F: GGATCTGCGTGCTCCATGGTTAG R: CCGTTCCTGCCAAGGTTGTATGTC	NC_018552.1:35289–36674	Internal light-harvesting complex proteins CP47	1386
CaPsbD	F: TGGTCACCGCTAACCGCTTTTG R: AGACCGACTACTCCAAGAGCACTC	NC_018552.1:34244–35305	Turnover proteins D2	1062
CaPsbS	F: AGGGAAAGGAGCATTGGCACAAC R: GCAGCAAAGAAGAAGAAGGCAACG	XM_016719676.1	Photosystem II 22 kDa protein	834
CaPsbP1	F: GCTGCTTCCACACAATGCTTCTTG R: TGGTTAGGCTTGAGGGTTGAAACG	XM_016701483.1	Oxygen-evolving enhancer protein 2	783
CaPsbP2	F: CTCGGGCAGCATTTGCTACCATAG R: CCTGAAATGAGTCGGCCACCAC	XM_016690370.1	*Capsicum annuum* photosynthetic NDH subunit	699

Note: F: Forward primer; R: Revers primer.

4.4. Statistical Analysis

The statistical analysis was performed with ANOVA using SPSS software (version 22.0, SPSS Institute Limited., California, USA) and EXCEL (Office, 2019). Significant differences ($p < 0.05$, $p < 0.01$) between the means of the different treatments were assessed using Duncan's multiple range tests. All figures were plotted using Origin Pro software. 9.0 (Origin Lab Institute, incorporated., California, USA).

5. Conclusions

In conclusion, the application of GB could be an effective way to improve the tolerance of pepper seedlings under low temperatures combined with low-light stress, which can involve photosynthetic and antioxidant aspects. The pre-spraying of GB alleviated the growth inhibition caused by LL stress by reducing photoinhibition, as indicated by the increased Fv/Fm and chl contents, as well as [Y(II)], Y(NPQ), Y(NO), NPQ, and qP, regulating the lipid peroxidation of membranes, thereby improving the ability to scavenge excess H_2O_2 and O_2^- through the increased activity of antioxidant enzymes and the expression of genes encoding antioxidant enzymes and the PSII reaction center proteins, and coordinating the activities of antioxidant enzymes in the mitochondria, chloroplasts, and cytosol. Overall, this study provides a better understanding of the mechanisms by which GB affects the growth of pepper plants under LL conditions and provides preliminary ideas for studying the response mechanisms of pepper seedlings under LL stress.

Author Contributions: N.L., J.L., J.X. and K.P.: conceptualization. N.L., J.L. and D.D.: methodology. N.L., D.D. and J.L.: software. N.L., J.L., D.D., Y.Y. and T.N.: formal analysis and investigation. N.L.: writing—original draft preparation. N.L., J.L. and J.X.: writing—review and editing. J.L. and J.X.: resources, supervision, project administration, and funding acquisition. All authors have read and agreed to the published version of the manuscript.

Funding: The work was funded by the Natural Science Foundation of Gansu Province (23JRRA1417), the Natural Science Foundation of Gansu Province (20JR10RA515), the Scientific Research Start-up

Funds for Openly Recruited Doctors (GAUKYQD-2018-35), the National Natural Science Foundation of China (32072657), and the Special Fund for Technical System of Melon and Vegetable Industry of Gansu Province (GARS-GC-1), Special Fund for Sci. and Tech. Innovation and Development Guided by Gansu Province (2018ZX-02). The authors would like to thank all of the authors, editors, and reviewers for their precious and valuable comments and suggestions, which have greatly improved this article.

Institutional Review Board Statement: Not applicable.

Informed Consent Statement: Not applicable.

Data Availability Statement: Data are contained within the article.

Conflicts of Interest: The authors declare no conflict of interest.

References

1. Zhu, J.K. Abiotic Stress Signaling and Responses in Plants. *Cell* **2016**, *167*, 313–324. [CrossRef] [PubMed]
2. Zhang, X.H.; Zhang, L.; Sun, Y.P.; Zheng, S.; Wang, J.; Zhang, T.G. Hydrogen peroxide is involved in strigolactone induced low temperature stress tolerance in rape seedlings (*Brassica rapa* L.). *Plant Physiol. Biochem.* **2020**, *157*, 402–415. [CrossRef]
3. Wang, Y.; Li, Y.; Wang, J.; Xiang, Z.; Xi, P.; Zhao, D. Physiological Changes and Differential Gene Expression of Tea Plants (*Camellia sinensis* (L.) Kuntze var. niaowangensis Q.H. Chen) Under Cold Stress. *DNA Cell Biol.* **2021**, *40*, 906–920. [CrossRef]
4. Li, M.; Wang, C.; Shi, J.; Zhang, Y.; Liu, T.; Qi, H. Abscisic acid and putrescine synergistically regulate the cold tolerance of melon seedlings. *Plant Physiol. Biochem.* **2021**, *166*, 1054–1064. [CrossRef] [PubMed]
5. Tang, C.; Xie, J.; Lv, J.; Li, J.; Zhang, J.; Wang, C.; Liang, G. Alleviating damage of photosystem and oxidative stress from chilling stress with exogenous zeaxanthin in pepper (*Capsicum annuum* L.) seedlings. *Plant Physiol. Biochem.* **2021**, *162*, 395–409. [CrossRef]
6. Guo, W.L.; Chen, R.G.; Gong, Z.H.; Yin, Y.X.; Li, D.W. Suppression Subtractive Hybridization Analysis of Genes Regulated by Application of Exogenous Abscisic Acid in Pepper Plant (*Capsicum annuum* L.) Leaves under Chilling Stress. *PLoS ONE* **2013**, *8*, e66667. [CrossRef]
7. Genzel, F.; Dicke, M.D.; Junker-Frohn, L.V.; Neuwohner, A.; Thiele, B.; Putz, A.; Usadel, B.; Wormit, A.; Wiese-Klinkenberg, A. Impact of Moderate Cold and Salt Stress on the Accumulation of Antioxidant Flavonoids in the Leaves of Two Capsicum Cultivars. *J. Agric. Food Chem.* **2021**, *69*, 6431–6443. [CrossRef] [PubMed]
8. Ma, J.; Wang, L.Y.; Dai, J.X.; Wang, Y.; Lin, D. The NAC-type transcription factor CaNAC46 regulates the salt and drought tolerance of transgenic *Arabidopsis thaliana*. *BMC Plant Biol.* **2021**, *21*, 11. [CrossRef]
9. Yadav, S.K. Cold stress tolerance mechanisms in plants. A review. *Agron. Sustain. Dev.* **2010**, *30*, 515–527. [CrossRef]
10. Casal, J.J.; Questa, J.I. Light and temperature cues: Multitasking receptors and transcriptional integrators. *New Phytol.* **2018**, *217*, 1029–1034. [CrossRef]
11. Ahmad, P.; Abd Allah, E.F.; Alyemeni, M.N.; Wijaya, L.; Alam, P.; Bhardwaj, R.; Siddique, K.H.M. Exogenous application of calcium to 24-epibrassinosteroid pre-treated tomato seedlings mitigates NaCl toxicity by modifying ascorbate-glutathione cycle and secondary metabolites. *Sci. Rep.* **2018**, *8*, 13515. [CrossRef]
12. Naus, J.; Smecko, S.; Spundova, M. Chloroplast avoidance movement as a sensitive indicator of relative water content during leaf desiccation in the dark. *Photosynth. Res.* **2016**, *129*, 217–225. [CrossRef]
13. Snider, J.L.; Thangthong, N.; Pilon, C.; Virk, G.; Tishchenko, V. OJIP-fluorescence parameters as rapid indicators of cotton (*Gossypium hirsutum* L.) seedling vigor under contrasting growth temperature regimes. *Plant Physiol. Biochem.* **2018**, *132*, 249–257. [CrossRef]
14. Heerden, P.D.R.v.; Villiers, O.T.d. Evaluation of the relative water content and the reduction of 2,3,5-triphenyltetrazoliumchloride as indicators of drought tolerance in spring wheat cultivars. *S. Afr. J. Plant Soil* **2013**, *13*, 131–135. [CrossRef]
15. Shan, Z.Y.; Luo, X.L.; Wei, M.G.; Huang, T.W.; Khan, A.; Zhu, Y.M. Physiological and proteomic analysis on long-term drought resistance of cassava (*Manihot esculenta* Crantz). *Sci. Rep.* **2018**, *8*, 17982. [CrossRef] [PubMed]
16. Kaufmann, H.; Blanke, M. Changes in carbohydrate levels and relative water content (RWC) to distinguish dormancy phases in sweet cherry. *J. Plant Physiol.* **2017**, *218*, 1–5. [CrossRef] [PubMed]
17. Syeed, S.; Sehar, Z.; Masood, A.; Anjum, N.A.; Khan, N.A. Control of Elevated Ion Accumulation, Oxidative Stress, and Lipid Peroxidation with Salicylic Acid-Induced Accumulation of Glycine Betaine in Salinity-Exposed *Vigna radiata* L. *Appl. Biochem. Biotechnol.* **2021**, *193*, 3301–3320. [CrossRef]
18. El-Amier, Y.; Elhindi, K.; El-Hendawy, S.; Al-Rashed, S.; Abd-ElGawad, A. Antioxidant System and Biomolecules Alteration in Pisum sativum under Heavy Metal Stress and Possible Alleviation by 5-Aminolevulinic Acid. *Molecules* **2019**, *24*, 4194. [CrossRef]
19. Iwaniuk, P.; Borusiewicz, A.; Lozowicka, B. Fluazinam and its mixtures induce diversified changes of crucial biochemical and antioxidant profile in leafy vegetable. *Sci. Hortic.* **2022**, *298*, 110988. [CrossRef]
20. Saja-Garbarz, D.; Libik-Konieczny, M.; Fellner, M.; Jurczyk, B.; Janowiak, F. Silicon-induced alterations in the expression of aquaporins and antioxidant system activity in well-watered and drought-stressed oilseed rape. *Plant Physiol. Biochem.* **2022**, *174*, 73–86. [CrossRef] [PubMed]

21. Mohammadi, H.; Moradi, F. Effects of growth regulators on enzymatic and non-enzymatic antioxidants in leaves of two contrasting wheat cultivars under water stress. *Braz. J. Bot.* **2016**, *39*, 495–505. [CrossRef]
22. Yu, F.; Chen, Y.; Yang, Z.; Wang, P.; Li, D.; Zhang, J. Effects of low temperature stress on antioxidant enzymes activity in the subcellular of two Sabrina species. *GuangXi Plants* **2014**, *34*, 686–693.
23. Geng, X.M.; Xiao, L.; Zhao, H.; Liu, P. Sub-cellular location of ROS-scavenging system in rhodoendron leaves under heat stress and H$_2$O$_2$ pretreatment. *J. Northwest Bot.* **2019**, *39*, 791–800.
24. Pinero, M.C.; Lorenzo, P.; Sanchez-Guerrero, M.C.; Medrano, E.; Lopez-Marin, J.; del Amor, F.M. Reducing extreme weather impacts in greenhouses: The effect of a new passive climate control system on nutritional quality of pepper fruits. *J. Sci. Food Agric.* **2021**, *102*, 2723–2730. [CrossRef] [PubMed]
25. Figueroa-Soto, C.G.; Valenzuela-Soto, E.M. Glycine betaine rather than acting only as an osmolyte also plays a role as regulator in cellular metabolism. *Biochimie* **2018**, *147*, 89–97. [CrossRef]
26. Li, J.; Zhang, Z.; Chong, K.; Xu, Y. Chilling tolerance in rice: Past and present. *J. Plant Physiol.* **2022**, *268*, 153576. [CrossRef]
27. Wu, J.X.; Nadeem, M.; Galagedara, L.; Thomas, R.; Cheema, M. Effects of Chilling Stress on Morphological, Physiological, and Biochemical Attributes of Silage Corn Genotypes during Seedling Establishment. *Plants* **2022**, *11*, 1217. [CrossRef]
28. Wei, D.D.; Zhang, T.P.; Wang, B.Q.; Zhang, H.L.; Ma, M.Y.; Li, S.F.; Chen, T.H.H.; Brestic, M.; Liu, Y.; Yang, X.H. Glycinebetaine mitigates tomato chilling stress by maintaining high-cyclic electron flow rate of photosystem I and stability of photosystem II. *Plant Cell Rep.* **2022**, *41*, 1087–1101. [CrossRef]
29. Park, E.J.; Jeknic, Z.; Chen, T.H. Exogenous application of glycinebetaine increases chilling tolerance in tomato plants. *Plant Cell Physiol.* **2006**, *47*, 706–714. [CrossRef]
30. Jia, W.; Ma, M.; Chen, J.; Wu, S. Plant Morphological, Physiological and Anatomical Adaption to Flooding Stress and the Underlying Molecular Mechanisms. *Int. J. Mol. Sci.* **2021**, *22*, 1088. [CrossRef]
31. Yamori, N.; Levine, C.P.; Mattson, N.S.; Yamori, W. Optimum root zone temperature of photosynthesis and plant growth depends on air temperature in lettuce plants. *Plant Mol. Biol.* **2022**, *110*, 385–395. [CrossRef]
32. Altaf, M.A.; Shahid, R.; Ren, M.X.; Naz, S.; Altaf, M.M.; Khan, L.U.; Tiwari, R.K.; Lal, M.K.; Shahid, M.A.; Kumar, R.; et al. Melatonin Improves Drought Stress Tolerance of Tomato by Modulating Plant Growth, Root Architecture, Photosynthesis, and Antioxidant Defense System. *Antioxidants* **2022**, *11*, 309. [CrossRef]
33. Giri, J. Glycinebetaine and abiotic stress tolerance in plants. *Plant Signal. Behav.* **2011**, *6*, 1746–1751. [CrossRef] [PubMed]
34. Gao, D.P.; Ran, C.; Zhang, Y.H.; Wang, X.L.; Lu, S.F.; Geng, Y.Q.; Guo, L.Y.; Shao, X.W. Effect of different concentrations of foliar iron fertilizer on chlorophyll fluorescence characteristics of iron-deficient rice seedlings under saline sodic conditions. *Plant Physiol. Biochem.* **2022**, *185*, 112–122. [CrossRef] [PubMed]
35. Tsai, Y.C.; Chen, K.C.; Cheng, T.S.; Lee, C.; Lin, S.H.; Tung, C.W. Chlorophyll fluorescence analysis in diverse rice varieties reveals the positive correlation between the seedlings salt tolerance and photosynthetic efficiency. *BMC Plant Biol.* **2019**, *19*, 403. [CrossRef] [PubMed]
36. Arikan, B.; Ozfidan-Konakci, C.; Alp, F.N.; Zengin, G.; Yildiztugay, E. Rosmarinic acid and hesperidin regulate gas exchange, chlorophyll fluorescence, antioxidant system and the fatty acid biosynthesis-related gene expression in *Arabidopsis thaliana* under heat stress. *Phytochemistry* **2022**, *198*, 113157. [CrossRef]
37. Shin, Y.K.; Bhandari, S.R.; Lee, J.G. Monitoring of Salinity, Temperature, and Drought Stress in Grafted Watermelon Seedlings Using Chlorophyll Fluorescence. *Front. Plant Sci.* **2021**, *12*, 786309. [CrossRef]
38. He, X.Y.; Richmond, M.E.A.; Williams, D.V.; Zheng, W.T.; Wu, F.B. Exogenous Glycinebetaine Reduces Cadmium Uptake and Mitigates Cadmium Toxicity in Two Tobacco Genotypes Differing in Cadmium Tolerance. *Int. J. Mol. Sci.* **2019**, *20*, 1612. [CrossRef]
39. Ma, Q.Q.; Wang, W.; Li, Y.H.; Li, D.Q.; Zou, Q. Alleviation of photoinhibition in drought-stressed wheat (*Triticum aestivum*) by foliar-applied glycinebetaine. *J. Plant Physiol.* **2006**, *163*, 165–175. [CrossRef]
40. Ran, X.; Wang, X.; Gao, X.; Liang, H.; Liu, B.; Huang, X. Effects of salt stress on the photosynthetic physiology and mineral ion absorption and distribution in white willow (*Salix alba* L.). *PLoS ONE* **2021**, *16*, e0260086. [CrossRef]
41. Alexandrina Stirbet, G. On the relation between the Kautsky effect (chlorophyll a fluorescence induction) and Photosystem II: Basics and applications of the OJIP fluorescence transient. *J. Photoch Photobio B* **2011**, *104*, 236–257. [CrossRef] [PubMed]
42. Benkov, M.A.; Yatsenko, A.M.; Tikhonov, A.N. Light acclimation of shade-tolerant and sun-resistant Tradescantia species: Photochemical activity of PSII and its sensitivity to heat treatment. *Photosynth. Res.* **2019**, *139*, 203–214. [CrossRef] [PubMed]
43. Neelam, S.; Subramanyam, R. Alteration of photochemistry and protein degradation of photosystem II from Chlamydomonas reinhardtii under high salt grown cells. *J. Photoch Photobio B* **2013**, *124*, 63–70. [CrossRef] [PubMed]
44. Chen, K.; Chen, L.; Fan, J.; Fu, J. Alleviation of heat damage to photosystem II by nitric oxide in tall fescue. *Photosynth. Res.* **2013**, *116*, 21–31. [CrossRef] [PubMed]
45. Yi, X.; Hargett, S.R.; Liu, H.; Frankel, L.K.; Bricker, T.M. The PsbP protein is required for photosystem II complex assembly/stability and photoautotrophy in *Arabidopsis thaliana*. *J. Biol. Chem.* **2007**, *282*, 24833–24841. [CrossRef]
46. Yi, X.; Hargett, S.R.; Frankel, L.K.; Bricker, T.M. The PsbP protein, but not the PsbQ protein, is required for normal thylakoid architecture in *Arabidopsis thaliana*. *FEBS Lett.* **2009**, *583*, 2142–2147. [CrossRef]
47. Endo, T.; Uebayashi, N.; Ishida, S.; Ikeuchi, M.; Sato, F. Light energy allocation at PSII under field light conditions: How much energy is lost in NPQ-associated dissipation? *Plant Physiol. Biochem.* **2014**, *81*, 115–120. [CrossRef]

48. Wang, X.Y.; Li, Y.P.; Wang, X.J.; Li, X.M.; Dong, S.K. Physiology and metabonomics reveal differences in drought resistance among soybean varieties. *Bot. Stud.* **2022**, *63*, 8. [CrossRef]
49. Weretilnyk, E.A.; Bednarek, S.; McCue, K.F.; Rhodes, D.; Hanson, A.D. Comparative biochemical and immunological studies of the glycine betaine synthesis pathway in diverse families of dicotyledons. *Planta* **1989**, *178*, 342–352. [CrossRef]
50. Zulfiqar, F.; Akram, N.A.; Ashraf, M. Osmoprotection in plants under abiotic stresses: New insights into a classical phenomenon. *Planta* **2020**, *251*, 3. [CrossRef]
51. Kaya, C.; Shabala, S. Melatonin improves drought stress tolerance of pepper (*Capsicum annuum*) plants via upregulating nitrogen metabolism. *Funct. Plant Biol.* **2023**. [CrossRef]
52. Nadarajah, K.K. ROS Homeostasis in Abiotic Stress Tolerance in Plants. *Int. J. Mol. Sci.* **2020**, *21*, 5208. [CrossRef] [PubMed]
53. Lee, K.P.; Kim, C.; Landgraf, F.; Apel, K. EXECUTER1- and EXECUTER2-dependent transfer of stress-related signals from the plastid to the nucleus of *Arabidopsis thaliana*. *Proc. Natl. Acad. Sci. USA* **2007**, *104*, 10270–10275. [CrossRef] [PubMed]
54. Janku, M.; Luhova, L.; Petrivalsky, M. On the Origin and Fate of Reactive Oxygen Species in Plant Cell Compartments. *Antioxidants* **2019**, *8*, 105. [CrossRef] [PubMed]
55. Alyemeni, M.N.; Ahanger, M.A.; Wijaya, L.; Alam, P.; Bhardwaj, R.; Ahmad, P. Correction to: Selenium mitigates cadmium-induced oxidative stress in tomato (*Solanum lycopersicum* L.) plants by modulating chlorophyll fluorescence, osmolyte accumulation, and antioxidant system. *Protoplasma* **2018**, *255*, 985–986. [CrossRef]
56. Yang, Y.; Yao, Y.; Li, J.; Zhang, J.; Zhang, X.; Hu, L.; Ding, D.; Bakpa, E.P.; Xie, J. Trehalose Alleviated Salt Stress in Tomato by Regulating ROS Metabolism, Photosynthesis, Osmolyte Synthesis, and Trehalose Metabolic Pathways. *Front. Plant Sci.* **2022**, *13*, 772948. [CrossRef]
57. Huang, X.M.; Song, S.Q.; Fu, J.R. Changes in water state and soluble protein contents of wampee axes during inducing desiccation tolerance by sucrose preculture. *Zhi Wu Sheng Li Yu Fen Zi Sheng Wu Xue Xue Bao* **2006**, *32*, 245–251.
58. You, L.; Song, Q.; Wu, Y.; Li, S.; Jiang, C.; Chang, L.; Yang, X.; Zhang, J. Accumulation of glycine betaine in transplastomic potato plants expressing choline oxidase confers improved drought tolerance. *Planta* **2019**, *249*, 1963–1975. [CrossRef]

 *plants*

Article

NPK Accumulation, Physiology, and Production of Sour Passion Fruit under Salt Stress Irrigated with Brackish Water in the Phenological Stages and K Fertilization

Geovani Soares de Lima [1,*], André Alisson Rodrigues da Silva [1], Rafaela Aparecida Frazão Torres [2], Lauriane Almeida dos Anjos Soares [2], Hans Raj Gheyi [1], Francisco Alves da Silva [2], Reginaldo Gomes Nobre [3], Carlos Alberto Vieira de Azevedo [1], Kilson Pinheiro Lopes [2], Lúcia Helena Garófalo Chaves [1] and Vera Lúcia Antunes de Lima [1]

[1] Post Graduate Program in Agricultural Engineering, Federal University of Campina Grande, Campina Grande 58430-380, Brazil
[2] Post Graduate Program in Tropical Horticulture, Federal University of Campina Grande, Pombal 58840-000, Brazil
[3] Post Graduate Program in Soil and Water Management, Federal Rural University of the Semi-Arid, Caraúbas 59780-000, Brazil
* Correspondence: geovani.soares@professor.ufcg.edu.br; Tel.: +55-83-2145-0622

Abstract: This research aimed to evaluate the effects of salt stress, varying the phenological stages, and K fertilization on NPK concentrations, physiology, and production of *Passiflora edulis* Sims. The research was carried out at the University Farm of São Domingos, Paraíba, Brazil, using a randomized block design with a 6 × 2 factorial arrangement. Six irrigation strategies were evaluated (use of low electrical conductivity water (0.3 dS m^{-1}) during all stages of development and application of high-salinity water (4.0 dS m^{-1}) in the following stages: vegetative, flowering, fruiting, successively in the vegetative/flowering, and vegetative/fruiting stages) and two potassium levels (207 and 345 g K$_2$O per plant), with four replications and three plants per plot. The leaf concentrations of N, P, and K in the sour passion fruit plants found in the present study were below the optimal levels reported in the literature, regardless of the development stage and the cultivation cycle. The relative water content, stomatal conductance, and photosynthesis were reduced by salt stress in the first cycle. However, in the second cycle, irrigation with 4.0 dS m^{-1} in the vegetative/flowering stages increased the CO$_2$ assimilation rate. Passion fruit is sensitive to salt stress in the vegetative/flowering stages of the first cycle. In the second cycle, salt stress in the fruiting stage resulted in higher production per plant.

Keywords: *Passiflora edulis*; water scarcity; osmotic regulation

Citation: de Lima, G.S.; da Silva, A.A.R.; Torres, R.A.F.; Soares, L.A.d.A.; Gheyi, H.R.; da Silva, F.A.; Nobre, R.G.; de Azevedo, C.A.V.; Lopes, K.P.; Chaves, L.H.G.; et al. NPK Accumulation, Physiology, and Production of Sour Passion Fruit under Salt Stress Irrigated with Brackish Water in the Phenological Stages and K Fertilization. *Plants* **2023**, *12*, 1573. https://doi.org/10.3390/plants12071573

Academic Editor: Anis Limami

Received: 4 March 2023
Revised: 25 March 2023
Accepted: 4 April 2023
Published: 6 April 2023

1. Introduction

Passiflora edulis Sims stands out as one of the main fruit species due to its high nutritive value, excellent organoleptic features, and significant economic potential as fresh fruit or in the agroindustry [1]. Although the northeastern semi-arid region has the potential for the increased cultivation of this fruit species, the occurrence of high temperatures and the irregular distribution of rainfall associated with intense evaporation throughout the year are limiting factors for irrigated fruit cultivation [2]. In this region, it is common to find water sources with high levels of dissolved salts, standing out as one of the abiotic stresses that promote osmotic and ionic changes in plants [3] when used to irrigate orchards.

Salts in the root zone cause osmotic stress, ion toxicity, nutrient imbalance, and water deficit, inducing a decrease in water and nutrient uptake by plants. Furthermore, excessive ion concentrations in the soil solution also damage photosynthetically active leaves and can result in chlorosis and early senescence, thus negatively affecting photosynthetic efficiency and production [4]. In this context, several studies have been carried out evaluating the effects of irrigation with saline water in the passion fruit crop, highlighting changes in gas exchange [5], photochemical efficiency [6], synthesis of photosynthetic pigments [7], and membrane instability [8], as well as a reduction in production components [9,10] and post-harvest quality [11]. However, these studies are restricted to the use of water with different levels of salinity throughout the plant development cycle, thus requiring further research, especially to assess the effects of salt stress on the accumulation of NPK, physiology, and production of sour passion fruit plants of the cultivar BRS GA1, one of the most commonly grown in northeast Brazil, at varying phenological stages.

It is important to highlight that the intensity of salt stress effects depends on the species, genotype, duration of exposure, fertilization, irrigation management, and development stage [12], in addition to stress tolerance mechanisms, including the maintenance of ion homeostasis and the osmotic balance and the elimination of reactive oxygen species [13]. Thus, the use of water with a high concentration of salts in the phenological phase(s) in which the crop expresses tolerance is a promising alternative for reducing the effects of salt stress [14,15]. Soares et al. [16] observed that salt stress in the initial stages leads to earlier flowering in cotton and does not compromise its production. Lima et al. [17] found that salt stress in the vegetative, flowering, and fruiting stages is a promising strategy for sesame cultivation in the semi-arid region of Brazil. Silva et al. [18], in a study evaluating the production and post-harvest quality of mini-watermelon fruits, saline water irrigation management strategies, and potassium fertilization, concluded that irrigation with water of 4.0 dS m^{-1} in the flowering and fruit maturation stages is a promising strategy for cultivation, since it does not compromise the production, and fertilization with 50% of the K$_2$O recommendation can be used in the cultivation of mini-watermelon without losses in production.

Thus, the use of saline water in the phenological stage(s) in which the crop has higher tolerance is a promising irrigation strategy, as it allows minimizing the effects of salt stress on plants and reduces changes in the physical-chemical attributes of the soil due to the formation of saline and sodic soils.

Another strategy capable of mitigating salt stress in plants is potassium fertilization [12–15]. Potassium is involved in almost all plant physiological processes that require water. These processes include stomatal regulation, photoassimilate transport, enzyme activation, and heliotropic movements of leaves. Potassium also assists in water transport, translocation of mineral compounds to the entire plant through the xylem, and maintenance of ion homeostasis and osmotic balance [19]. Under salt stress conditions, Munir et al. [20] observed that high K$^+$ levels improved the concentrations of antioxidant enzymes and the morphophysiological attributes of the plants. From this perspective, potassium fertilization could be an alternative means to maintain ion homeostasis, compensate for the negative charges of the macromolecules, maintain electroneutrality, and establish cell turgor and volume [21].

In this context, several studies have been carried out evaluating the effects of irrigation with saline water in the passion fruit crop, highlighting changes in gas exchange [5,22], photochemical efficiency [6], synthesis of photosynthetic pigments [7], and membrane instability [8], as well as a reduction in production components [9,10] and post-harvest quality [11]. However, these studies are restricted to the use of water with different levels of electrical conductivity throughout the plant development cycle, thus requiring further research, especially to assess the concentrations of NPK, physiology, and production of *Passiflora edulis* Sims, cultivar BRS GA1, grown under salt stress in the phenological stages and fertilization with K.

The hypothesis of this study is that the sensitivity and/or tolerance of passion fruit to salt stress varies with the development stage of the crop and that the intensity of salt stress effects on NPK accumulation, physiology and production can be attenuated by potassium fertilization through its function in osmoregulation, enzymatic activation, and oxidative protection. Thus, it is imperative to conduct new studies aimed at identifying the stage of the development cycle of BRS GA1 sour passion fruit in which it is sensitive and/or tolerant to water salinity, as well as a dose of potassium capable of mitigating the intensity of salt stress as an alternative for production in areas with qualitative and quantitative scarcity of water resources, a situation found in semi-arid areas of northeastern Brazil.

In this scenario, the aim of this study was to evaluate the NPK concentrations, physiology, and production of *Passiflora edulis* Sims under salt stress in the phenological stages and fertilization with K.

2. Results

The ISBW × KD significantly affected the leaf concentrations of nitrogen and potassium of *Passiflora edulis* Sims in the first and second cycle. The brackish water irrigation strategies significantly influenced all studied variables in the first cycle. In the second cycle, the ISBW had a significant effect on all variables except for the leaf P concentration. The potassium levels, in turn, significantly interfered with the nitrogen and potassium concentrations in the first cycle; in the second cycle, the KD significantly affected the K concentrations in leaves (Table 1).

Table 1. Summary of the analysis of variance referring to the leaf concentrations of nitrogen (N), phosphorus (P), and potassium (K) in *Passiflora edulis* Sims under brackish water irrigation strategies (ISBW) and potassium fertilization (KD).

	Mean Squares					
	ISBW	**KD**	**ISBW × KD**	**Blocks**	**Residual**	**CV (%)**
			1st cycle			
N	81.35 **	22.77 *	41.48 *	6.22 ^{ns}	6.30	13.67
P	0.42 *	0.34 ^{ns}	0.25 ^{ns}	0.49 *	0.16	18.26
K	112.46 **	100.31 **	10.75 *	9.00 ^{ns}	4.55	15.35
			2nd cycle			
N	26.14 **	3.34 ^{ns}	15.26 *	1.65 ^{ns}	4.09	14.21
P	0.15 ^{ns}	0.16 ^{ns}	0.02 ^{ns}	0.07 ^{ns}	0.05	14.71
K	102.09 **	78.23 *	19.48 *	8.14 ^{ns}	5.10	24.77
DF	5	1	5	3	33	-

DF—degree of freedom; CV (%)—coefficient of variation; * significant at 0.05 probability level; ** significant at 0.01 probability level; ^{ns} not significant.

In the first cycle (Figure 1A), the leaf nitrogen concentrations in plants subjected to the T3 and T5 strategies were higher than those found in the other strategies (T1, T2, and T4) when receiving 345 g of K_2O. On the other hand, fertilization with 207 g of K_2O resulted in the lowest N concentrations in plants cultivated under the T4 and T6 strategies compared to those subjected to salt stress in T2 and T5 strategies. When decomposing the effect of the potassium levels in each strategy, the N concentrations of plants subjected to the ECw of 4.0 dS m^{-1} in the vegetative stage and fertilization with 207 g of K_2O were superior to those of plants that received 345 g K_2O.

Despite the increase in the N concentrations, especially in plants fertilized with 345 g K_2O and irrigated with high-salinity water in T3, T5, and T6, the values obtained (24.07 and 19.21 g kg^{-1}) are considered insufficient for the adequate nutrition of *Passiflora edulis* Sims, which should range from 40 to 50 g kg^{-1} [23]. Carvalho [24] evaluated the effects of nitrogen fertilization, irrigation, and sampling time on the leaf nutrient concentrations of *Passiflora edulis f. flavicarpa Deg.* and observed that the leaf N concentrations were below the range considered adequate for the crop during the flowering and fruiting peaks. However, the plants did not show any nutritional deficiency symptoms.

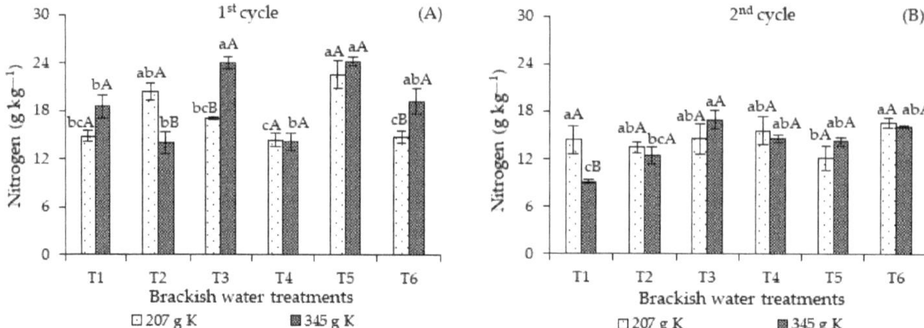

Figure 1. Leaf nitrogen concentrations of *Passiflora edulis* Sims as a function of the interaction ISBW × KD in the first (**A**) and second (**B**) cycle. Standard error of the mean (n = 4); Means followed by different lowercase letters indicate a significant difference between the irrigation strategies with brackish water for the same KD ($p \leq 0.05$), and different uppercase letters indicate a significant difference between potassium levels for the same irrigation strategy. T1—irrigation with low-salinity water during the entire cultivation cycle; salt stress in the following stages: T2—vegetative; T3—flowering; T4—fruiting; T5—vegetative and flowering; T6—vegetative and fruiting.

For the second cycle (Figure 1B), plants under salt stress in T3, T4, T5, and T6 and fertilized with 345 g K_2O obtained higher N concentrations than plants that received T1 during the entire cultivation cycle and high-salinity water in T2. On the other hand, treatments T2 and T4 and fertilization with 207 g K_2O resulted in a higher N concentrations, differing significantly only from the T5 strategy. When analyzing the effects of the potassium levels in each irrigation strategy, the N concentrations of plants fertilized with 207 g K_2O differed significantly only when irrigated with the lowest ECw water (T1) during the entire cycle. There were no significant differences in the remaining irrigation strategies regardless of the potassium level. When comparing the two cycles, the leaf N concentrations of the plants decreased markedly in the second cycle compared to the first.

Concerning the effects of salt stress on the phosphorus concentrations of the first cycle, plants irrigated with high ECw successively in the vegetative and flowering stages differed significantly only from those cultivated under salt stress in the vegetative stage (Figure 2). There were no significant differences when comparing the concentrations of plants subjected strategies T1, T3, T4, T5, and T6.

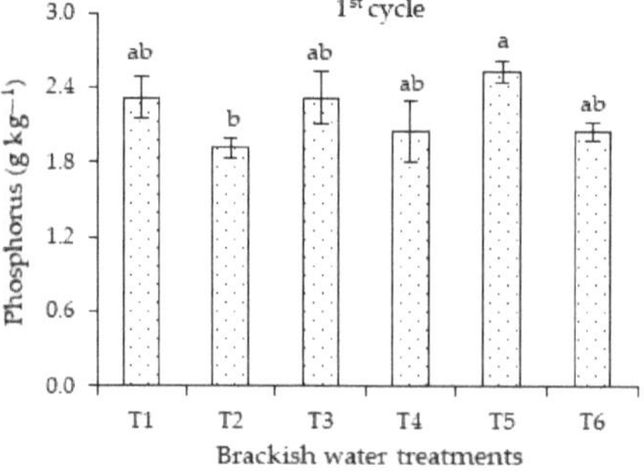

Figure 2. Leaf concentrations of phosphorus of *Passiflora edulis* Sims as a function of salt stress, varying the phenological stages in the first cycle. Standard error of the mean (n = 4); Means followed by different letters indicate a significant difference between treatments. T1—irrigation with low-salinity water during the entire cultivation cycle; salt stress in the following stages: T2—vegetative; T3—flowering; T4—fruiting; T5—vegetative and flowering; T6—vegetative and fruiting.

The potassium concentrations of sour passion fruit plants fertilized with 207 g of K_2O and continuously irrigated with high-salinity water in T5 decreased significantly compared to those found in plants subjected to the other irrigation strategies (T1, T2, T3, T4, and T6) in the first cultivation cycle (Figure 3A). Also, there were no significant differences when comparing the control treatment (T1) with the T1, T3, T4, and T6 strategies. Furthermore, fertilization with 345 g of K_2O promoted the highest leaf concentrations of potassium in plants subjected to the T1, T4, and T6 strategies. There were significant differences in the potassium concentrations of plants subjected to T2, T3, and T5 compared to the other strategies, T1, T4, and T6. When analyzing KD in each strategy, a clear superiority of the leaf K concentrations in plants that received 345 g K_2O in T1, T4, and T6 was observed.

Regardless of the irrigation strategy and the cultivation cycles, the leaf concentrations of N, P, and K obtained in this study are below the range considered adequate [23]. However, it should be noted that fertilization management followed the recommendations of [25] and the plants during the crop cycle did not show any deficiency symptoms. Soares et al. [26] studied the effects of salinity on the early development of sour passion fruit and also observed that potassium uptake was reduced even at low salinity. According to these authors, this reduction occurred as a function of the loss in the selective absorption capacity of ions by the plasmalemma.

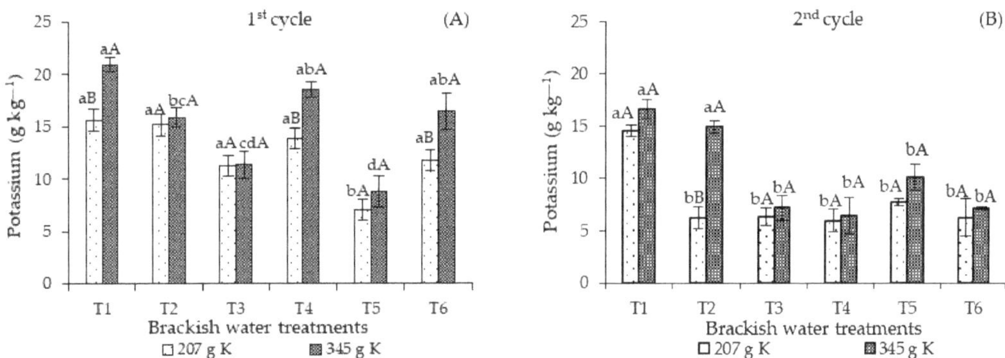

Figure 3. Leaf K concentrations of *Passiflora edulis* Sims as a function of the interaction ISBW × KD in the first (**A**) and second (**B**) cycle. Standard error of the mean (n = 4); Means followed by different lowercase letters indicate a significant difference between irrigation strategies with brackish water for the same KD ($p \leq 0.05$), and different uppercase letters indicate a significant difference between potassium levels for the same irrigation strategy. T1—irrigation with low-salinity water during the entire cultivation cycle; salt stress in the following stages: T2—vegetative; T3—flowering; T4—fruiting; T5—vegetative and flowering; T6—vegetative and fruiting.

The ISBW × KD interaction also influenced the leaf K concentrations in the second cycle (Figure 3B). Fertilization with 207 g of K_2O promoted the highest leaf K concentrations in plants subjected to T1 during the entire cultivation cycle, differing significantly from the other strategies. On the other hand, fertilization with 345 g K_2O resulted in the highest K concentrations in plants cultivated under the T1 strategy and salt stress in the vegetative stage (T2), surpassing the values observed in the other irrigation strategies with brackish water. When analyzing the effects of potassium levels in each irrigation strategy (Figure 4B), significant differences were observed only in the K concentrations of T2, with plants fertilized with 345 g K_2O attaining the highest values.

There was a significant effect of the irrigation strategies with brackish water on all variables measured in the first and second cycle (Table 2). However, the KD and the ISBW × KD interaction did not significantly influence any of the variables measured in the first and second cycle.

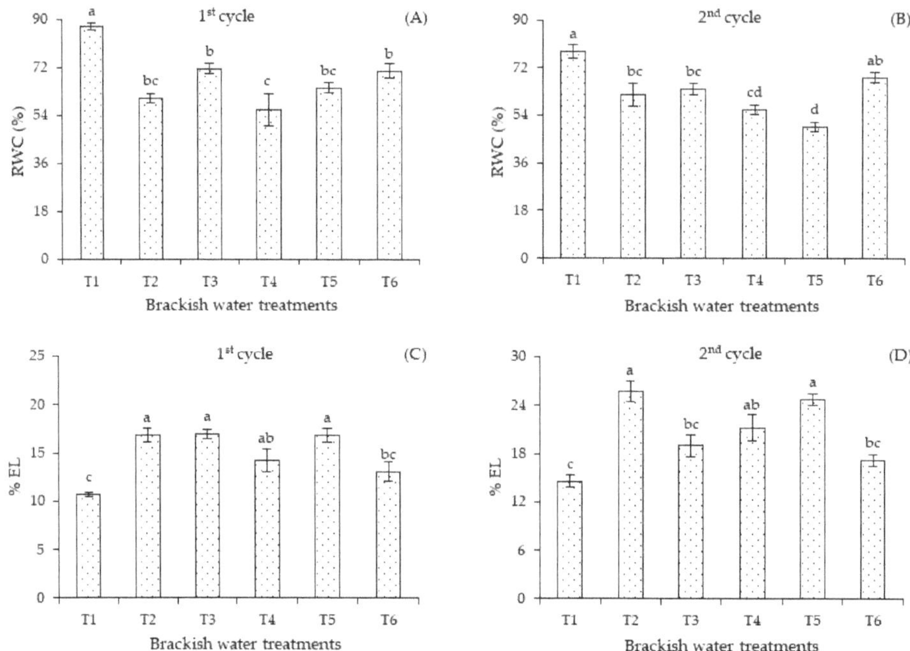

Figure 4. Relative water content—RWC (**A**,**B**) and electrolyte leakage—%EL (**C**,**D**) in *Passiflora edulis* Sims as a function of salt stress, varying the phenological stages in the first (240 DAT) and second (445 DAT) cycles. Standard error of the mean ($n = 8$); Means followed by different letters indicate a significant difference between treatments. T1—irrigation with low-salinity water during the entire cultivation cycle; salt stress in the following stages: T2—vegetative; T3—flowering; T4—fruiting; T5—vegetative and flowering; T6—vegetative and fruiting.

Table 2. Analysis of variance for relative water content (RWC), electrolyte leakage (%EL), stomatal conductance (*gs*), intercellular concentration of carbon dioxide (*Ci*), photosynthesis (*A*), and production per plant (PROD) in *Passiflora edulis* Sims under brackish water irrigation strategies (ISBW) (ISBW) and potassium fertilization (KD).

	Mean Squares					
	ISBW	**KD**	**ISBW × KD**	**Blocks**	**Residual**	**CV (%)**
	1st cycle					
RWC	969.94 **	91.52 ns	52.08 ns	22.16 ns	82.43	13.29
%EL	52.92 **	12.56 ns	5.74 ns	2.77 ns	4.73	14.72
gs	0.14 **	0.01 ns	0.004 ns	0.004 ns	0.01	32.76
Ci	5944.07 *	0.02 ns	3059.97 ns	213.57 ns	1787.48	22.04
A	418.15 **	9.62 ns	21.38 ns	5.10 ns	12.71	20.46
PROD	1,989,844.2 **	13,612,052.5 ns	5,063,282.6 ns	1,225,254.3 ns	2866,438.2	25.63
	2nd cycle					
RWC	764.30 **	186.48 ns	18.42 ns	69.36 ns	48.55	11.06
%EL	149.28 **	14.64	18.00	12.67	8.97	14.67
gs	0.10 **	0.004 ns	0.005 ns	0.01 ns	0.006	28.91
Ci	3213.20 *	74.32 ns	135.61 ns	1161.56 ns	763.06 ns	12.15
A	177.71 **	2.55 ns	8.27 ns	15.46 ns	4.41 ns	16.26
PROD	1,989,844.2 **	13,612,052.5 ns	5,063,282.6 ns	1,225,254.3 ns	2,866,438.2	27.48
DF	5	1	5	3	33	-

DF—degree of freedom; CV (%)—coefficient of variation; * significant at 0.05 probability level; ** significant at 0.01 probability level; ns not significant.

In the first cycle, the relative water content (Figure 4A) of plants subjected to T1 surpassed the values of plants that received high-salinity water in the other irrigation strategies (T2, T3, T4, T5, and T6). When comparing the RWC of plants subjected to water stress in the different phenological stages, significant differences were observed between T3 and T6 compared to T4. However, there were no significant differences in the RWC of plants subjected to salt stress in T2, T3, T5, and T6.

In the second cycle, the relative water content (Figure 4B) of plants cultivated under T1 differed significantly from that of plants that received salt stress in the T2, T3, T4, and T5 strategies. Plants subjected to continuous salt stress in T5 and T4 obtained the lowest RWC values, possibly due to water uptake restrictions.

Electrolyte leakage of plants subjected to salt stress in T2, T3, and T5 differed significantly compared to those grown under the T1 and T6 strategies in the first cultivation cycle (Figure 4C). No significant effect was observed when comparing the T2, T3, T4, and T6 strategies. In the second cultivation cycle (Figure 4D), irrigation with 4.0 dS m^{-1} promoted the highest electrolyte leakage of plants under the T2 and T5 strategies, differing significantly from those that received T1, T3, and T6. Similar to the first cycle (Figure 4A), irrigation with ECw of 1.3 dS m^{-1} during the entire cycle (T1) and the ECw of 4.0 dS m^{-1} in T6 resulted in a lower %EL. The reduction in %EL in the T6 strategy indicates the recovery potential of plants subjected to salt stress since, from the end of the vegetative stage to the beginning of the fruiting stage, the plants were irrigated for 19 days with the ECw of 1.3 dS m^{-1}. Lima et al. [27] studied *Passiflora edulis* Sims plants irrigated with different cationic natures and observed that the water with the ECw of 3.0 dS m^{-1} containing Na$^+$ and Na$^+$ + Ca^{2+} increased electrolyte leakage in the leaf tissues.

Stomatal conductance was negatively affected by irrigation with water of 4.0 dS m^{-1} in the two cultivation cycles, regardless of the plant development stage (Figure 5). In the first cycle (Figure 5A), the *gs* was superior in plants under T1 compared to plants subjected to salt stress, regardless of the irrigation strategy. In the second cycle (Figure 5B), plants under T1 also had the highest *gs*, differing significantly from those subjected to T2, T3, T4, T5, and T6. When comparing the stomatal conductance of plants irrigated with 4.0 dS m^{-1}, it was observed that the salt stress applied in T5 resulted in the lowest value compared to the T3, T4, and T6 strategies.

Intercellular concentration of carbon dioxide (Figure 5C) under the treatment T1 in the first cycle did not differ significantly from the values found in T3, T4, T5, and T6. There were significant differences only between plants subjected to the T1 and T2 strategies. In the second cycle (Figure 5D), plants grown with 4.0 dS m^{-1} in T2, T3, and T5 obtained higher intercellular concentrations of carbon dioxide compared to treatment T1. However, there were no significant differences in the *Ci* of plants grown under T4 and T5 compared to the T1 strategy.

The photosynthesis of the plants was influenced by the irrigation strategies with brackish water (Figure 5E,F). In the first cycle (Figure 5E), irrigation with the treatment T1 promoted the highest photosynthesis compared to plants grown under other strategies. Continuous irrigation with 4.0 dS m^{-1} in the vegetative and flowering stages resulted in the lowest photosynthesis in both cycles.

In the second cycle (Figure 5F), the photosynthesis of plants under the treatment T5 did not differ significantly from the values of those receiving T1. This situation indicates a possible recovery of plants that received T2 and T4. The application of treatments T2 and T5 compromised the photosynthesis. In plants cultivated under irrigation with water of high electrical conductivity, the stomata usually close and therefore limit the entry of CO_2 into the substomatal chamber.

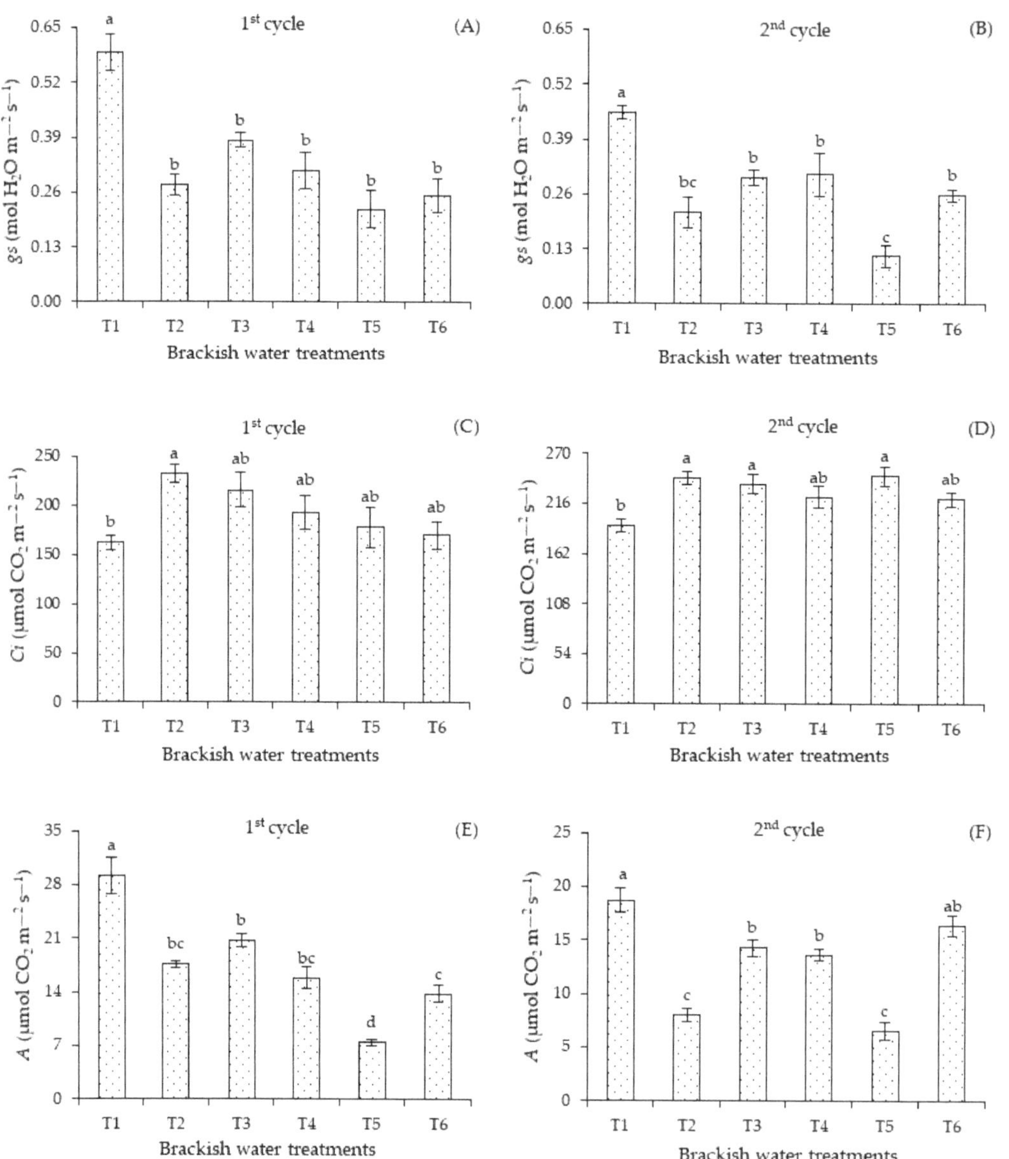

Figure 5. Stomatal conductance—gs (**A,B**), intercellular concentration of carbon dioxide—Ci (**C,D**), and photosynthesis—A (**E,F**) of *Passiflora edulis* Sims as a function of salt stress, varying the phenological stages in the first (240 DAT) and second (445 DAT) cycle. Standard error of the mean (n = 8); Means followed by different letters indicate a significant difference between treatments. T1—irrigation with low-salinity water during the entire cultivation cycle; salt stress in the following stages: T2—vegetative; T3—flowering; T4—fruiting; T5—vegetative and flowering; T6—vegetative and fruiting.

The irrigation management strategies with brackish water influenced the production per plant of sour passion fruit. In the first cycle (Figure 6A), plants subjected to the T3 and T5 strategies attained the lowest PROD values (5.54 and 4.16 kg per plant). In the second cycle (Figure 6B), plants under strategies T2, T3, T5, and T6 (4.32, 4.72, 3.28, and 4.37 kg per plant) did not differ significantly among themselves and showed the lowest PROD values, on average 34.63% lower than that obtained with strategies T1 and T4 (7.00 and 5.75 kg per plant), which were statistically equal. The lowest PROD values observed in plants under strategy T5 reflect the changes observed in the gas exchange parameters, especially stomatal conductance, photosynthesis, and ion homeostasis. Another reason for the reduced production is the decrease in N, P, and K accumulations, which were considered insufficient for the nutrition of sour passion fruit [23], regardless of the irrigation strategy and the cultivation cycle.

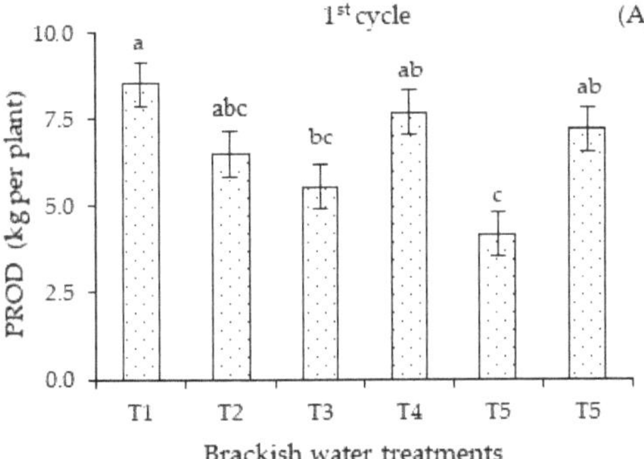
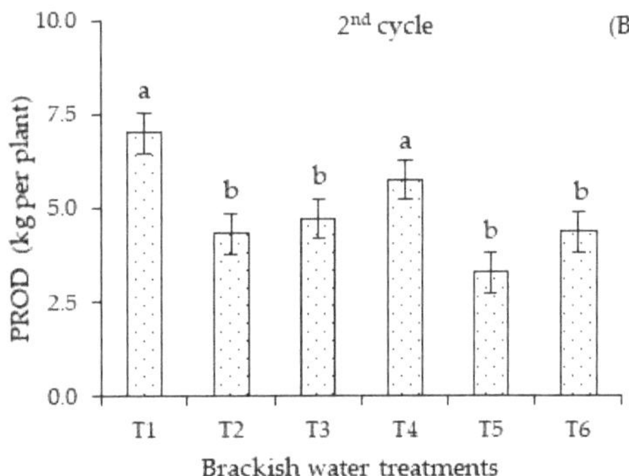

Figure 6. Production per plant—PROD of *Passiflora edulis* Sims as a function of salt stress, varying the phenological stages in the first (**A**) and second (**B**) cycles. Standard error of the mean ($n = 8$); Means followed by different letters indicate a significant difference between treatments. T1—irrigation with low-salinity water during the entire cultivation cycle; salt stress in the following stages: T2—vegetative; T3—flowering; T4—fruiting; T5—vegetative and flowering; T6—vegetative and fruiting.

The reduction in production per plant (Figure 6A,B) in the first cycle under strategies T3 and T5 and in the second cycle under T2, T3, T5, and T6 indicates that *Passiflora edulis* Sims is sensitive to salt stress during the vegetative and flowering stages. Plant sensitivity to salinity normally varies with the development stages.

In this research, the highest values of polar diameter of fruits in the first cycle were obtained in plants under the T1, T4, and T6 strategies, but did not differ significantly from those observed in plants grown with 4.0 dS m^{-1} water in T2 and T3. On the other hand, irrigation with high-salinity water in T5 resulted in lower values of fruit polar diameter compared to those found under T1, T4, and T6 strategies. In the second cycle, the fruits produced under the T2 and T5 strategies had lower polar diameter values compared to those of plants irrigated with ECw of 1.3 dS m^{-1} during the entire cycle. Regarding the equatorial diameter, passion fruit plants subjected to the T1, T2, T4, and T6 strategies had the highest values in the first cycle, while irrigation with ECw of 4.0 dS m^{-1} in T3 and T5 resulted in fruits with smaller equatorial diameter. In the second cycle, there was no significant effect of irrigation strategies with brackish water and K doses on the equatorial diameter of the fruits.

Regarding the chemical characteristics of the fruits, as highlighted by Lima et al. [3], irrigation with water of 4.0 dS m^{-1} in T5 and fertilization with 345 g K$_2$O per plant per year increases the total titratable acidity and reduces the hydrogen potential and soluble solids in the passion fruit pulp, regardless of the adopted strategy. On the other hand, fertilization with 345 g K$_2$O and irrigation with water of 4.0 dS m^{-1} in T3 increases the flavonoid contents and the ratio of soluble solids to total titratable acidity, while the salt stress in the fruiting stage promotes an increase in anthocyanin contents in the passion fruit pulp.

3. Discussion

In the semi-arid region of northeastern Brazil, the salt stress caused by the high concentrations of salts found in surface and subsurface water sources negatively interferes with plant growth and development. Salt stress induces osmotic, ionic, and oxidative stresses and changes in gene expression, hindering normal physiological processes and nutrient absorption and hence reducing plant production [28]. Thus, irrigation with high-salinity water at varying phenological stages is an alternative means to minimize the deleterious effects on plants through the maintenance of ionic homeostasis and modulation of physiology and through the reduction in negative impacts and physical-chemical attributes of the soil [29]. In addition, for performing osmoregulatory function, potassium also contributes to the ionic and osmotic homeostasis of plants.

Plants fertilized with 345 g K_2O had higher N concentrations than those that received 207 g K_2O under irrigation with 4.0 dS m^{-1} in the T3 and T6 strategies. Potassium plays a key role in the nitrogen metabolism of plants and can increase the synthesis of amino acids and proteins. Furthermore, this macronutrient contributes to the balance of cations and anions in the cytoplasm [30]. According to [31], plants that accumulate higher K contents tend to restrict the absorption and transport of toxic ions (Na^+ and Cl^-) from irrigation water.

This reduction in the second cycle could be related to the time of exposure of the plants to salt stress, which is considered one of the main factors responsible for the low N availability in the soil [32]. The results obtained for the N concentrations are lower than those found by [33], who evaluated the macronutrient concentrations of *Passiflora edulis f. flavicarpa Deg.* at the beginning of flowering under irrigation with saline water in soils with and without mineral fertilization and obtained the mean value of 50.81 g kg^{-1}.

The increase in the leaf K concentrations is an important mechanism of plants under salt stress since this macronutrient plays a relevant role in stomatal regulation, especially under restricted conditions of water, and ensures the turgidity of guard cells through a reduction in the osmotic potential [34].

The phosphorus concentrations obtained in this study (1.91 to 2.53 g kg^{-1}) are considered inadequate to meet the nutrient requirement of plants, i.e., they are below the range indicated by [23] (between 4 and 5 g kg^{-1}). Carvalho et al. [35], in a study on the effects of potassium fertilization, irrigation depths, and times of the year on the concentrations of macronutrients, micronutrients, and Na^+ of *Passiflora edulis f. flavicarpa*, also observed that the leaf concentrations of P and K in all seasons were below the range considered adequate for the cycle.

The higher reduction in the potassium concentrations of the second cultivation cycle could be related to the higher Na^+ concentration compared to K^+ in the root zone, with Na^+ inhibiting the K^+ uptake by plants due to the competition between Na^+ and K^+ at the uptake sites of the cell membranes. Sodium can replace potassium at the binding sites and, as a result, inhibit regular plant metabolism [13].

The leaf contents of N, P, and K, regardless of the phenological stage and the cultivation cycle, obtained in this study are below the range considered adequate by Malavolta et al. [23]. However, it is important to highlight that fertilization management was recommended by [25] and the plants at no stage of their cycle showed symptoms of nutritional deficiency. Soares et al. [26], in a study evaluating the effects of salinity on the early development of passion fruit, also found that potassium absorption was drastically reduced, even at low salinity levels. According to these authors, this reduction occurred due to the loss of capacity for selective ion absorption by the plasmalemma.

The reduction in the water content in plants subjected to salt stress (T1, T2, T3, T4, T5, and T6) is a result of the negative effect of osmotic stress on water availability in the soil and consequently reduced water uptake by the plants, which affects their general water status [36]. The reduction in the relative water content in the leaf blade in the different phenological stages influenced the gas exchange, limiting stomatal conductance and CO_2

assimilation rate. Usually stomatal closure is a strategy to minimize water losses to the atmosphere and reduce excessive absorption of toxic ions (Na^+ and Cl^-).

Although the increase in electrolyte leakage occurred in plants under high salinity (4.0 dS m^{-1}) compared to those that received ECw of 1.3 dS m^{-1}, the values obtained were less than 50%, which is an indication that there was no significant damage to the cell membrane of the sour passion fruit, as the tissue is considered injured when the percentage of electrolyte leakage exceeds 50% [37].

Under salt stress conditions, plants normally lose water from their tissues, which can have rapid and significant effects on cell expansion and division, stomatal opening, and accumulation of abscisic acid [38]. On the other hand, salt stress induces the formation of reactive oxygen species (ROS), which cause oxidative damage and lipid peroxidation of the membrane, decreasing membrane fluidity and selectivity [4]. Thus, with the overproduction of ROS in the cells, the stability or integrity of the membrane is interrupted, resulting in electrolyte leakage in plants under stress conditions [36].

The partial stomatal closure in plants cultivated under salt stress decreases CO_2 absorption and affects the functions of the photosynthetic apparatus [4]. The control of stomatal opening and closure is one of the adaptative mechanisms used to avoid the loss of cell turgor due to the limited water supply [39]. As observed in this study, Lima et al. [40] found that irrigation with water of 3.2 dS m^{-1} limited the stomatal conductance of plants subjected to salt stress in the vegetative, vegetative/flowering, flowering, fruiting, and vegetative/fruiting stages. According to these authors, the lower values of stomatal conductance in plants under salt stress occur due to reductions in leaf turgor and atmospheric vapor pressure, which are ways to reduce the delay between water absorption by roots and transpiration, consequently leading to partial closure of the stomata to avoid excessive dehydration of the guard cells.

The increase in the Ci of plants grown under salt stress indicates damage to the photosynthetic apparatus, e.g., decrease in the carboxylation efficiency of RuBisCO, caused mainly by salt accumulation in the leaf tissues [41].

The increase in the intercellular concentration of carbon dioxide in plants subjected to T5 did not influence photosynthesis, suggesting that this process was inhibited by the action of factors of non-stomatal origin. The non-stomatal regulation induced by salt stress is related to the activities of the photosynthetic enzymes of the Calvin cycle, disturbances in chlorophyll synthesis, and damage to the photosynthetic apparatus [42].

The reduction in the concentration of CO_2 in the substomatal spaces is reflected in the rate of CO_2 assimilation. Another factor contributing to the decrease in CO_2 assimilation is the inhibition of the RuBisCO enzyme activity, which prevents the conversion of absorbed CO_2 into photoassimilates. In many cycles, notably, salt stress interferes with traits of gas exchange under moderate to severe levels of salinity, thus reducing stomatal conductance and the rate of CO_2 assimilation due to the limitations that occur in the total water potential caused by excess salts [43].

However, most plants, especially those of economic importance, are more sensitive to salinity during the early phenological stages [16], which may have contributed to the observed lower production per plant. Further, the higher energy expenditure to maintain the metabolic activities under salt stress conditions may have caused the formation of fruits with lower mass [44], thus justifying the reduced PROD in plants under these strategies. The reduction in production mainly in plants subjected to salt stress in the flowering and flowering/fruiting stages stands out as a challenge for irrigated cultivation, since the water requirement in these stages is greater when compared to the vegetative stage; with this, more salts are incorporated into the soil, compromising the production of plants [45].

Unlike the results obtained in this study, Souto et al. [31] concluded that there was an increase in the photosynthesis of yellow passion fruit plants grafted on *P. cincinnata*, but this was not reflected in fruit yield. In the present research, the reduction in photosynthesis observed in plants subjected to salt stress in T5 in both crop cycles decreased production per plant. Soares et al. [16], in a study with cotton genotypes, observed that the successive application of saline water in the flowering and boll formation stages caused a drastic reduction in the physiological aspects of the cycle, with recovery of the plants after the interruption of stress.

In general, it is possible to observe that sour passion fruit showed contrasting responses in NPK accumulation, physiology, and production as a function of the different phenological stages and production cycles, proving that the sensitivity and/or tolerance of plants depends on several factors such as intensity and duration of stress, irrigation and fertilization management practices, and soil and climatic conditions of the region.

4. Materials and Methods

4.1. Characterization of Study

The research was carried out in the field at the Center of Agricultural Sciences and Technology (CCTA), in Pombal, PB, located at 06°48′50″ S, 37°56′31″ W, at an altitude of 190 m.

4.2. Details of Sources of Variation and Period of Application of Treatments

The experiment consisted of six irrigation strategies with brackish water—ISBW (T1—irrigation with low-salinity water during the entire cultivation cycle as a control; irrigation with high-salinity water at the respective stages: T2—vegetative; T3—flowering; T4—fruiting; at the successive stages, vegetative and flowering—T5; vegetative and fruiting—T6 and two potassium levels corresponding to 207 and 345 g of K_2O per plant per year of the potassium recommendation [25], distributed in randomized blocks in a 6 × 2 factorial arrangement with four replications, with each plot consisting of three usable plants. The experimental layout is shown in Figure 7. The area had a row of plants externally to the usable plots.

Figure 7. Layout of the experimental area.

Two salinity levels of irrigation water were used in the experiment, one corresponding to moderate salinity (1.3 dS m^{-1}—T1) and the other to high electrical conductivity (4.0 dS m^{-1}—T2, T3, T4, T5, and T6), applied during the following cycle development stages in the first cycle: T1—irrigation with low-salinity water during the entire cultivation cycle (1–253 days after transplanting—DAT); irrigation with high-salinity water in T2—from transplanting until the formation of the floral primordia (50–113 DAT); T3—from the formation of the floral primordia to the total development of the flower bud (anthesis) (114–198 DAT); T4—from the fecundation of the flower bud to the formation of fruits with interspersed yellow spots (199–253 DAT); T5—in the vegetative and flowering stages

(50–198 DAT); T6—in the vegetative (50–113 DAT) and fruiting stages (199–253 DAT). The procedure was similar in the second cultivation cycle, T1—irrigation with low-salinity water during the entire cultivation cycle (254–475 DAT); salt stress in T2 (254–340 DAT); T3—flowering (341–360 DAT); T4—fruiting (361–475 DAT); T5—vegetative and flowering stages (254–360 DAT); T6—vegetative (254–340 DAT) and fruiting stages (361–475 DAT).

4.3. Crop Management

In this research, the BRS GA1 genotype was used, as it has a great potential for exploitation in Brazilian orchards due to its high yield (42 t ha^{-1}) and tolerance to Anthracnose. The seedlings were produced by sowing two seeds in plastic bags filled with a substrate composed of soil, sand, and cattle manure in the ratio of 84:15:1 (v/v). From the moment the plants started to produce tendrils, transplanting to the field was carried out (61 DAS).

Before transplanting the seedlings, plowing and harrowing were carried out in the area. The sour passion fruit was cultivated in an Entisol, with the following physical and chemical properties determined according to [46]: Hydrogen potential (1:2.5 soil/water) = 7.82, organic matter = 0.81 dag kg^{-1}, P = 10.60 mg kg^{-1}, P = 10.60 mg kg^{-1}; K$^+$, Na$^+$, Ca^{2+}, Mg^{2+} and Al^{3+} + H$^+$ equivalent to 0.30, 0.81, 2.44, 1.81, and 0 cmol$_c$ kg^{-1}, respectively; Particle-size fraction: Sand, Silt and Clay = 820.90, 170.10, and 9.00 g kg^{-1}, respectively; Moisture content (dag kg^{-1}) at field capacity (33.42 kPa) and permanent wilting point (1519.5 kPa) = 12.87 and 5.29, respectively.

The sour passion fruit plant was cultivated in holes measuring 40 cm $\times$ 40 cm $\times$ 40 cm. In the basal fertilization, 20 L of bovine manure and 50 g of single superphosphate were used, following the recommendation of [25]. Nitrogen and potassium were applied in the top dressing, using urea and potassium chloride, respectively. Plants subjected to the dose of 345 g K$_2$O per plant received 65 g in the vegetative stage and 280 g during the flowering and fruiting stages. The other plants received 39 and 168 g K$_2$O per plant in the same stages of development.

Every 15 days after plant emergence the micronutrients were applied using 1 g L^{-1} of Dripsol Micro$^®$ (Candeias, Bahia, Brazil), containing 1.1, 0.85, 0.5, 3.4, 3.2, 0.05, and 4.2% of Mg^{2+}, B, Cu, Fe, Mn, Mo, and Zn, respectively. The applications were carried out through the leaves. The plants were cultivated with a spacing of 3 m $\times$ 3 m, adopting the vertical trellis system at a height of 1.80 m. A nylon ribbon was used to guide the plants. When plants exceeded the trellis height by 10 cm, the apical bud was cut to induce the formation of secondary branches. After the production of the secondary branches, two were conducted, one on each side up to a length of 1.5 m, and later were conducted up to 30 cm from the ground level. Tendrils and unwanted branches were eliminated throughout the crop cycle to promote the full development of the crop.

At 254 DAT the second production cycle began. At this time, the canopy was pruned [3]. This procedure reduces the problems caused by pests and diseases, improves the phytosanitary status of plants, and facilitates crop management, especially fertilization and irrigation. The pruning of tertiary and quaternary branches was performed at 40 cm from the wire.

The 1.3 dS m^{-1} water was obtained from a CCTA/UFCG well, with Ca^{2+}, Mg^{2+}, Na$^+$, K$^+$, HCO$_3{}^-$, CO$_3{}^{2-}$, and Cl$^-$ concentrations of 0.85, 0.40, 5.81, 0.40, 5.09, 0, and 4.07 mmol$_c$ L^{-1}, respectively, and hydrogen potential of 6.69. Water with electrical conductivity of 4.0 dS m^{-1} was prepared by adding NaCl, based on the relationship between ECw and salt concentration [47].

Brackish water irrigation management began at 50 DAT. A drip irrigation system was adopted, using 32-mm-diameter PVC tubes in the main line and 16-mm-diameter low-density polyethylene tubes in the lateral lines, with emitters operating at a flow rate of 10 L h^{-1}. Each plant had two pressure-compensating drippers (model GA 10 Grapa), each at 15 cm from the stem. The plants were irrigated daily at 7:00 a.m. by supplying water according to the strategy adopted. The irrigation depth was estimated based on the crop evapotranspiration [48], obtained using Equation (1):

$$ETc = ETo \times Kc \tag{1}$$

where:
ETc—crop evapotranspiration, mm day^{-1};
ETo—reference evapotranspiration of Penman-Monteith, mm day^{-1}; and
Kc—crop coefficient, dimensionless.

Climatic data obtained from the Meteorological Station were used to quantify the reference evapotranspiration (ETo) using the crop coefficients recommended by [49]. It is important to point out that irrigation with water of high and low electrical conductivity, varying the phenological stages, reduces salt stress in plants and impacts on the salt accumulation in soils. Additionally, at 30-day intervals, a leaching fraction of 0.15 was applied to reduce the salt content in the soil. The water depths applied in the different irrigation strategies are presented in Table 3.

Table 3. Water depth values applied to BRS GA1 yellow passion fruit in different strategies of irrigation with brackish water in the phenological stages.

ISBW	DAT	1st Cycle	
		Water Depth (mm)	
		1.3 dS m^{-1}	4.0 dS m^{-1}
T1	1–253	1256	0
T2	50–113	877	253
T3	114–198	847	273
T4	199–253	1079	119
T5	50–198	468	525
T6	50–113/199–253	699	371
	2nd cycle		
T1	254–475	1043	0
T2	254–340	628	277
T3	341–360	924	79
T4	361–475	533	340
T5	254–360	510	356
T6	254–340/361–475	119	616

ISBW—irrigation strategies with brackish water; T1—irrigation with low-salinity water during the entire cultivation cycle; salt stress in the following stages: T2—vegetative; T3—flowering; T4—fruiting; T5—vegetative and flowering; T6—vegetative and fruiting.

4.4. Variables Analyzed

The mineral composition (N, P, K) was evaluated by selecting the fourth leaf from the apex of the intermediate branches of each experimental unit according to [33]. The collection was performed during the transition from flowering to fruiting in the first (199 DAT) and second (361 DAT) production cycles. After drying, the samples were ground and subjected to chemical analyses according to the methodology of [23]. The P and K concentrations were determined by nitric acid digestion, whereas the N concentration was determined by sulfuric acid digestion.

The relative water content (RWC), electrolyte leakage (EL) in the leaf blade, and gas exchange were also determined at 199 DAT (1st cycle) and 361 DAT (2nd cycle). The relative water content (RWC) in the leaf blade was evaluated using 8 leaf discs with areas of 113 mm^2, collected from leaves located in the middle third of the secondary branches. Immediately after, the discs were weighed to obtain the fresh mass (FM). The samples were placed in plastic bags, immersed in 50 mL of distilled water, and stored for 24 h. After this period, the excess water was removed with paper towels, and the samples were weighed to obtain the turgid mass (TM). Subsequently, the samples were dried in the oven (temperature $\approx 65\,^{\circ}\text{C} \pm 3\,^{\circ}\text{C}$, until reaching constant weight) to obtain the dry mass (DM). The RWC was determined according to [50] using Equation (2):

$$\text{RWC} = \frac{(\text{FM} - \text{DM})}{(\text{TM} - \text{DM})} \times 100 \tag{2}$$

where:
RWC—relative water content (%);
FM—leaf fresh mass (g);
TM—leaf turgid mass (g); and
DM—leaf dry mass (g).

To determine electrolyte leakage in the leaf blade, leaves were collected from the middle third of the secondary branches; 8 leaf discs with areas of $113\,\text{mm}^2$ were immediately removed, washed with distilled water to eliminate other electrolytes adhered to the leaves, and then placed in a beaker with 50 mL of double-distilled water, which was closed with aluminum foil. The samples remained at a temperature of $25\,^{\circ}\text{C}$ for 24 h and then the initial electrical conductivity (Ci) was measured; subsequently, the beakers were taken to an oven with forced air circulation and subjected to a temperature of $80\,^{\circ}\text{C}$ for 150 min and then cooled to determine the final electrical conductivity (Cf). Electrolyte leakage in the leaf blade was obtained according to [51] using Equation (3):

$$\%\text{EL} = \frac{\text{Ci}}{\text{Cf}} \times 100 \tag{3}$$

where:
%EL—electrolyte leakage in the leaf blade;
Ci—initial electrical conductivity (dS m^{-1}); and
Cf—final electrical conductivity (dS m^{-1}).

Gas exchange was evaluated using an intermediate and intact leaf of the productive branch to determine stomatal conductance $(\text{gs-mol H}_2\text{O m}^{-2}\,\text{s}^{-1})$, photosynthesis (A) $(\mu\text{mol CO}_2\,\text{m}^{-2}\,\text{s}^{-1})$, and intercellular concentration of carbon dioxide (Ci) $(\mu\text{mol CO}_2\,\text{m}^{-2}\,\text{s}^{-1})$ using the portable photosynthesis meter "LCPro+" from ADC BioScientific Ltd. (Hoddesdon, England). The readings were performed from 7:00 to 10:00 a.m. using the third fully expanded leaf counting from the apical bud under natural conditions of air temperature, CO_2 concentration, and using an artificial radiation source established through the photosynthetic light response curve and determining the photosynthetic light saturation point [52].

Mature fruits (with a yellow peel color) were harvested from 199 to 253 DAT in the first cycle and from 361 to 445 DAT in the second cycle. The first cycle comprised the period from transplanting to the end of the harvest (1–253 DAT). At the end of harvest, a cleaning pruning was carried out to renew productive branches, eliminating dead, old, diseased and/or unproductive branches, reducing the problems caused by pests and diseases and improving the phytosanitary state of the plants. The second cycle began after pruning, keeping the same plants used in the first production cycle (254–475 DAT). The cultivation of passion fruit in the second production cycle had the purpose of validating the changes that occurred in the first cycle in the accumulation of NPK, in physiology, and production, considering that the responses of plants to stress vary according to the time of year and the irrigation management, fertilization, and climatic conditions. After harvest, the production per plant (PROD, kg per plant) was determined and fruits harvested per treatment were counted, allowing the calculation of average fruit weight.

4.5. Statistical Procedures

The data obtained were evaluated through analysis of variance by the F test after the data normality test (Shapiro-Wilk). Then, an analysis of variance was performed, with the Tukey test $(p \leq 0.05)$ used for irrigation strategies and K doses. These analyses were performed using the statistical software SISVAR ESAL version 5.7 [53].

5. Conclusions

Relative water content, stomatal conductance, and photosynthesis were reduced by salt stress during the first cultivation cycle. However, in the second cycle, irrigation with 4.0 dS m^{-1} water in the fruiting stage and successively in the vegetative and fruiting stages resulted in the highest production per plant and in elevation of photosynthesis, respectively. In the present study, the concentrations of N, P, and K in passion fruit leaves are below the optimal levels established in the literature, regardless of the development stage and crop cycle. Changes in ionic homeostasis by inhibition of NPK absorption caused a reduction in the production of passion fruit under the semi-arid conditions of Northeast Brazil, leading to a yield lower than the production potential of the BRS GA1 cultivar. Despite the importance of potassium in osmoregulation and its role in various physiological processes, such as stomatal regulation, photosynthetic fixation of CO_2, and transport and use of photoassimilates, in the present study, no significant effects of potassium doses were observed on the relative water content and electrolyte leakage in the leaf blade as well as in the gas exchange of sour passion fruit in both cultivation cycles. The hypothesis formulated was confirmed only for potassium concentrations, as fertilization with the recommended dose of 100% increased the levels of K in plants cultivated under irrigation water salinity of 1.3 dS m^{-1} throughout the growth cycle and under water salinity of 4.0 dS m^{-1} in the fruiting and vegetative/fruiting stages in the first cycle and in the vegetative stage in the second cycle. To expand knowledge about *Passiflora edulis* Sims under salt stress, future studies should elucidate the effects of salinity on the activity of antioxidant enzymes and the accumulation of organic solutes and their role in osmotic adjustment, in addition to ionic homeostasis through the determination of levels of macro and micronutrients and the accumulation of sodium and chloride in passion fruit plants.

Author Contributions: Conceptualization, R.A.F.T. and G.S.d.L.; methodology, A.A.R.d.S. and R.A.F.T.; software, A.A.R.d.S.; validation, R.A.F.T., F.A.d.S., G.S.d.L. and L.A.d.A.S.; formal analysis, R.G.N.; L.A.d.A.S. and V.L.A.d.L.; investigation, G.S.d.L. and R.A.F.T.; resources, G.S.d.L. and L.A.d.A.S.; data collection, A.A.R.d.S.; writing—original draft preparation, G.S.d.L.; writing—review and editing, A.A.R.d.S., G.S.d.L., C.A.V.d.A., V.L.A.d.L., K.P.L., L.H.G.C. and H.R.G.; supervision, H.R.G.; project administration, G.S.d.L. and L.A.d.A.S. All authors have read and agreed to the published version of the manuscript.

Funding: National Council for Scientific and Technological Development—CNPq (Proc. 429732/2018-0)), Coordination for the Improvement of Higher Education Personnel—CAPES (financial code—001), and Universidade Federal de Campina Grande.

Acknowledgments: The authors thank the Graduate Program in Agricultural Engineering of the Federal University of Campina Grande, the National Council for Scientific and Technological Development (CNPq), and the Coordination for the Improvement of Higher Education Personnel (CAPES) for the support in carrying out this research.

Conflicts of Interest: The authors declare no conflict of interest.

References

1. Santos, V.A.; Ramos, J.D.; Laredo, R.R.; dos Silva, F.O.R.; Chagas, E.A.; Pasqual, M. Produção e qualidade de frutos de maracujazeiro-amarelo provenientes do cultivo com mudas em diferentes idades. *Rev. Ciênc. Agroveter.* **2017**, *16*, 33–40. [CrossRef]
2. Nobre, R.G.; de Lima, G.S.; Gheyi, H.R.; de Medeiros, E.P.; dos Soares, L.A.A.; Alves, A.N. Teor de óleo e produtividade da mamoneira de acordo com a adubação nitrogenada e irrigação com água salina. *Pesq. Agropec. Bras.* **2012**, *47*, 991–999. [CrossRef]
3. De Lima, G.S.; Pinheiro, F.W.A.; Gheyi, H.R.; dos Soares, L.A.A.; Soares, M.D.M.; da Silva, F.A.; de Azevedo, C.A.V.; de Lima, V.L.A. Postharvest quality of sour passion fruit under irrigation strategies with brackish water and potassium application in two crop cycles. *Wat. Air Soil Poll.* **2022**, *233*, e452. [CrossRef]
4. Hannachi, S.; Steppe, K.; Eloudi, M.; Mechi, L.; Bahrini, I.; Labeke, M.-C.V. Salt stress induced changes in photosynthesis and metabolic profiles of one tolerant ('Bonica') and one sensitive ('Black Beauty') eggplant cultivars (*Solanum melongena* L.). *Plants* **2022**, *11*, 590. [CrossRef]
5. De Lima, G.S.; Fernandes, C.G.J.; dos Soares, L.A.A.; Gheyi, H.R.; Fernandes, P.D. Gas exchange, chloroplast pigments and growth of passion fruit cultivated with saline water and potassium fertilization. *Rev. Caatinga.* **2020**, *33*, 184–194. [CrossRef]

6. Andrade, E.M.G.; de Lima, G.S.; de Lima, V.L.A.; da Silva, S.S.; Dias, A.S.; Gheyi, H.R. Hydrogen peroxide as attenuator of salt stress effects on the physiology and biomass of yellow passion fruit. *Rev. Bras. Eng. Agríc. Ambient.* **2022**, *26*, 571–578. [CrossRef]

7. Ramos, J.G.; de Lima, V.L.A.; de Lima, G.S.; da Paiva, F.J.S.; de Pereira, M.O.; Nunes, K.G. Hydrogen peroxide as salt stress attenuator in sour passion fruit. *Rev. Caatinga* **2022**, *35*, 412–422. [CrossRef]

8. De Lima, G.S.; das Soares, M.G.S.; dos Soares, L.A.A.; Gheyi, H.R.; Pinheiro, F.W.A.; da Silva, J.B. Potassium and irrigation water salinity on the formation of sour passion fruit seedlings. *Rev. Bras. Eng. Agríc. Ambient.* **2021**, *25*, 393–401. [CrossRef]

9. Da Paiva, F.J.S.; de Lima, G.S.; de Lima, V.L.A.; Nunes, K.G.; Fernandes, P.D. Gas exchange and production of passion fruit as affected by cationic nature of irrigation water. *Rev. Caatinga* **2021**, *34*, 926–936. [CrossRef]

10. Nunes, J.C.; de Lima Neto, A.J.; Cavalcante, L.F.; Pereira, W.E.; Gheyi, H.R.; de Lima, G.S.; de Oliveira, F.F.; da Nunes, J.A.S. Leaching of salts and production of sour passion fruit irrigated with low- and high-salinity water. *Rev. Bras. Eng. Agríc. Ambient.* **2023**, *27*, 393–399. [CrossRef]

11. Andrade, E.M.G.; de Lima, G.S.; de Lima, V.L.A.; da Silva, S.S.; Gheyi, H.R.; Araújo, A.C.; Gomes, J.P.; dos Soares, L.A.A. Production and postharvest quality of yellow passion fruit cultivated with saline water and hydrogen peroxide. *AIMS Agric. Food* **2019**, *4*, 907–920. [CrossRef]

12. Da Silva, S.S.; de Lima, G.S.; de Lima, V.L.A.; de Soares, L.A.A.; Gheyi, H.R.; Fernandes, P.D. Quantum yield, photosynthetic pigments and biomass of mini-watermelon under irrigation strategies and potassium. *Rev. Caatinga* **2021**, *34*, 659–669. [CrossRef]

13. Guo, H.; Huang, Z.; Li, M.; Hou, Z. Growth, ionic homeostasis, and physiological responses of cotton under different salt and alkali stresses. *Sci. Rep.* **2020**, *10*, e21844. [CrossRef] [PubMed]

14. De Silva, S.S.; de Lima, G.S.; de Lima, V.L.A.; Gheyi, H.R.; dos Soares, L.A.A.; Oliveira, J.P.M.; de Araújo, A.C.; Gomes, J.P. Production and quality of watermelon fruits under salinity management strategies and nitrogen fertilization. *Semin. Ciênc. Agrár.* **2020**, *41*, 2923–2936. [CrossRef]

15. Pinheiro, F.W.A.; de Lima, G.S.; Gheyi, H.R.; dos Soares, L.A.A.; Nobre, R.G.; Fernandes, P.D. Brackish water irrigation strategies and potassium fertilization in the cultivation of yellow passion fruit. *Ciênc. Agrotec.* **2022**, *46*, e022621. [CrossRef]

16. Dos Soares, L.A.A.; Fernandes, P.D.; de Lima, G.S.; Suassuna, J.F.; Pereira, R.F. Gas exchanges and production of colored cotton irrigated with saline water at different phenological stages. *Rev. Ciênc. Agron.* **2018**, *49*, 239–248. [CrossRef]

17. De Lima, G.S.; de Lacerda, C.N.; dos Soares, L.A.A.; Gheyi, H.R.; Araújo, R.H.C.R. Production characteristics of sesame genotypes under different strategies of saline water application. *Rev. Caatinga* **2021**, *33*, 490–499. [CrossRef]

18. Da Silva, S.S.; de Lima, G.S.; de Lima, V.L.A.; Gheyi, H.R.; dos Soares, L.A.A.; Oliveira, J.P.M. Production and post-harvest quality of mini-watermelon crop under irrigation management strategies and potassium fertilization. *Rev. Bras. Eng. Agríc. Ambient.* **2022**, *26*, 51–58. [CrossRef]

19. Hasanuzzaman, M.; Bhuyan, M.H.M.B.; Nahar, K.; Hossain, M.S.; Mahmud, J.A.; Hossen, M.S.; Masud, A.A.C.; Fujita, M. Potassium: A vital regulator of plant responses and tolerance to abiotic stresses. *Agronomy* **2018**, *8*, 31. [CrossRef]

20. Munir, A.; Shehzad, M.T.; Qadir, A.A.; Murtaza, G.; Khalid, H.I. Use of potassium fertilization to ameliorate the adverse effects of saline-sodic stress condition (ECw: SARw Levels) in rice (*Oryza sativa* L.). *Commun. Soil Sci. Plant Anal.* **2019**, *50*, 1975–1985. [CrossRef]

21. Mulet, J.M.; Campos, F.; Yenush, L. Editorial: Ion homeostasis in plant stress and development. *Front. Plant Sci.* **2020**, *11*, e618273. [CrossRef] [PubMed]

22. Andrade, E.M.G.; de Lima, G.S.; de Lima, V.L.A.; da Silva, S.S.; Gheyi, H.R.; da Silva, A.A.R. Gas exchanges and growth of passion fruit under saline water irrigation and H_2O_2 application. *Rev. Bras. Eng. Agríc. Ambient.* **2019**, *23*, 945–951. [CrossRef]

23. Malavolta, E.; Vitti, G.C.; Oliveira, S.A. *Avaliação do Estado Nutricional das Plantas: Princípios e Aplicações*; POTAFOS: Piracicaba, Brazil, 1997; 201p.

24. De Carvalho, A.J.C.; Monnerat, P.H.; Martins, D.P.; Bernardo, S.; da Silva, J.A. Teores foliares de nutrientes no maracujazeiro amarelo em função de adubação nitrogenada, irrigação e épocas de amostragem. *Sci. Agric.* **2002**, *59*, 121–127. [CrossRef]

25. De Costa, A.F.S.; da Costa, A.N.; Ventura, J.A.; Fanton, C.J.; de Lima, I.M.; Caetano, L.C.S.; de Santana, E.N. *Recomendações Técnicas para o Cultivo do Maracujazeiro*; Incaper: Vitória, Brazil, 2008.

26. Soares, F.A.L.; Gheyi, H.R.; Viana, S.B.A.; Uyeda, C.A.; Fernandes, P.D. Water salinity and initial development of yellow passion fruit. *Sci. Agric.* **2002**, *59*, 491–497. [CrossRef]

27. De Lima, G.S.; de Souza, W.B.B.; dos Soares, L.A.A.; Pinheiro, F.W.A.; Gheyi, H.A.R.; Oliveira, V.K.N. Dano celular e pigmentos fotossintéticos do maracujazeiro-azedo em função da natureza catiônica da água. *Irriga* **2020**, *25*, 663–669. [CrossRef]

28. Deng, X.; Ma, Y.; Shuang, C.; Jin, Z.; Shi, C.; Liu, J.; Lin, J.; Yan, X. Castor plant adaptation to salinity stress during early seedling stage by physiological and transcriptomic methods. *Agronomy* **2023**, *13*, 693. [CrossRef]

29. dos Soares, L.A.A.; Fernandes, P.D.; de Lima, G.S.; Brito, M.E.B.; do Nascimento, R.; Arriel, N.H.C. Physiology and production of naturally-colored cotton under irrigation strategies using salinized water. *Pesqui. Agropecuária Bras.* **2018**, *53*, 746–755. [CrossRef]

30. Xu, X.; Du, X.; Wang, F.; Sha, J.; Chen, Q.; Tian, G.; Zhu, Z.; Ge, S.; Jiang, Y. Effects of potassium levels on plant growth, accumulation and distribution of carbon, and nitrate metabolism in apple dwarf rootstock seedlings. *Front. Plant Sci.* **2020**, *11*, e904. [CrossRef]

31. De Souto, A.G.L.; Cavalcante, L.F.; de Melo, E.N.; Cavalcante, Í.H.L.; da Silva, R.Í.L.; de Lima, G.S.; Gheyi, H.R.; Pereira, W.E.; de Paiva Neto, V.B.; de Oliveira, C.J.A.; et al. Salinity and mulching effects on nutrition and production of grafted sour passion fruit. *Plants* **2023**, *12*, 1035. [CrossRef]

32. Da Gomes, M.A.C.; Suzuki, M.S.; da Cunha, M.; Tullii, C.F. Effect of salt stress on nutrient concentration, photosynthetic pigments, proline and foliar morphology of *Salvinia auriculata* Aubl. *Acta Limnol.* **2011**, *23*, 164–176. [CrossRef]
33. Do Nascimento, J.A.M.; Cavalcante, L.F.; Dantas, S.A.G.; da Silva, S.A. Estado nutricional de maracujazeiro-amarelo irrigado com água salina e adubação organomineral. *Rev. Bras. Frutic.* **2011**, *33*, 729–735. [CrossRef]
34. Da Silva, A.R.A.; Bezerra, F.M.L.; de Lacerda, C.F.; de Miranda, R.S.; Marques, E.C. Ion accumulation in young plants of the 'green dwarf' coconut under water and salt stress. *Rev. Ciênc. Agron.* **2018**, *49*, 249–258. [CrossRef]
35. De Carvalho, A.J.C.; Martins, D.P.; Monnerat, P.H.; Bernardo, S.; da Silva, J.A. Teores de nutrientes foliares no maracujazeiro-amarelo associados à estação fenológica, adubação potássica e lâminas de irrigação. *Rev. Bras. Frutic.* **2001**, *23*, 403–408. [CrossRef]
36. Singh, P.; Kumar, V.; Sharma, J.; Saini, S.; Sharma, P.; Kumar, S.; Sinhmar, Y.; Kumar, D.; Sharma, A. Silicon supplementation alleviates the salinity stress in wheat plants by enhancing the plant water status, photosynthetic pigments, proline content and antioxidant enzyme activities. *Plants* **2022**, *11*, 2525. [CrossRef] [PubMed]
37. Sullivan, C.Y. Sorghum in the seventies: Mechanism of heat and drought resistance in grain sorghum and methods of measurement. *Sorghum Seventies* **1972**, *1*, 247–264.
38. Negrão, S.; Schmockel, S.M.; Tester, M. Evaluating physiological responses of plants to salinity stress. *Ann. Bot.* **2017**, *119*, mcw191. [CrossRef] [PubMed]
39. Hnilickova, H.; Kraus, K.; Vachova, P.; Hnilicka, F. Salinity stress affects photosynthesis, malondialdehyde formation, and proline content in *Portulaca oleracea* L. *Plants* **2021**, *10*, 845. [CrossRef]
40. De Lima, G.S.; da Silva, J.B.; Pinheiro, F.W.A.; dos Soares, L.A.A.; Gheyi, H.R. Potassium does not attenuate salt stress in yellow passion fruit under irrigation management strategies. *Rev. Caatinga* **2020**, *33*, 1082–1091. [CrossRef]
41. Dias, A.S.; de Lima, G.S.; Gheyi, H.R.; de Melo, A.S.; Silva, P.C.C.; dos Soares, L.A.A.; da Paiva, F.J.S.; da Silva, S.S. Effect of combined potassium-phosphorus fertilization on gas exchange, antioxidant activity and fruit production of West Indian cherry under salt stress. *Arid. Land Res. Manag.* **2021**, *36*, 163–180. [CrossRef]
42. Pan, T.; Liu, M.; Kreslavski, V.D.; Zharmukhamedov, S.K.; Nie, C.; Yu, M.; Kuznetsov, V.V.; Allakhverdiev, S.I.; Shabala, S. Non-stomatal limitation of photosynthesis by soil salinity. *Crit. Rev. Environ. Sci. Technol.* **2020**, *51*, 791–825. [CrossRef]
43. Zahra, N.; Hinai, M.S.A.; Hafeez, M.B.; Rehman, A.; Wahid, A.; Siddique, K.H.M.; Farooq, M. Regulation of photosynthesis under salt stress and associated tolerance mechanisms. *Plant Physiol. Biochem.* **2022**, *178*, 55–69. [CrossRef]
44. De Lima, G.S.; Félix, C.M.; da Silva, S.S.; dos Soares, L.A.A.; Gheyi, H.R.; Soares, M.D.M.; do Sousa, P.F.N.; Fernandes, P.D. Gas exchange, growth, and production of mini-watermelon under saline water irrigation and phosphate fertilization. *Semin. Ciênc. Agrár.* **2020**, *41*, 3039–3052. [CrossRef]
45. Dos Soares, L.A.A.; Fernandes, P.D.; de Lima, G.S.; Gheyi, H.R.; Nobre, R.G.; da Sá, F.V.S.; Moreira, R.C.L. Saline water irrigation strategies in two production cycles of naturally colored cotton. *Irrig. Sci.* **2020**, *38*, 401–413. [CrossRef]
46. Teixeira, P.C.; Donagemma, G.K.; Fontana, A.; Teixeira, W.G. *Manual de Métodos de Análise de Solo*, 3rd ed.; Embrapa: Brasília, Brazil, 2017; 573p.
47. Richards, L.A. *Diagnosis and Improvement of Saline and Alkali Soils*; Agriculture Handbook 60; U.S. Department of Agriculture: Washington, DC, USA, 1954; 160p.
48. Bernardo, S.; Soares, A.A.; Mantovani, E.C. *Manual de Irrigação*, 8th ed.; UFV: Viçosa, Brazil, 2013; 625p.
49. De Freire, J.L.O.; Cavalcante, L.F.; Rebequi, A.M.; Dias, T.J.; de Souto, A.G.L. Necessidade hídrica do maracujazeiro amarelo cultivado sob estresse salino, biofertilização e cobertura do solo. *Rev. Caatinga* **2011**, *24*, 82–91.
50. Weatherley, P.E. Studies in the water relations of the cotton plant. I—The field measurements of water deficits in leaves. *New Phytol.* **1950**, *49*, 81–97. [CrossRef]
51. Scotti-Campos, P.; Pham-Thi, A.T.; Semedo, J.N.; Pais, I.P.; Ramalho, J.C.; do Matos, M.C. Physiological responses and membrane integrity in three Vigna genotypes with contrasting drought tolerance. *Emir. J. Food Agric.* **2013**, *25*, 1002–1013. [CrossRef]
52. Fernandes, E.A.; dos Soares, L.A.A.; de Lima, G.S.; de Silva Neta, A.M.S.; Roque, I.A.; da Silva, F.A.; Fernandes, P.D.; de Lacerda, C.N. Cell damage, gas exchange, and growth of *Annona squamosa* L. under saline water irrigation and potassium fertilization. *Semin. Ciênc. Agrár.* **2021**, *42*, 999–1018. [CrossRef]
53. Ferreira, D.F. SISVAR: A computer analysis system to fixed effects split plot type designs. *Rev. Bras. Biom.* **2019**, *37*, 529–535. [CrossRef]

Article

Osmoregulatory and Antioxidants Modulation by Salicylic Acid and Methionine in Cowpea Plants under the Water Restriction

Auta Paulina da Silva Oliveira [1], Yuri Lima Melo [1], Rayanne Silva de Alencar [1], Pedro Roberto Almeida Viégas [2], Guilherme Felix Dias [1], Rener Luciano de Souza Ferraz [3], Francisco Vanies da Silva Sá [4], José Dantas Neto [5], Ivomberg Dourado Magalhães [6], Hans Raj Gheyi [5], Claudivan Feitosa de Lacerda [7] and Alberto Soares de Melo [1,*]

[1]   Center of Biological and Health Sciences, Universidade Estadual da Paraíba, Campus I, Campina Grande 58429-500, PB, Brazil

[2]   Department of Agronomy, Federal University of Sergipe, São Cristóvão 49100-000, SE, Brazil

[3]   Academic Unit of Development Technology, Federal University of Campina Grande, Sumé 58540-000, PB, Brazil

[4]   Department of Agronomic and Forest Science, Federal Rural University of the Semi-Arid, Mossoró 59625-900, RN, Brazil

[5]   Center of Tecnologia and Natural Resources, Universidade Federal de Campina Grande, Campina Grande 58429-900, PB, Brazil

[6]   Department of Agronomy, Federal University of Alagoas, Rio Largo 57309-005, AL, Brazil

[7]   Department of Agronomy, Federal University of Ceará, Fortaleza 60020-903, CE, Brazil

*   Correspondence: alberto.melo@servidor.uepb.edu.br

Abstract: Global climate changes have intensified water stress in arid and semi-arid regions, reducing plant growth and yield. In this scenario, the present study aimed to evaluate the mitigating action of salicylic acid and methionine in cowpea cultivars under water restriction conditions. An experiment was conducted in a completely randomized design with treatments set up in a 2×5 factorial arrangement corresponding to two cowpea cultivars (BRS Novaera and BRS Pajeú) and five treatments of water replenishment, salicylic acid, and methionine. After eight days, water stress decreased the Ψw, leaf area, and fresh mass and increased the total soluble sugars and catalase activity in the two cultivars. After sixteen days, water stress increased the activity of the superoxide dismutase and ascorbate peroxidase enzymes and decreased the total soluble sugars content and catalase activity of BRS Pajeú plants. This stress response was intensified in the BRS Pajeú plants sprayed with salicylic acid and the BRS Novaera plants with salicylic acid or methionine. BRS Pajeú is more tolerant to water stress than BRS Novaera; therefore, the regulations induced by the isolated application of salicylic acid and methionine were more intense in BRS Novaera, stimulating the tolerance mechanism of this cultivar to water stress.

Keywords: *Vigna unguiculata* (L.) Walp; antioxidant enzymes; drought tolerance

Citation: Oliveira, A.P.d.S.; Melo, Y.L.; de Alencar, R.S.; Viégas, P.R.A.; Dias, G.F.; Ferraz, R.L.d.S.; Sá, F.V.d.S.; Dantas Neto, J.; Magalhães, I.D.; Gheyi, H.R.; et al. Osmoregulatory and Antioxidants Modulation by Salicylic Acid and Methionine in Cowpea Plants under the Water Restriction. *Plants* **2023**, *12*, 1341. https://doi.org/10.3390/plants12061341

Academic Editor: Ismael Aranda

Received: 27 February 2023
Revised: 8 March 2023
Accepted: 11 March 2023
Published: 16 March 2023

1. Introduction

Global climate changes significantly disturb the space-time distribution of rainfall, causing water deficit in agroecosystems and reducing crop growth and yield. This reality has highlighted the need for genetically improved genotypes with traits of agricultural interest and capability of adapting to different environments under adverse conditions, e.g., cowpea (*Vigna unguiculata* (L.) Walp.) [1]. Cowpea performs a critical nutritional role due to its use as human food, animal forage, and green manure. It also contributes to social and economic development in arid and semi-arid regions. In those regions, water limitation in the soil-plant-atmosphere continuum inhibits germination and plant establishment and reduces vegetative growth and grain production [2].

The tolerant ones had the highest leaf water potential when comparing tolerant and sensitive cowpea genotypes to water stress under water restriction conditions. In contrast, leaf gas exchange and chlorophyll fluorescence decreased more rapidly in the sensitive ones [3]. These differences in the physiological and biochemical responses highlight the need for studies that explore the aptitude of multiple genotypes [4]. The divergences between genotypes regarding physiological and biochemical variations and gene expression [5] suggest that the use of elicitors, i.e., low molecular weight molecules that modulate the cell metabolism, can induce responses that result in higher tolerance to water stress [6]. However, applying elicitors stimulates a cascade of biochemical reactions that modify the secondary metabolism of plant species [7].

Among elicitors, salicylic acid (SA) is an essential plant growth regulator that, when applied exogenously, intensifies physiological and molecular processes, including changes in gene expression, increasing protein synthesis, and activating specific enzymes [8]. SA applications from 0.1 to 1 mM increase the germination percentage and the activity of SOD, CAT, and APX in cowpea plants under drought conditions [9,10]. Araújo et al. [10] observed that applying 1 mM of SA in cowpea plants under water restriction prevented membrane damage, increased the proline content, regulated early growth, and increased the levels of chlorophyll *a*, *b*, and carotenoids. In another study, the yield of cowpea plants under water restriction conditions reached values close to 2.64–2.73 Mg ha^{-1} in the first and second crop cycles after applying 0.3 g L^{-1} of SA [8].

Methionine is an essential amino acid that participates in several physiological functions in plants. Its limitation compromises plant survival since this amino acid acts as an effective regulator in the growth and development of plants subjected to water restriction [11]. Some studies show that the exogenous foliar application of methionine is practical and positively impacts the integrity of photosynthetic pigments, the accumulation of compatible osmolytes, the removal of reactive oxygen species (ROS), and the improvement in cowpea growth and yield [12].

We hypothesized that foliar application of salicylic acid and methionine could modulate cowpea plants' osmotic and antioxidant metabolism under water restriction. From this perspective, since the effects of the exogenous applications of elicitors on cowpea plants grown under stress still require explanation, the present study investigated the effects of the application of salicylic acid and methionine on the water status, growth, osmotic adjustment, oxidative stress, and leaf gas exchange indicators of two cowpea cultivars subjected to water restriction.

2. Results

Eight days after treatment (DAT) application, the first four PCs represented 78% of the total data variance (s^2). PC$_1$ (34% s^2) was formed by the linear combination of the water potential (Ψw) with the leaf area (LA), shoot fresh matter (FM), the total content of soluble sugars (TSS), and catalase activity (CAT). PC$_2$ (23% s^2) is the combination between the relative leaf water content (RWC) and the content of proline (PRO), total soluble proteins (TSPs), hydrogen peroxide (H$_2$O$_2$), and ascorbate peroxidase activity (APX). PC$_3$ (13.5% s^2) is the combination of Ψw with superoxide dismutase activity (SOD), whereas PC$_4$ (8% s^2) is the combination of RWC with the content of total free amino acids (TFAAs). There was a significant difference ($p < 0.01$) between genotypes (G), treatments (T), and the G $\times$ T interaction in the four PCs (Table 1).

Sixteen DAT, the first four PCs explained 80% of s^2. PC$_1$ (38% s^2) originated from the combination between RWC, LA, FM, PRO, TFAA, the CO$_2$ assimilation rate (A), transpiration (E), stomatal conductance (gs), and the internal CO$_2$ concentration (Ci); PC$_2$ (20% s^2) combined TSPs, APX, CAT, and SOD; PC$_3$ (13% s^2) is the combination between Ψw, TSS, and H$_2$O$_2$; finally, PC$_4$ (9% s^2) is the combination between TFAAs and SOD (Table 1).

After eight days, the PC1 separated treatments (left–right), and PC2 separated genotypes (lower–upper). In PC1, water stress decreased the Ψw, LA, and FM and increased the TSS and CAT activity, a process that occurred more markedly in the G2 plants (BRS Novaera) sprayed with SA + MET (T5) or when not receiving spraying (T2), and in the G1 plants (BRS Pajeú) sprayed with SA (T3) (Figure 1A,B).

Table 1. Correlation between original variables and principal components, eigenvalues, explained and cumulative variances, and the probability of significance of the test of hypothesis after eight and 16 days of the application of treatments.

Variables	Principal Components							
	8 Days after Treatments				16 Days after Treatments			
	PC_1	PC_2	PC_3	PC_4	PC_1	PC_2	PC_3	PC_4
Ψw—Leaf water potential	0.73 *	−0.12	0.58 *	−0.00	−0.03	−0.31	−0.90 *	−0.11
RWC—Relative leaf water content	0.42	−0.61 *	0.12	0.55 *	−0.58 *	−0.15	−0.36	−0.35
LA—Leaf surface area	0.88 *	0.39	−0.01	0.03	−0.92 *	0.08	−0.03	−0.13
FM—Shoot fresh matter	0.83 *	0.48	0.01	0.12	−0.85 *	0.25	−0.06	0.02
TSS—Total soluble sugars	−0.77 *	−0.08	0.30	−0.08	0.16	−0.50	0.58 *	−0.43
PRO—Proline content	−0.58	0.59 *	−0.40	−0.2	0.65 *	0.23	0.22	−0.01
TFAAs—Total free amino acids	−0.55	0.22	−0.26	0.65 *	0.53 *	0.39	−0.05	−0.61 *
TSPs—Total soluble proteins	−0.21	0.76 *	0.29	0.28	−0.42	0.74 *	−0.17	0.00
H_2O_2—Hydrogen peroxide	−0.21	−0.63 *	0.18	−0.15	−0.32	−0.52	0.70 *	−0.00
APX—Ascorbate peroxidase activity	−0.09	−0.78 *	−0.32	0.07	−0.13	−0.82 *	−0.27	−0.33
CAT—Catalase activity	−0.69 *	−0.02	0.42	0.26	−0.31	0.76 *	0.17	0.23
SOD –Superoxide dismutase activity	−0.35	0.16	0.74 *	−0.13	−0.17	−0.66 *	0.06	0.58 *
A—Net CO_2 assimilation rate	ne	ne	ne	ne	−0.94 *	−0.12	−0.04	0.11
E—Water vapor transpiration rate	ne	ne	ne	ne	−0.94 *	−0.09	0.15	−0.06
gs—Stomatal conductance to water vapor	ne	ne	ne	ne	−0.96 *	−0.05	0.02	−0.08
Ci—Internal carbon dioxide concentration	ne	ne	ne	ne	−0.52 *	0.32	0.42	−0.37
λ—Eigenvalues	4.08	2.78	1.62	0.97	6.01	3.27	2.15	1.37
S^2 (%)—Explained variance	34.02	23.16	13.51	8.07	37.56	20.45	13.42	8.53
S^2 (%)—Cumulative variance	34.02	57.18	70.69	78.77	37.56	58.00	71.42	79.96
MANOVA—Multivariate analysis of variance	*p*-value							
Hotelling's T-squared test for genotypes—G	<0.01	<0.01	<0.01	<0.01	<0.01	<0.01	<0.01	<0.01
Hotelling's T-squared test for treatments—T	<0.01	<0.01	<0.01	<0.01	<0.01	<0.01	<0.01	<0.01
Hotelling's T-squared test for the G × T interaction	<0.01	<0.01	<0.01	<0.01	<0.01	<0.01	<0.01	<0.01

*: Variable with Pearson's correlation coefficient (r ≥ 0.55) considered for PC; ne: not evaluated eight days after the application of treatments; PC: principal component.

In PC2, G2 showed the highest PRO accumulation and TSP content and the lowest RWC, H_2O_2, and APX activity compared to G1. When analyzing the effect of T within each G, the G2 plants under water stress and no spraying of elicitors (T2) and those sprayed with SA (T3) showed the highest PRO and TSP contents, similar to control plants (T1). In contrast, the plants sprayed with MET (T4) or SA + MET (T5) increased the RWC, H_2O_2, and APX, unlike T1 and T2. In G1, the spraying with SA, MET, and SA + MET increased the RWC, H_2O_2, and APX and decreased the PRO and TSPs compared to the control.

In PC3, water stress decreased the Ψw and SOD activity in G2 compared to control. However, MET spraying increased these indicators. On the other hand, the G1 plants showed increased Ψw and SOD activity under water stress, whereas spraying with SA, MET, or SA + MET maintained these indicators at levels similar to the control treatment.

In PC4, water stress decreased the RWC and TFAAs of G1 plants, with SA or MET spraying increasing these indicators in stressed plants to levels similar to the control treatment. However, when SA + MET were applied together, the RWC and TFAA levels increased but did not reach levels similar to the control treatment. G2 plants under water stress and SA spraying showed lower RWC and TFAA values, whereas MET increased these indicators to levels similar to the control treatment, and the joint application of SA + MET increased these levels to values above those of the control treatment (Figure 1C,D).

Sixteen DAT, in PC1, water stress decreased the gas exchange variables (A, E, gs, and Ci), increased the content of osmoprotectants (TFAAs and PRO), and decreased the water content (RWC) and growth (LA and FM) in the two genotypes. This process is intensified by SA or SA + MET spraying in G1 and SA in G2. Regardless of treatment, the process described above was more pronounced in G2 (Figure 2A,B and Table 2).

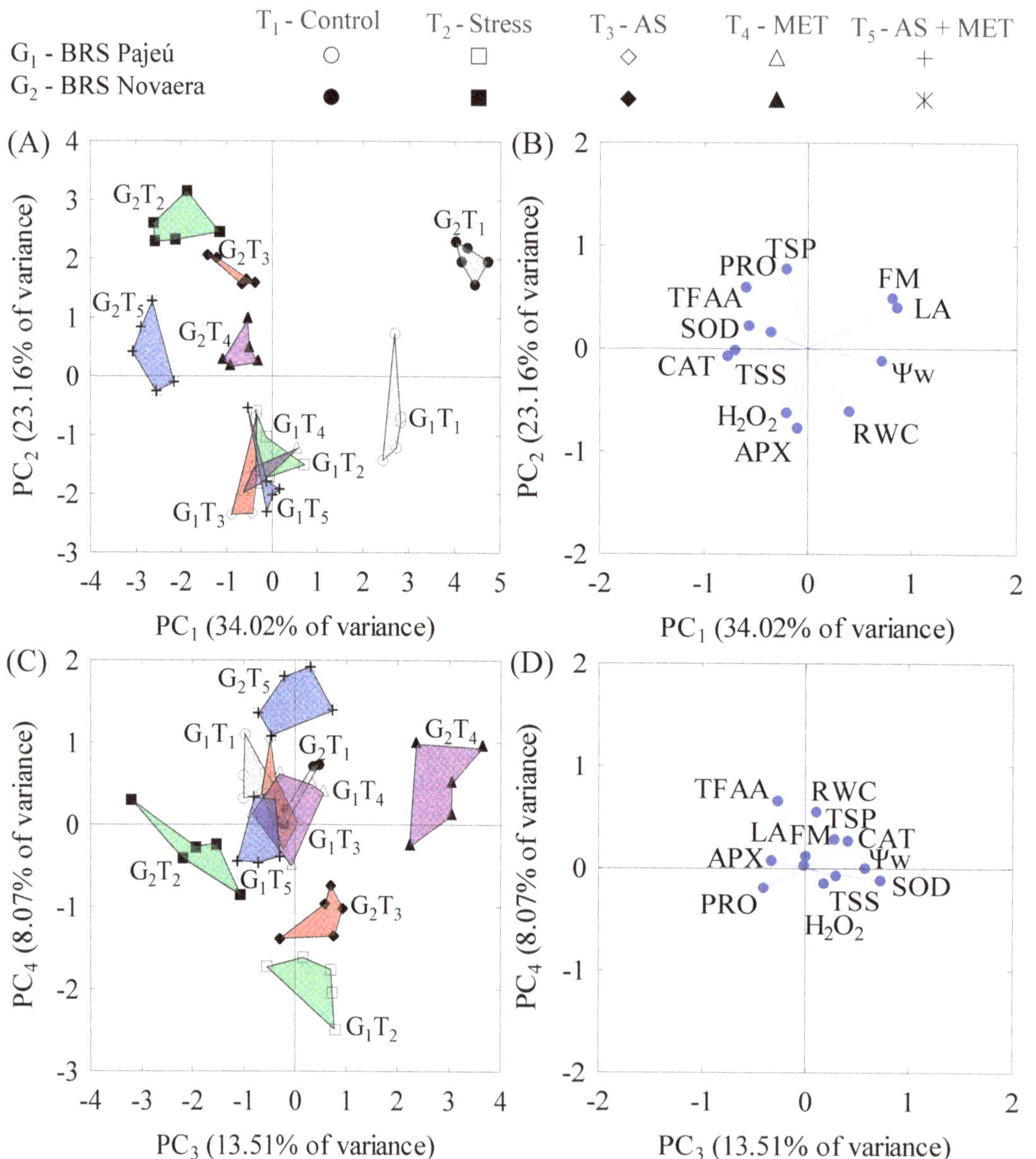

Figure 1. Two-dimensional projection of the interactions between genotypes and treatments (**A**,**C**) and between the correlation coefficients of the variables (**B**,**D**) with the first four principal components (PCs 1, 2, 3, and 4) eight days after the application of treatments.

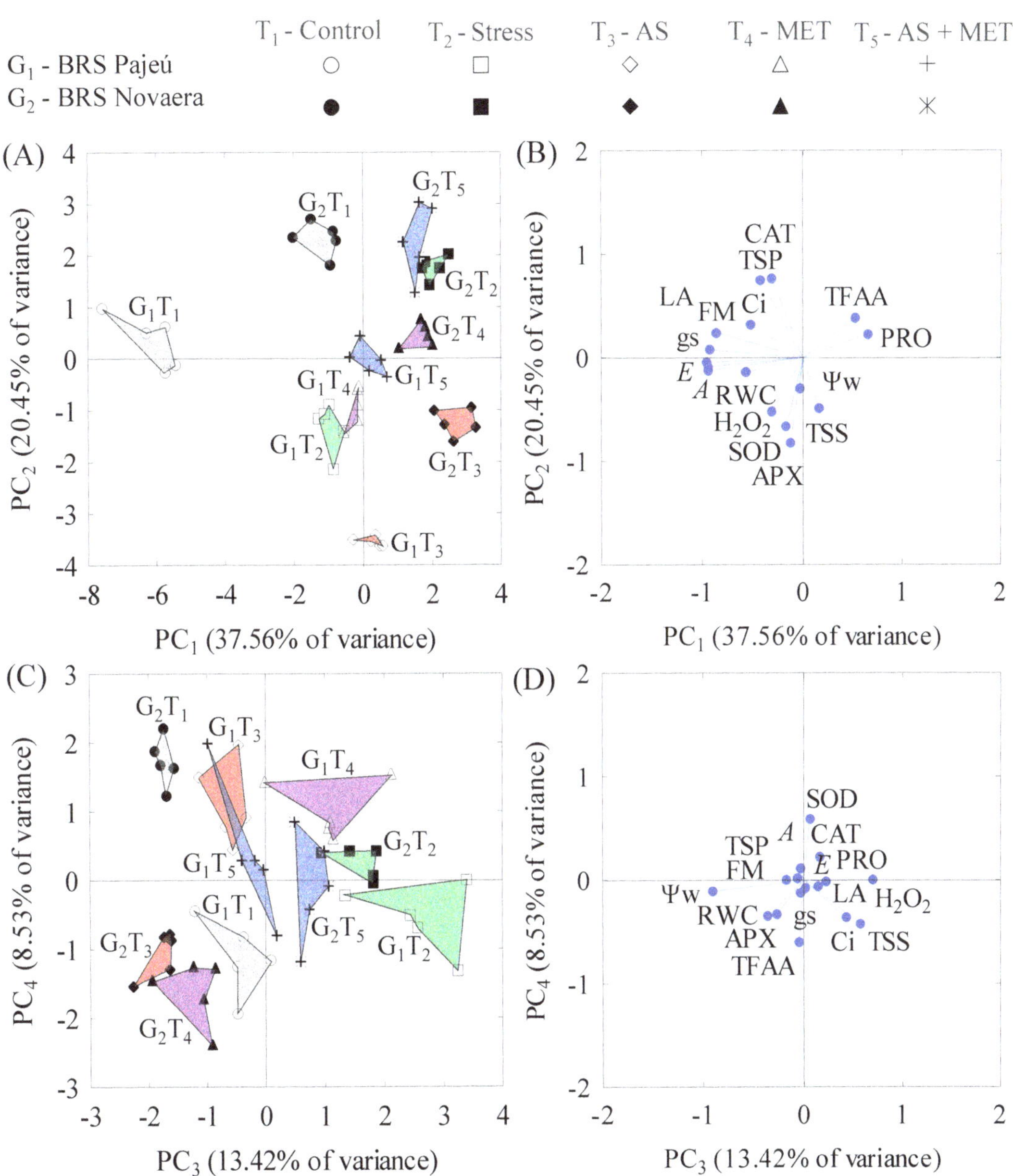

Figure 2. Two-dimensional projection of the interactions between genotypes and treatments (**A**,**C**) and between the correlation coefficients of the variables (**B**,**D**) and the first four principal components (PCs 1, 2, 3, and 4) 16 days after application of treatments.

Table 2. Means of the original variables and scores of the principal components eight and sixteen days after application of treatments.

| Var | *p*-Value | Means of the Combinations between Genotypes and Treatments after 8 Days | | | | | | | | | |
| | | G1—BRS Pajeú | | | | | G2—BRS Novaera | | | | |
		T1	T2	T3	T4	T5	T1	T2	T3	T4	T5
Ψw	<0.01	−0.47Ba	−0.52Aa	−0.53Aa	−0.57Ba	−0.56Aa	−0.38Aa	−0.74Bc	−0.56Ab	−0.43Aa	−0.65Bbc
RWC	<0.01	93.43Aa	82.84Ac	86.98Abc	88.37Aabc	89.63Aab	86.53Ba	74.97Bb	74.64Bb	87.75Aa	86.00Aa
LA	<0.01	156.80Ba	98.91Ab	77.92Bc	75.04Ac	84.62Abc	260.29Aa	79.55Bc	116.93Ab	85.08Ac	73.44Bc
FM	<0.01	11.74Ba	5.89Bb	5.13Bbc	5.55Bbc	4.66Ac	15.39Aa	6.84Ac	8.22Ab	7.03Ac	5.34Ad
TSS	<0.01	12.05Ad	20.01Aa	16.78Bbc	14.59Bc	19.05Bab	10.94Ad	15.91Bc	20.01Ab	19.17Ab	24.21Aa
PRO	<0.01	8.61Ab	18.64Ba	10.17Bab	6.76Ab	12.03Bab	6.95Ac	60.71Aa	32.82Ab	10.23Ac	28.89Ab
TFAAs	<0.01	0.48Ab	0.35Bc	0.66Aa	0.52Ab	0.42Bbc	0.45Acd	0.77Ab	0.40Bd	0.54Ac	0.94Aa
TSPs	<0.03	5.92Bab	4.01Bb	5.78Bab	6.73Ba	4.79Bb	7.85Ab	8.32Aab	10.03Aa	8.95Aab	8.50Aab
H_2O_2	<0.01	8.12Ac	13.37Ab	15.39Aa	16.41Aa	7.59Bc	6.98Bb	6.89Bb	9.50Ba	9.77Ba	9.35Aa
APX	<0.06	65.70Aa	49.61Aa	54.99Aa	49.07Aa	70.82Aa	13.46Bb	26.00Bab	34.86Bab	26.00Bab	47.71Ba
CAT	<0.50	1.26Ac	1.53Bbc	2.32Aa	2.59Aa	2.25Aab	1.31Ac	2.16Ab	2.49Aab	3.01Aa	2.66Aab
SOD	<0.01	7.69Ab	18.16Aa	4.72Bb	4.67Bb	5.58Bb	4.18Ac	9.10Bbc	13.16Ab	29.07Aa	14.54Ab
		Means comparison test for the scores of the principal components									
PC1	<0.01	1.35Ba	−0.01Ab	−0.23Ab	−0.13Ab	−0.05Ab	2.14Aa	−1.02Bc	−0.41Ab	−0.33Ab	−1.31Bc
PC2	<0.01	−0.41Ba	−0.79Bab	−1.04Bb	−1.01Bb	−1.04Bb	1.19Aa	1.53Aa	1.06Aa	0.26Ab	0.25Ab
PC3	<0.01	−0.61Bb	0.30Aa	−0.26Bab	−0.04Bab	−0.52Ab	0.02Ab	−1.54Bc	0.44Ab	2.26Aa	−0.05Ab
PC4	<0.01	0.61Aa	−1.97Bc	0.34Aab	0.24Aab	−0.14Bb	0.33Abc	−0.31Ac	−1.12Bd	0.48Ab	1.53Aa
		Means of the combinations between genotypes and treatments after 16 days									
Ψw	<0.01	−0.28Aa	−0.42Ac	−0.25Ba	−0.36Bb	−0.26Aa	−0.28Ab	−0.42Ac	−0.20Aa	−0.26Ab	−0.37Bc
RWC	<0.01	81.95Aa	73.67Ab	71.18Bb	74.28Bb	71.79Ab	74.40Ba	69.90Bb	74.49Aa	77.86Aa	64.83Bc
LA	<0.01	267.48Aa	149.73Ab	129.45Ab	92.17Bc	125.22Ab	144.89Ba	75.38Bc	58.85Bc	111.86Ab	130.74Aab
FM	<0.01	15.24Aa	10.60Ab	7.95Ac	6.30Ac	7.78Ac	12.52Ba	7.20Bbc	5.73Bc	7.07Abc	8.33Ab
TSS	<0.01	14.30Ac	20.20Aa	15.70Abc	16.64Ab	16.16Abc	11.34Bb	14.49Ba	16.02Aa	16.12Aa	15.86Aa
PRO	<0.01	9.04Bc	14.23Bbc	14.61Babc	17.46Aab	20.40Aa	14.14Ac	30.43Aa	18.85Abc	20.70Ab	16.90Abc
TFAAs	<0.01	0.77Aab	0.91Aa	0.71Bb	0.74Bab	0.81Bab	0.87Ab	0.96Aab	1.07Aa	1.08Aa	1.05Aab
TSPs	<0.27	9.69Aa	6.40Ab	3.61Bc	5.36Bb	8.62Aa	10.27Aa	6.58Abc	5.21Ac	7.10Ab	9.16Aa
H_2O_2	<0.01	8.12Ab	11.84Aa	8.54Ab	8.27Ab	7.61Ab	5.18Bb	7.82Ba	5.81Bb	5.81Bb	5.33Bb
APX	<0.01	90.00Ab	73.81Abc	131.58Aa	59.14Bcd	54.25Ad	20.32Bc	12.86Bc	98.98Ba	86.13Aa	41.46Bb
CAT	<0.01	3.03Aa	2.69Aa	1.59Ab	3.14Aa	2.47Ba	3.88Aa	2.87Ab	1.71Ac	2.72Ab	3.43Aab
SOD	<0.01	12.28Ac	16.74Abc	25.21Aa	20.10Ab	13.27Ac	14.34Aa	7.94Bbc	12.49Bab	6.91Bc	12.10Aab
A	<0.01	23.28Aa	12.17Abc	11.48Ac	13.06Ab	13.38Ab	14.05Ba	7.33Bb	5.65Bc	6.08Bbc	3.86Bd
E	<0.01	3.11Aa	1.65Ab	1.16Ac	1.76Ab	1.20Ac	1.39Ba	0.98Bb	0.58Bc	0.71Bbc	0.53Bc
gs	<0.01	157.85Aa	65.50Ab	55.30Ab	69.45Ab	57.05Ab	68.25Ba	39.50Bb	26.00Bb	30.00Bb	22.00Bb
Ci	<0.01	132.05Aa	83.30Ab	56.75Ab	83.70Ab	70.05Bb	52.45Bc	87.10Aab	58.59Ac	69.80Abc	111.70Aa
		Means comparison test for the scores of the principal components									
PC1	<0.01	−2.51Bc	−0.38Bb	0.11Ba	−0.09Bab	0.08Ba	−0.49Ac	0.84Aab	1.11Aa	0.69Ab	0.65Ab
PC2	<0.01	0.18Ba	−0.74Bb	−1.97Bc	−0.52Bb	−0.02Ba	1.28Aa	0.97Aa	−0.69Ac	0.25Ab	1.26Aa
PC3	<0.01	−0.33Ac	1.78Aa	−0.43Ac	0.74Ab	−0.19Bc	−1.17Bb	1.08Ba	−1.20Bb	−0.81Bb	0.54Aa
PC4	<0.01	−0.97Bc	−0.48Bbc	0.95Aa	0.88Aa	0.32Aab	1.47Aa	0.21Ab	−0.91Bcd	−1.39Bd	−0.08Abc

Var: variables; T1: control (100% of soil field capacity); T2: stress (50% of soil field capacity); T3: stress + salicylic acid (SA); T4: stress + methionine application (MET); T5: stress + SA + MET. Ψw: water potential (MPa); RWC: relative water content (%); LA: leaf area (cm^2); FM: fresh mass (g); TSS: total soluble sugars in leaf blades (mg g^{-1} FM); PRO: free proline in leaf blades (μmol g^{-1} FM); TFAAs: total free amino acids in leaf blades (μmol g^{-1} FM); TSPs: total soluble proteins in leaf blades (mg g^{-1} FM); H_2O_2: hydrogen peroxide in leaf blades (μmol g^{-1} FM); APX: ascorbate peroxidase activity in leaf blades (nmol of ascorbate min^{-1} mg^{-1} of protein); CAT: catalase activity in leaf blades (μmol of H_2O_2 min^{-1} mg^{-1} of protein); SOD: superoxide dismutase activity in leaf blades (U min^{-1} mg^{-1} of protein); A: net photosynthesis (μmol CO_2 m^{-1} s^{-1}); E: transpiration (mmol H_2O m^{-1} s^{-1}); gs: stomatal conductance (mmol H_2O m^{-1} s^{-1}); Ci: internal CO_2 concentration (ppm); PC: principal component. Genotype means followed by the same uppercase letters do not differ by Student's *t*-test (*p* > 0.05), and treatment means followed by the same lowercase letters in the rows do not differ by Tukey's test (*p* > 0.05).

In PC2, water stress increased the activity of the SOD and APX enzymes and decreased the TSP content and CAT activity of G1 plants. This stress response was intensified in the G1 plants sprayed with SA and the G2 plants with SA or MET. However, foliar spraying with SA + MET significantly inhibited this response in the two genotypes and maintained enzyme activity and the TSP content at levels similar to the control plants (Figure 2A,B and Table 2).

In PC3, the two genotypes subjected to stress showed increased H_2O_2 contents, triggering TSS accumulation and Ψw reduction. In G1, the foliar spraying with SA or SA + MET reversed this process and made these indicators similar to control plants. In G2, the stress effect was reversed by the isolated spraying with SA or MET. However, the effect was maintained when these mitigators were applied together (Figure 2C,D and Table 2).

In PC4, spraying with SA or MET increased SOD activity and decreased the TFAA content of G1, differing from the control plants and those under stress and no spraying, which did not differ concerning these indicators. On the other hand, stress imposition in G2 increased TFAA accumulation and decreased SOD activity, a process intensified by SA or MET spraying (Figure 2C,D and Table 2).

Table 2 shows the mean values of the original variables, the scores of the principal components, and the means comparison tests for these scores.

In cowpea under water stress after 8 days of treatment (8 DAT), only 'BRS Novaera' had the leaf water potential decreased by 94% compared to the control. In the same period, the 'BRS Pajeú' under water stress showed a higher water potential (-0.52 MPa) similar to the control (-0.47 MPa) and higher than the 'BRS Novaera' cultivar (-0.74 MPa). Sixteen DAT, the leaf water potential of the 'BRS Pajeú' and 'BRS Novaera' cultivars decreased by 50% under water stress compared to the control. The relative leaf water content (RWC) of the two cowpea cultivars was decreased by water stress. Eight DAT, the RWC of 'BRS Novaera' decreased by 13.74% under water stress compared to the control. Sixteen DAT, 'BRS Pajeú' showed a 20% decrease in RWC under water stress compared to the control (Table 2).

Cowpea cultivars decreased leaf area (LA) under water stress compared to the control. Eight DAT, 'BRS Pajeú' had higher FA (98.91 cm^2 per plant) than 'BRS Novaera' (79.55 cm^2 per plant) under water stress. Sixteen DAT, the LA of 'BRS Pajeú' decreased by 40% and that of 'BRS Novaera' by 48% under water stress compared to the control. Water stress decreased fresh cowpea biomass (FM) compared to the control (Table 2). Eight DAT, 'BRS Novaera' (6.84 g FM) was superior to 'BRS Pajeú' (5.89 g FM) in water stress. Sixteen DAT, the reduction trend of FM in water stress was maintained, being 30.44% in 'BRS Pajeú' and 67% in 'BRS Novaera' compared to the control (Table 2).

In the cowpea under water stress conditions, total soluble sugars (TSS) increased by 66% and 41.2% in 'BRS Pajeú' and 41.4% and 27.7% in 'BRS Novaera' compared to the control for the 8 and 16 DAT, respectively (Table 2). In the isolated application of salicylic acid (SA, 1.5 mM), TSS decreased by 16.2% and 22.25% in 'BRS Pajeú' and increased by 26 and 11% in 'BRS Novaera' compared to water stress for 8 and 16 DAT, respectively. In the isolated application of methionine (MET, 6 mM), only 'BRS Novaera' increased TSS by 20.5% and 11.24% compared to water stress for 8 and 16 DAT, respectively. The proline (PRO) of 'BRS Pajeú' increased by 116% and 57.4% under water stress compared to the control at 8 and 16 DAT (Table 2). In the 'BRS Novaera', PRO increased by 774.78% and 115.28% compared to the control at 8 and 16 DAT (Table 2). The isolated application of SA decreased the PRO levels in 'BRS Novaera' by 46% and 38% compared to water stress at 8 and 16 DAT. The isolated application of MET (6 mM) reduced the PRO levels of 'BRS Novaera' by 83.14% compared to water stress at 8 DAT (Table 2).

The total free amino acids (TFAAs) of the 'BRS Novaera' under water stress increased 68% compared to the control at 8 DAT (Table 2). At 16 DAT, TFAAs increased by 20% in 'BRS Pajeú' and 11% in 'BRS Novaera' under water stress compared to the control. The isolated application of SA in 'BRS Pajeú' increased the TFAAs by 91% and 22% compared to water stress at 8 and 16 DAT. However, in 'BRS Novaera' with SA application, the TFAAs decreased by 46% compared to water stress at 8 DAT, and at 16 DAT, the TFAAs increased by 12% compared to water stress. The isolated application of MET (6 mM) increased the TFAA content by 50% in 'BRS Pajeú' at 8 DAT and 14% in 'BRS Novaera' at 16 DAT compared to water stress. At 8 DAT, the joint application of SA and MET increased the TFAA content in 'BRS Novaera' by 27% compared to water stress (Table 2).

In the cowpea under water stress conditions, the total soluble protein (TSP) content decreased by 32.26% in 'BRS Pajeú' and increased by 5.99% in 'BRS Novaera' at 8 DAT compared to the control. At 16 DAT, the TSPs of plants under water stress decreased by 34 and 36% for 'BRS Pajeú' and 'BRS Novaera' compared to the control. The isolated application of SA (1.5 mM) compared to water stress increased TSPs by 43 and 21% for 'BRS Pajeú' and 'BRS Novaera' at 8 DAT. At 16 DAT, the isolated application of SA (1.5 mM) decreased TSPs by 43.66% and 21% for 'BRS Pajeú' and 'BRS Novaera' compared to water stress (Table 2).

In plants under water stress at 8 DAT, H_2O_2 was increased by 65% in 'BRS Pajeú' compared to the control (Table 2). However, at 16 DAT, H_2O_2 increased by 46% in 'BRS Pajeú' and 51% in 'BRS Novaera' compared to the control. At 8 DAT, the isolated application of SA increased H_2O_2 by 15% in 'BRS Pajeú' and 38% in 'BRS Novaera' compared to water stress. However, in the isolated application of SA at 16 DAT, the 'BRS Pajeú' and 'BRS Novaera' cultivars decreased the H_2O_2 content by 26 and 28% compared to water stress. In the isolated application of MET compared to water stress at 8 DAT, H_2O_2 increased by 23% in 'BRS Pajéu' and 42% in 'BRS Novaera' (42%). At 16 DAT with MET application, H_2O_2 increased by 30% only in 'BRS Pajeú' compared to water stress (Table 2).

The ascorbate peroxidase activity (APX) of plants under water stress at 8 DAT decreased by 24.5% in 'BRS Pajeú' and increased by 93% in 'BRS Novaera' compared to the control. At 16 DAT, APX activity decreased by 18% and 37% for 'BRS Pajeú' and 'BRS Novaera' compared to the control. The foliar application of SA (1.5 mM) at 8 DAT increased APX activity by 11% and 34% for 'BRS Pajeú' and 'BRS Novaera' compared to water stress. At 16 DAT, SA increased APX activity by 78% in 'BRS Pajéu' and 670% in 'BRS Novaera' compared to water stress. Foliar application of MET (6 mM) at 16 DAT increased APX activity by 570% only for 'BRS Novaera', compared to water stress. The joint application of SA and MET increased APX activity by 222% for 'BRS Novaera' at 16 DAT compared to water stress (Table 2).

The catalase (CAT) activity of plants under water stress at 8 DAT increased by 20.63% for 'BRS Pajeú' and 66.15% for 'BRS Novaera' compared to the control. However, at 16 DAT, the CAT activity of plants under water stress decreased by 26% in 'BRS Novaera' compared to the control. The isolated application of SA (1.5 mM) at 8 DAT increased CAT activity by 53% for 'BRS Pajeú' and 15% for 'BRS Novaera', compared to the water stress. However, the isolated application of SA (1.5 mM) at 16 DAT decreased CAT activity by 41% for 'BRS Pajeú' and 40% for 'BRS Novaera' compared to the water stress. At 8 DAT, MET foliar application (6 mM) increased CAT activity by 70% in 'BRS Pajeú' and 39% in 'BRS Novaera' compared to water stress. However, at 16 DAT, MET foliar application increased 16.35% CAT activity in 'BRS Pajeú' compared to the water stress. In the joint application of SA and MET compared to water stress, CAT activity increased by 46% for 'BRS Pajeú' and 23% for 'BRS Novaera' at 8 DAT. However, at 16 DAT, CAT activity increased by 19% only for 'BRS Novaera' compared to water stress (Table 2).

In plants under water stress at 8 DAT, superoxide dismutase (SOD) activity increased by 136% for 'BRS Pajeú' and 118% for 'BRS Novaera' compared to the control. At 16 DAT, the SOD activity of plants under water stress decreased by 45% for 'BRS Novaera' compared to the control (Table 2). At 8 DAT, the SA application compared to water stress decreased SOD activity in 'BRS Pajeú' by 73%. At 16 DAT, the SA application compared to water stress decreased increased SOD activity by 51% for 'BRS Pajeú' and 57% for 'BRS Novaera'. Foliar application of MET (6 mM) at 8 DAT compared to water stress increased SOD activity by 220% in 'BRS Novaera' and decreased SOD activity by 74% in 'BRS Pajeú' (Table 2).

Leaf gas exchange was evaluated only at 16 DAT, and all leaf gas exchange variables of 'BRS Pajeú' decreased under water stress, with a decrease of 47% in photosynthesis, 47% in transpiration, 59% in stomatal conductance, and 37% in carbon internal compared to the control. In 'BRS Novaera' under water stress, internal carbon increased by 66% compared to the control, while photosynthesis, transpiration, and stomatal conductance decreased by 47%, 29%, and 42% compared to the control. The foliar application of SA (1.5 mM) decreased photosynthesis by 7.35% and 24%, transpiration by 30% and 41%, and stomatal conductance by 16% and 34.17% for 'BRS Pajeú' and 'BRS Novaera' compared to water stress, respectively. However, only the internal carbon of 'BRS Novaera' decreased by 33% compared to water stress. The foliar application of MET (6 mM) increased photosynthesis by 5.53%, transpiration by 6.36%, and stomatal conductance of 'BRS Pajéu' by 6.03% compared to water stress. However, in 'BRS Novaera', the application of MET decreased photosynthesis by 17.68%, transpiration by 28.57%, and stomatal conductance by 24.05% compared to water stress. In the joint application of SA and MET, only the internal carbon of 'BRS Novaera' increased by 28.24% compared to water stress (Table 2).

3. Discussion

Some cowpea cultivars (e.g., BRS Novaera) show reductions in the leaf water potential under water restriction conditions, damaging different growth indicators, including leaf area and fresh mass [13]. One of the effects most related to the reduction in growth indicators caused by water restriction is the loss of cell turgor, which restricts cell division and elongation and causes physiological changes that include disturbances in photosynthesis [14].

The probable damage caused to the photosynthetic apparatus might have favored an increase in the levels of reactive oxygen species in the cultivar BRS Novaera as soon as the first eight days of stress, justifying the increase in the activity of antioxidant enzymes such as CAT. The increase in the activity of this enzyme is due to the drought tolerance mechanism attributed to some cowpea cultivars [6]. In the cultivar BRS Novaera, stress imposition for eight weeks increased the TSS content in cowpea leaves, which is considered an expected reaction and could be related to starch degradation, favoring the action of the osmolyte as an osmotic adjuster or metabolic signaling molecule frequently involved with drought tolerance [15,16].

In addition to TSS, other compatible osmolytes, e.g., PRO and TSPs, can contribute to the osmotic adjustment process of different species under stress conditions, including cowpea [17,18]. In addition to the cultivar BRS Novaera, an increase in PRO levels after SA application (T3) was already observed in other cowpea cultivars by Andrade et al. [19]. SA is related to several regulatory functions of plant metabolism and activates defense mechanisms against water deficit, including osmotic adjustment [20]. Compatible solutes are typically hydrophilic and can replace water molecules on the surface of proteins and membranes, which increases osmotic pressure and the potential water gradient between soil and roots at the first moment, thus enabling a continuous influx of water by osmosis throughout the plant [17].

In the cowpea cultivars BRS Pajeú (G1) and BRS Novaera (G2), the application of plant elicitors eight days after the beginning of the water deficit increased the RWC, H_2O_2, and APX activity. The action mechanism of SA suggests that this acid is also responsible for increasing the concentrations of ROS, such as H_2O_2, during the first days of stress through a signaling process that leads to the activation of the cellular detoxification mechanism, thus promoting stress tolerance [21]. Applying both SA and MET can intensify the activity of antioxidant enzymes under water deficit conditions, maintaining membrane stability and increasing the plant water status [11,12]. Specifically, in the cultivar BRS Pajeú (G1), the application of elicitors might have directly contributed to water status maintenance processes unrelated to the accumulation of compatible osmolytes, e.g., PRO and TSPs, even during the first evaluation period.

Since the enzyme activity observed results from both synthesis and degradation, the net SOD activity in the cultivar BRS Novaera (G2) reduced after 16 days under water deficit. According to Liang et al. [22], a decrease in the SOD synthesis or an increase in the SOD proteolysis occurs in plants under water deficit due to disturbances in the photosynthetic mechanism. Furthermore, H_2O_2 accumulation under drought conditions can also decrease SOD activity. After eight days, MET application in the cultivar BRS Novaera (G2) and the remaining elicitors in BRS Pajeú (G1) increased the SOD levels and contributed to the recovery of the plant water status. Both MET and SA effectively removed the superoxide ion, primarily through increased SOD activity [9,11]. The increase in this enzyme's activity contributes to maintaining membrane integrity by reducing lipid peroxidation [22].

In the present study, the water deficit decreased the TFAA levels in both cowpea cultivars after eight days. According to Goufo et al. [17], cowpea can regulate its nitrogen metabolism, converting amino acids into proteins and vice versa, depending on the current needs of the plant. Therefore, the results of the present study suggest that reduced TFAA levels could be related to increased TSP concentrations, especially in the cultivar BRS Novaera (G2). The application of elicitors increased the TFAA levels of the cultivars BRS Pajeú (G1) and BRS Novaera (G2) under water restriction conditions in the first eight days. Due to the apparent plasticity of cowpea plants in controlling their N metabolism, increases in the TFAA levels are relevant not only for protein biosynthesis but also for influencing different physiological processes, e.g., growth and development. Gorni et al. [23] indicate that TFAA levels contribute to intracellular pH control, metabolic energy generation, and plant water stress tolerance.

After 16 days of stress, the plasticity of cowpea plants in regulating their morphophysiological attributes became clear. However, the photosynthetic metabolism of the species is sensitive to water deficit since the reduction in net photosynthesis occurs by stomatal closure, limiting the CO_2 supply for RuBisCO as soon as the stress is imposed [6]. In general, cowpea cultivars under water deficit show a rapid reduction in stomatal conductance, which signals the closing of stomata, followed by low photosynthetic and transpiration rates [3]. Nevertheless, stomatal closure is one of the first responses of the species, working as an efficient adaptative mechanism to control transpiration. The increase in the osmoregulatory levels such as PRO and TFAAs, in turn, could work as an osmotic adjustment in some cowpea cultivars under water restriction [3,13,17,24], which, in the present study, also happened after 16 days of stress. The active mechanism works through the synthesis and accumulation of organic solutes in the cytoplasm, and the result is the reduction in the water potential of the plant, providing a water potential gradient favorable to water uptake and the maintenance of cell turgor [17,25].

The prolonged water stress of 16 days (16 DAT) might have intensified the disturbances in photosynthetic processes and affected the growth indicators of cowpea, e.g., dry matter production and leaf area expansion [19,26] and pod weight and yield [27]. Although SA is frequently associated with beneficial effects on photosynthetic processes even under water deficit conditions, the second application of elicitors in both cultivars of the present study seems to have negatively affected the gas exchange indicators. The second application of 0.21 g L^{-1} SA in the 8-day interval may have generated momentary photosynthetic disturbances during the evaluated period. Kumar et al. [28] reported the harmful effects on the photosynthetic activity and the inhibition of the nitrate uptake system in *Trifloium alexandrinum* (L.) after the application of a high level of salicylic acid (100 μg mL^{-1}).

The increased ROS levels observed after 16 days of stress, especially in the cultivar BRS Pajeú (G1), could result in lipid peroxidation, protein oxidation, inhibited enzyme activity, oxidative damage to RNA and DNA, and cell death [11,29]. In the present study, the modulation in the biochemical mechanism of cowpea plants under water restriction becomes more evident over time. After 16 days under stress, the increase in the levels of antioxidant enzymes in both cultivars suggests the intensification of the antioxidant metabolism mediated by SOD, which converts O^{2-} into H_2O_2 so that this oxidizing agent can be more easily converted into H_2O, preferably by the enzyme APX at the second moment, reducing the oxidative damage to plant cells [19]. In the referred period, the intensification of SOD and APX activity after SA application in both cultivars and MET application in the cultivar BRS Novaera (G2), under water restriction, was also observed by Andrade et al. [19] and Merwad et al. [12] in other cowpea cultivars.

After 16 days under water deficit, stress continued to restrict the water supply to leaf tissues, causing oxidative damage to cowpea plants, represented by the increased H_2O_2 levels. As a result, high H_2O_2 levels can interrupt the photosynthetic machinery [30]. However, changes to the gene expression and protein levels are also triggered during the stress period [31], inducing increases in cowpea [32]. The application of elicitors in both cultivars, especially SA, enabled plants to resume the water supply and balance the oxidative metabolism by increasing the leaf water potential and reducing the H_2O_2 levels [11]. Salicylic acid protects the cell membranes and their carrier proteins, which maintain their structure and function against the toxic and disruptive effects of reactive oxygen species released during stress [33]. At a higher intensity in the cultivar BRS Novaera (G2), MET application after 16 DAT effectively reduced the H_2O_2 contents in stressed plants, which may have contributed to reducing the lipid peroxidation levels and, consequently, maintaining the damage to cell membranes [11,12].

The divergences observed between the cowpea cultivars BRS Pajeú (G1) and BRS Novaera (G2) after 16 DAT for the SOD and TFAA contents under stress conditions in the absence and presence of elicitors highlight the evidence that the responses of plants of the same species to stress conditions depend on each genotype [4,34]. In the cultivar BRS Pajeú (G1), water restriction may have reduced nitrogen assimilation even after the second application of elicitors [12] since this nutrient is an indispensable intermediate in the nitrogen metabolism and the biosynthesis of amino acids. After SA application, the intensified reduction in TFAA levels indicates the participation of the elicitor in inducing the expression of 11 new cowpea proteins in plants subjected to water stress, which is related to the improvement in growth and production [8]. On the other hand, in the cultivar BRS Novaera (G2) under water restriction for 16 days, the increase in the TFAA levels could be related to the osmotic adjustment process, in which the osmoprotectant mediated by these molecules reduced the damage to cell membranes and, consequently, the levels of ROS, possibly justifying the reduction in antioxidant activity. SA application intensified the synthesis of TFAAs and highlighted the importance of these molecules in different anti-stress metabolic pathways since they act as signaling molecules and precursors for synthesizing plant hormones or secondary metabolites of the defense mechanism [23].

4. Materials and Methods

4.1. Location of the Research Area and Experimental Design

The experiment was conducted from October to December 2019 at the Forest Garden (Horto Florestal) area (an extension of the Integrated Research Complex of Três Marias, Campus I) of the State University of Paraíba, located in Campina Grande—PB ($07°13'50''$ S, $35°52'52''$ W, at an elevation of 551 m). The climate is classified as *BSh*, according to Köppen and Geiger, with a mean temperature of 23.3 °C and a mean annual rainfall of 503 mm [35].

We subjected two cowpea cultivars (BRS Novaera and BRS Pajeú) to five treatments, described as follows: control (T_1: 100% of soil field capacity); stress (T_2: 50% of soil field capacity); T_3: stress + salicylic acid application − SA (0.21 g L^{-1}); T_4: stress + methionine application − MET (0.89 g L^{-1}); T_5: stress + SA (0.21 g L^{-1}) + MET (0.89 g L^{-1}). The SA and MET concentrations were established based on previous studies conducted by Dutra et al. [9] and Merwad et al. [12]. The experiment was set up in a completely randomized design with five replications (n = 5), totaling fifty experimental units composed of two plants per pot.

4.2. Conduction of the Experiment

Pots with a capacity of 3.6 dm^3 were filled with clayey-sandy soil whose physical and fertility characteristics were as follows: sand: 659 g kg^{-1}; silt: 101 g kg^{-1}; clay: 240 g kg^{-1}; apparent density: 1.38 kg dm^{-3}; particle density: 2.63 kg dm^{-3}; total porosity: 0.48 m^3 m^{-3}; calcium: 2.38 cmol$_c$ dm^{-3}; magnesium: 1.66 cmol$_c$ dm^{-3}; sodium: 0.23 cmol$_c$ dm^{-3}; potassium: 0.14 cmol$_c$ dm^{-3}; hydrogen + aluminum: 5.69 cmol$_c$ dm^{-3}; and organic matter: 20.38 g kg^{-1}; pH: 4.8.

The soil was corrected to increase the base saturation percentage to 70% using 4.8 mg $CaCO_3$ (pure for analysis) per pot. After correction, the soil was moist for 30 days before sowing. We applied 20 kg ha^{-1} of P_2O_5, 30 kg ha^{-1} of N, and 35 kg ha^{-1} of K_2O [36], corresponding to 36 mg of P_2O_5, 54 mg of N, and 63 mg of K_2O per pot. The fertilizers used were urea (45% N), monoammonium phosphate (12% N and 65% P_2O_5), and potassium chloride (60% K_2O). Monoammonium phosphate was applied before sowing. Urea and potassium chloride were applied in equal portions at $30°$, $47°$, and $52°$ days after sowing. The seeds used in the experiment were processed to remove those with physical damage and malformations. Subsequently, the seeds were exposed to the preventive fungicide

Captan® (Captana 800 g kg^{-1} a.i) at 0.11 g per 100 g^{-1} of seeds and were left to rest for 24 h. Soil saturation was performed before sowing for 48 h, followed by excess water drainage to maintain the soil at field capacity on the sowing [37]. The pots were arranged in five rows with ten pots each and spaced 0.8 m between tows and 0.6 m between pots.

Sowing was performed by equidistantly distributing five seeds per pot at a mean depth of two centimeters. The plants were thinned to two seedlings per pot 17 days after sowing. Water replenishment was performed daily and manually from sowing to the beginning of treatments. This procedure was performed according to the water volume lost by evapotranspiration in each pot, with irrigation using 100 and 50% of the water volume lost by evapotranspiration in the control and other treatments (stress, stress + SA, stress + MET, and stress + SA + MET), respectively.

Water replenishment was performed daily by weighing the pots and calculating the water volume to be replenished based on the difference between the maximum storage–AM (water capacity available in the pot) and the current storage—AT, according to Casaroli and van Lier [38], using the following equation:

$$I = AM - AT \tag{1}$$

where:

I—necessary irrigation, L pot^{-1};
AM—maximum storage, L pot^{-1};
AT—current storage, L pot^{-1}.

When the plants reached the developmental stage V8 (55 days after sowing—DAS), the application of treatments began with water restriction and the first application of elicitor substances at a level of 20 mL per plant, totaling 40 mL per pot for each solution (SA, MET, and SA + MET). The elicitors were applied via spraying until the runoff point of the solution in the abaxial and adaxial regions of the leaves.

After eight days of treatment (8 DAT), one plant from each pot was collected for destructive analysis using the following parameters: water status (water potential—Ψw; relative water content—RWC), growth (leaf area—LA and total fresh mass—FM), osmoregulators (total soluble sugars—TSS, proline—PRO, total free amino acids—TFAAs, total soluble proteins—TSPs), and the antioxidant mechanism (hydrogen peroxide—H_2O_2, superoxide dismutase—SOD, ascorbate peroxidase—APX, and catalase—CAT). The treatments were reapplied in the plants that remained in the pots 8 DAT. After eight days of the reapplication, the gas exchange variables were measured at 16 DAT, followed by sample collection to measure these variables.

4.3. Growth Analysis and Plant Water Status

The Ψw measurements in the plant petioles were performed with a Scholander 3005F01 pressure chamber (Soil Moisture Corp., Santa Barbara, CA, USA) from 3:00 to 5:00 a.m., with values expressed as MPa [39]. For the RWC analysis, three fresh leaf disks were removed with a copper cutter and weighed (DFM), immersed in 10 mL of distilled water for 24 h, and weighed again to obtain the turgid mass (DTM). Then, the material was dried in a forced air oven at 80 °C for 24 h to obtain its dry mass (DDM). The RWC was calculated using the equation proposed by Smart and Bingham [40]:

$$RWC(\%) = \frac{(DFM - DDM)}{(DTM - DDM)} \times 100 \tag{2}$$

where: RWC (%) = relative water content. DFM = fresh disk mass. DTM = turgid disk mass. DDM = dry disk mass.

The leaf area (LA) was determined using a planimeter (LI-3100, USA) and expressed as cm^2. The fresh leaf mass (FM) was obtained by weighing the plant material in an analytical balance.

4.4. Analysis of Compatible Osmotic Concentrations

TSS quantification in leaf blades was performed by the phenol-sulfuric acid method described by Dubois et al. [41], with readings in a spectrophotometer at 490 nm of absorbance and expressed as mg TSS g^{-1} of FM. PRO quantification in leaf blades was performed by the colorimetric method proposed by Bates et al. [42], with readings at 520 nm. The concentrations were based on the L-proline standard curve and were expressed as μmol of PRO g^{-1} of FM.

The TFAA concentration in leaf blades was determined according to the method described by Peoples et al. [43], with readings at 570 nm quantified according to the glutamine standard curve and expressed as μmol TFAA g^{-1} FM. The TSP concentration was determined according to Bradford [44], with readings at 595 nm and data described as mg TSP g^{-1} FM using the albumin standard curve as a reference.

Analysis of Antioxidant Mechanism Components

H_2O_2 quantification in leaf blades was performed following the method of Velikova et al. [45], with readings at 390 nm and concentrations expressed as μmol H_2O_2 g^{-1} FM. SOD activity in leaf blades was determined based on the photoreduction inhibition capacity of nitro blue tetrazolium chloride (NBT) by the enzyme of the plant extract, following the method of Beauchamp and Fridovich [46] and expressed as U min^{-1} mg^{-1} of protein, with readings at 560 nm of absorbance. CAT activity in leaf blades was quantified according to Kar and Mishra [47], with readings at 240 nm expressed as μmol of H_2O_2 min^{-1} mg^{-1} of protein.

APX activity in leaf blades was determined by the method proposed by Nakano and Asada [48], calculated based on ascorbate consumption by monitoring the decrease in absorbance in 10 readings at 290 nm in a quartz cuvette, expressed as nmol of ascorbate min^{-1} mg^{-1} of protein.

4.5. Leaf Gas Exchange

The gas exchange variables were measured in the middle region of the plant, represented by net photosynthesis—A (μmol CO_2 m^{-1} s^{-1}), transpiration—E (mmol H_2O m^{-1} s^{-1}), stomatal conductance—gs (mmol H_2O m^{-1} s^{-1}), and internal CO_2 concentration—Ci (ppm) using an infrared gas analyzer—IRGA (Infra-red Gas Analyzer)—GFS 3000 FL. All measurements were performed in the morning, between 8:00 and 11:00 a.m., in a leaf area of 8 cm^2 using an artificial radiation source with an intensity of 1200 μmol m^{-2} s^{-1} under control cuvette conditions at a temperature of 26.4 °C ($\pm$ 1), relative humidity of 60% ($\pm$1), and CO_2 concentration of 400 μmol mol^{-1}.

4.6. Statistical Analysis

The data on the original variables were standardized to obtain the zero mean and unit variance ($\bar{x} = 0$ and $s^2 = 1$) and evaluated by principal component analysis (PCA) and multivariate analysis (MANOVA) by Hotelling's test ($p \leq 0.05$). The means of the scores of each principal component (PC) for the cultivation conditions were compared by Tukey's test ($p \leq 0.05$), and the means of the cultivars were compared by Student's *t*-test ($p \leq 0.05$).

5. Conclusions

In the present study, exogenous applications of salicylic acid and methionine in cowpea plants under water restriction modulated the osmoregulation metabolism of free amino acids, proline, soluble proteins, and free carbohydrates, in addition to the antioxidant activity of the SOD, APX, and CAT enzymes, improving the water status after eight days of application of treatments. Furthermore, after 16 days, the cowpea cultivars under stress showed reduced water loss by transpiration, which may have regulated the photosynthetic processes.

The regulations in the cultivar BRS Novaera induced by salicylic acid and methionine were more expressive and beneficial to intensifying the tolerance mechanism of the species to water restriction. Therefore, both elicitors act as modulators of the species' metabolism and effectively mitigate the effects of water stress. Further research is required to identify the influence of elicitors in the reproductive phase and their consequent representativeness in the production indicators of the species, including under field conditions.

Author Contributions: Conceptualization, A.S.d.M.; methodology, A.P.d.S.O., Y.L.M. and A.S.d.M.; validation, P.R.A.V., A.S.d.M. and F.V.d.S.S.; investigation, A.P.d.S.O., G.F.D., R.S.d.A. and Y.L.M.; resources, R.L.d.S.F. and A.S.d.M.; writing—original draft preparation, A.S.d.M., A.P.d.S.O., R.L.d.S.F., P.R.A.V. and Y.L.M. writing—review and editing, R.L.d.S.F., J.D.N., A.S.d.M., F.V.d.S.S., H.R.G., I.D.M., C.F.d.L. and Y.L.M.; supervision, A.S.d.M., Y.L.M., R.L.d.S.F. and A.P.d.S.O.; funding acquisition, A.S.d.M. All authors have read and agreed to the published version of the manuscript.

Funding: This research was funded by Conselho Nacional de Desenvolvimento Científico e Tecnológico—CNPq, research project No. 306155/2019-2, Coordenação de Aperfeiçoamento de Pessoal de Nível Superior—Brasil (CAPES)—Brazil: finance code 001 and n.8881.707827/2022-01, and the State University of Paraíba (UEPB-PRPGP)—finance.

Institutional Review Board Statement: Not applicable.

Informed Consent Statement: Not applicable.

Data Availability Statement: Not applicable.

Acknowledgments: The Coordination for the Improvement of Higher Education Personnel—Brazil (CAPES)—finance code 001. To the Conselho Nacional de Desenvolvimento Científico e Tecnológico—CNPq, for granting financial aid (Proc. 306155/2019-2).

Conflicts of Interest: The authors declare no conflict of interest.

References

1. Martey, E.; Etwire, P.M.; Mockshell, J. Climate-smart cowpea adoption and welfare effects of comprehensive agricultural training programs. *Technol. Soc.* **2021**, *64*, 101468. [CrossRef] [PubMed]
2. Tavares, D.S.; Fernandes, T.E.K.; Rita, Y.L.; Rocha, D.C.; Sant'anna-Santos, B.F.; Gomes, M.P. Germinative metabolism and seedling growth of cowpea (*Vigna unguiculata*) under salt and osmotic stress. *S. Afr. J. Bot.* **2021**, *139*, 399–408. [CrossRef]
3. Rivas, R.; Falcão, H.M.; Ribeiro, R.V.; Machado, E.C.; Pimentel, C.; Santos, M.G. Drought tolerance in cowpea species is driven by less sensitivity of leaf gas exchange to water deficit and rapid recovery of photosynthesis after rehydration. *S. Afr. J. Bot.* **2016**, *103*, 101–107. [CrossRef]
4. Gomes, A.M.; Rodrigues, A.P.; António, C.; Rodrigues, A.M.; Leitão, A.E.; Batista-Santos, P.; Nhantumbo, N.; Massinga, R.; Ribeiro-Barros, A.I.; Ramalho, J.C. Drought response of cowpea (*Vigna unguiculata* (L.) Walp.) landraces at leaf physiological and metabolite profile levels. *Environ. Exp. Bot.* **2020**, *175*, e104060. [CrossRef]
5. Carvalho, M.; Castro, I.; Moutinho-Pereira, J.; Correia, C.; Egea-Cortines, M.; Matos, M.; Lino-Neto, T. Evaluating stress responses in cowpea under drought stress. *J. Plant Physiol.* **2019**, *241*, e153001. [CrossRef]
6. Melo, A.S.; Melo, Y.L.; Lacerda, C.F.; Viégas, P.R.; Ferraz, R.L.S.; Gheyi, H.R. Water restriction in cowpea plants [*Vigna unguiculata* (L.) Walp.]: Metabolic changes and tolerance induction. *Rev. Bras. Eng. Agrícola Ambient.* **2022**, *26*, 190–197. [CrossRef]
7. Chakraborty, N.; Sarkar, A.; Acharya, K. Elicitor-mediated amelioration of abiotic stress in plants. In *Molecular Plant Abiotic Stress: Biology and Biotechnology*; John Wiley & Sons, Ltd.: Hoboken, NJ, USA, 2019; pp. 105–122. [CrossRef]
8. Nassef, D.M. Impact of irrigation water deficit and foliar application with salicylic acid on the productivity of two cowpea cultivars. *Egypt. J. Hortic.* **2017**, *44*, 75–90. [CrossRef]
9. Dutra, W.F.; Melo, A.S.; Suassuna, J.F.; Dutra, A.F.; Silva, D.C.; Maia, J.M. Antioxidative responses of cowpea cultivars to water deficit and salicylic acid treatment. *Agron. J.* **2017**, *109*, 895–905. [CrossRef]
10. Araújo, E.D.; Melo, A.S.; Rocha, M.S.; Carneiro, R.F.; Rocha, M.M. Germination and initial growth of cowpea cultivars under osmotic stress and salicylic acid. *Rev. Caatinga* **2018**, *31*, 80–89. [CrossRef]
11. Mehak, G.; Akram, N.A.; Ashraf, M.; Kaushik, P.; El-Sheikh, M.A.; Ahmad, P. Methionine-induced regulation of growth, secondary metabolites and oxidative defense system in sunflower (*Helianthus annuus* L.) plants subjected to water deficit stress. *PLoS ONE* **2021**, *16*, e0259585. [CrossRef]
12. Merwad, A.R.M.; Desoky, E.S.M.; Rady, M.M. Response of water deficit-stressed *Vigna unguiculata* performances to silicon, proline or methionine foliar application. *Sci. Hortic.* **2018**, *228*, 132–144. [CrossRef]
13. Silva, D.C.; Melo, A.S.; Melo, Y.L.; Andrade, W.L.; Lima, L.M.; Santos, A.R. Silicon foliar application attenuates the effects of water suppression on cowpea cultivars. *Ciência Agrotecnologia* **2019**, *43*, 1–10. [CrossRef]

14. Freitas, R.M.O.; Dombroski, J.L.D.; Freitas, F.C.L.; Nogueira, N.W.; Pinto, J.R.S. Physiological responses of cowpea under water stress and rewatering in no-tillage and conventional tillage systems. *Rev. Caatinga* **2017**, *30*, 559–567. [CrossRef]
15. Mohamed, H.I.; Latif, H.H. Improvement of drought tolerance of soybean plants by using methyl jasmonate. *Physiol. Mol. Biol. Plants* **2017**, *23*, 545–556. [CrossRef]
16. Sadeghipour, O. Drought tolerance of cowpea enhanced by exogenous application of methyl jasmonate. *Int. J. Mod. Agric.* **2018**, *7*, 51–57.
17. Goufo, P.; Moutinho-Pereira, J.M.; Jorge, T.F.; Correia, C.M.; Oliveira, M.R.; Rosa, E.A.S.; Trindade, H. Cowpea (*Vigna unguiculata* L. Walp.) metabolomics: Osmoprotection as a physiological strategy for drought stress resistance and improved yield. *Front. Plant Sci.* **2017**, *8*, 586. [CrossRef]
18. Zhang, W.; Yu, X.; LI, M.; Lang, D.; Zhang, X.; Xie, Z. Silicon promotes growth and root yield of Glycyrrhiza uralensis under salt and drought stresses through enhancing osmotic adjustment and regulating antioxidant metabolism. *Crop Prot.* **2018**, *107*, 1–11. [CrossRef]
19. Andrade, W.L.d.; Melo, A.S.d.; Melo, Y.L.; Sá, F.V.d.S.; Rocha, M.M.; Oliveira, A.P.S.; Fernandes Júnior, P.I. *Bradyrhizobium* inoculation plus foliar application of salicylic acid mitigates water deficit effects on cowpea. *J. Plant Growth Regul.* **2021**, *40*, 656–667. [CrossRef]
20. Afshari, M.; Shekari, F.; Azimkhani, R.; Habibi, H.; Fotokian, M.H. Effects of foliar application of salicylic acid on growth and physiological attributes of cowpea under water stress conditions. *Iran Agric. Res.* **2013**, *32*, 55–70. [CrossRef]
21. Horváth, E.; Szalai, G.; Janda, T. Induction of abiotic stress tolerance by salicylic acid signaling. *J. Plant Growth Regul.* **2007**, *26*, 290–300. [CrossRef]
22. Liang, G.; Bu, J.; Zhang, S.; Jing, G.; Zhang, G.; Liu, X. Effects of drought stress on the photosynthetic physiological parameters of Populus x euramericana "Neva". *J. For. Res.* **2019**, *30*, 409–416. [CrossRef]
23. Gorni, P.H.; Pacheco, A.C.; Moro, A.L.; Silva, J.F.A.; Moreli, R.R.; Miranda, G.R.; Silva, R.M.G. Salicylic acid foliar application increases biomass, nutrient assimilation, primary metabolites and essential oil content in *Achillea millefolium* L. *Sci. Hortic.* **2020**, *270*, e109436. [CrossRef]
24. Costa, R.C.L.; Lobato, A.K.S.; Silveira, J.A.G.; Laughinghouse, H.D. ABA-mediated proline synthesis in cowpea leaves exposed to water deficiency and rehydration. *Turk. J. Agric. For.* **2011**, *35*, 309–317. [CrossRef]
25. Turner, N.C. Turgor maintenance by osmotic adjustment, an adaptive mechanism for coping with plant water deficits. *Plant Cell Environ.* **2017**, *40*, 1–3. [CrossRef]
26. Melo, A.S.; Silva, A.R.F.; Dutra, A.F.; Dutra, W.F.; Brito, M.E.B.; SÁ, F.V.S. Photosynthetic efficiency and production of cowpea cultivars under deficit irrigation. *Rev. Ambiente Água* **2018**, *13*, 1–8. [CrossRef]
27. Dutra, A.F.; Melo, A.S.; Filgueiras, L.M.B.; Silva, Á.R.F.; Oliveira, I.M.; Brito, M.E.B. Parâmetros fisiológicos e componentes de produção de feijão-caupi cultivado sob deficiência hídrica. *Rev. Bras. Ciências Agrárias* **2015**, *10*, 189–197. [CrossRef]
28. Kumar, B.; Ram, H.; Sarlach, R.S. Enhancing seed yield and quality of Egyptian clover (*Trifolium alexandrinum* L.) with foliar application of bio-regulators. *Field Crops Res.* **2013**, *146*, 25–30. [CrossRef]
29. Tamburino, R.; Vitale, M.; Ruggiero, A.; Sassi, M.; Sannino, L.; Arena, S.; Scotti, N. Chloroplast proteome response to drought stress and recovery in tomato (*Solanum lycopersicum* L.). *BMC Plant Biol.* **2017**, *17*, 40. Available online: https://bmcplantbiol.biomedcentral.com/articles/10.1186/s12870-017-0971-0 (accessed on 13 February 2023). [CrossRef] [PubMed]
30. Ojuederie, O.B.; Olanrewaju, O.S.; Babalola, O.O. Plant growth promoting rhizobacterial mitigation of drought stress in crop plants: Implications for sustainable agriculture. *Agronomy* **2019**, *9*, 712. [CrossRef]
31. Caverzan, A.; Bonifacio, A.; Carvalho, F.E.; Andrade, C.M.; Passaia, G.; Schünemann, M.; Margis-Pinheiro, M. The knockdown of chloroplastic ascorbate peroxidases reveals its regulatory role in the photosynthesis and protection under photo-oxidative stress in rice. *Plant Sci.* **2014**, *214*, 74–87. [CrossRef]
32. Mishra, S.; Sahu, G.; Shaw, B.P. Insight into the cellular and physiological regulatory modulations of Class-I TCP9 to enhance drought and salinity stress tolerance in cowpea. *Physiol. Plant.* **2022**, *174*, e13542. [CrossRef] [PubMed]
33. Kumar, B.; Sarlach, R.S. Forage cowpea (*Vigna unguiculata*) seed yield and seed quality response to foliar application of bio-regulators. *Int. J. Agric. Environ. Biotechnol.* **2015**, *8*, 891. [CrossRef]
34. Iwuagwu, M.O.; Ogbonnaya, C.I.; Onyike, N.B. Physiological response of cowpea [*Vigna unguiculata* (L.) Walp.] to drought: The osmotic adjustment resistance strategy. *Acad. J. Sci.* **2017**, *7*, 329–344.
35. Francisco, P.R.M.; Medeiros, R.D.; Santos, D.; Matos, R.D. Classificação climática de Köppen e Thornthwaite para o estado da Paraíba. *Rev. Bras. Geogr. Física* **2015**, *8*, 1006–1016. [CrossRef]
36. Melo, F.B.; Cardoso, M.J.; Salviano, A.A.C. Soil fertility and fertilization. In *Cowpea: Technological Advances*, 1st ed.; Freire-Filho, F.R., Lima, J.A.A., Ribeiro, V.Q., Eds.; Embrapa Meio Norte: Brasília, Brasil, 2005; Volume 1, pp. 229–242.
37. Feitosa, S.S.; Albuquerque, M.B.; Oliveira, A.P.; Pereira, W.E.; Brito-Neto, J.F. Fisiologia do *Sesamum indicum* L. sob estresse hídrico e aplicação de ácido salicílico. *Irriga* **2016**, *21*, 711–723. [CrossRef]
38. Casaroli, D.; Jong Van Lier, Q.D. Critérios para determinação da capacidade de vaso. *Rev. Bras. Ciência Solo* **2008**, *32*, 59–66. [CrossRef]
39. Scholander, P.F.; Hammel, H.T.; Hemmingsen, E.A.; Bradstreet, E.D. Sap pressure in vascular plants. *Science* **1965**, *148*, 339–346. [CrossRef]
40. Smart, R.E.; Bingham, G.E. Rapid estimates of relative water content. *Plant Physiol.* **1974**, *53*, 258–260. [CrossRef]

41. Dubois, M.; Gilles, K.A.; Hamilton, J.K.; Rebers, P.A.; Smith, F. Colorimetric method for determination of sugars and related substances. *Anal. Chem.* **1956**, *28*, 350–356. [CrossRef]

42. Bates, L.S.; Waldren, R.P.; Teare, I.D. Rapid determination of free proline for water-stress studies. *Plant Soil* **1973**, *39*, 205–207. [CrossRef]

43. Peoples, M.B.; Faizah, A.W.; Rerkasem, B.; Herridge, D.F. *Methods for Evaluating Nitrogen Fixation by Nodulated Legumes in the Field*; Australian Centre for International Agricultural Research: Canberra, Australia, 1989; 79p.

44. Bradford, M.M. A rapid and sensitive method for the quantitation of microgram quantities of protein utilizing the principle of protein-dye binding. *Anal. Biochem.* **1976**, *72*, 248–254. [CrossRef] [PubMed]

45. Velikova, V.; Yordanovi, I.; Edreva, A. Oxidative stress and some antioxidante systems in acid raintreated bean plants. *Plant Sci.* **2000**, *151*, 59–66. [CrossRef]

46. Beauchamp, C.; Fridovich, I. Superoxide dismutase: Improved assays and an assay applicable to acrylamide gels. *Anal. Biochem.* **1971**, *44*, 276–287. [CrossRef]

47. Kar, M.; Mishra, D. Catalase, peroxidase, and polyphenoloxidase activities during rice leaf senescence. *Plant Physiol.* **1976**, *57*, 315–319. [CrossRef] [PubMed]

48. Nakano, Y.; Asada, K. Hydrogen peroxide is scavenged by ascorbate-specific peroxidases in spinach chloroplast. *Plant Cell Physiol.* **1981**, *22*, 867–880. [CrossRef]

Article

Bacilli Rhizobacteria as Biostimulants of Growth and Production of Sesame Cultivars under Water Deficit

Giliard Bruno Primo de Lima [1], Erika Fernandes Gomes [2], Geisenilma Maria Gonçalves da Rocha [3], Francisco de Assis Silva [4], Pedro Dantas Fernandes [4], Alexandre Paulo Machado [5], Paulo Ivan Fernandes-Junior [6], Alberto Soares de Melo [1], Nair Helena Castro Arriel [3], Tarcisio Marcos de Souza Gondim [3] and Liziane Maria de Lima [3,*]

[1] Agricultural Sciences, UEPB, Campina Grande 58429-500, PB, Brazil
[2] Biological Sciences, UEPB, Campina Grande 58429-500, PB, Brazil
[3] Embrapa Algodão, Campina Grande 58428-095, PB, Brazil
[4] Agricultural Engineering, UFCG, Campina Grande 58429-900, PB, Brazil
[5] Department of Basic Health Sciences, Federal University of Mato Grosso, Campo Grande 78060-900, MT, Brazil
[6] Embrapa Semiárido, Petrolina 56302-970, PE, Brazil
* Correspondence: liziane.lima@embrapa.br

Abstract: A strategy using bacilli was adopted aiming to investigate the mitigation of the effects of water deficit in sesame. An experiment was carried out in a greenhouse with 2 sesame cultivars (BRS Seda and BRS Anahí) and 4 inoculants (pant001, ESA 13, ESA 402, and ESA 441). On the 30th day of the cycle, irrigation was suspended for eight days, and the plants were subjected to physiological analysis using an infrared gas analyzer (IRGA). On the 8th day of water suspension, leaves were collected for analysis: superoxide dismutase, catalase, ascorbate peroxidase, proline, nitrogen, chlorophyll, and carotenoids. At the end of the crop cycle, data on biomass and vegetative growth characteristics were collected. Data were submitted for variance analysis and comparison of means by the Tukey and Shapiro–Wilk tests. A positive effect of inoculants was observed for all characteristics evaluated, contributing to improvements in plant physiology, induction of biochemical responses, vegetative development, and productivity. ESA 13 established better interaction with the BRS Anahí cultivar and ESA 402 with BRS Seda, with an increase of 49% and 34%, respectively, for the mass of one thousand seeds. Thus, biological indicators are identified regarding the potential of inoculants for application in sesame cultivation.

Keywords: abiotic stress; water restriction; rhizobacteria; osmoprotection system; leaf gas exchange; sesame

Citation: Lima, G.B.P.d.; Gomes, E.F.; Rocha, G.M.G.d.; Silva, F.d.A.; Fernandes, P.D.; Machado, A.P.; Fernandes-Junior, P.I.; Melo, A.S.d.; Arriel, N.H.C.; Gondim, T.M.d.S.; et al. Bacilli Rhizobacteria as Biostimulants of Growth and Production of Sesame Cultivars under Water Deficit. *Plants* **2023**, *12*, 1337. https://doi.org/10.3390/plants12061337

Academic Editor: Daniela Businelli

Received: 6 February 2023
Revised: 28 February 2023
Accepted: 8 March 2023
Published: 16 March 2023

1. Introduction

A broadly defined group of rhizobacteria has been associated with plants, as promoters of plant growth and root development by several mechanisms, such as biological nitrogen fixation [1,2], in the release and uptake of insoluble nutrients (e.g., iron and phosphorus), stimulation of phytohormone synthesis [3], suppression of pathogens through the production of antibiotics and siderophores [2,4], among others, establishing beneficial associations with their host plants.

In drylands (arid and semi-arid regions), the association of plants (both crops and native species) with stimulating rhizobacteria is important to the plant's establishment and development, since in addition to providing nutrients and stimulating growth, these microbes also help plants cope with drought stress [5–8]. These bacteria could be used as inoculants in agriculture since biofertilizers with high concentrations of microorganisms act on the growth and development of plants, increasing crop production, preserving soil life, reducing production costs, without causing damage to water resources or emissions of pollutants, and acting upon soil bioremediation [9,10]. In addition, due to the mitigating action on the negative effects caused by biotic and abiotic stresses [11,12], inoculants can be used as a strategy to promote various agricultural crops in arid and semi-arid environments. In drylands, the effects of inoculants are already observed in several crops, for example, maize (*Zea mays*) [13], sorghum (*Sorghum bicolor*) [14], pear-millet (*Pennissetum glaucum*) [15], and sesame (*Sesamum indicum*) [16,17], indicating the potential of selected bacteria for agricultural applications in the field.

Sesame (*Sesamum indicum* L.) is an agricultural food crop that is oleaginous and has seeds with excellent quality oil as well as antioxidants from the presence of sesamol, sesamolin, and sesamin [18]. This crop is used in the production of various foods, in the medicinal and pharmaceutical industries, and also consumed in nature and in animal feed. It is easy to grow and tolerant to certain periods of drought, which makes it an ideal alternative for small and medium rural producers [19]. However, to achieve maximum yield, it requires well-distributed rainfall during the different phases of its cycle [20]. In this context, the application of stimulating bacteria could benefit plant growth under water deficit conditions.

Water deficit is limiting for water relations at the cellular level or for the whole organism of plants, causing economic losses in agriculture [21]. Rhizobacteria can act on the regulation of plant physiology and produce compounds that act as mediators for abiotic stress in plants by inducing the expression of specific genes [22–24]. Drought is a serious problem for agricultural crops, and to overcome this limitation, technologies are emerging, such as the use of new inoculants [5,7], aiming to promote increased production, especially in environments where climatic conditions are limiting, such as arid and semi-arid environments.

Based on this perspective, this study aimed to investigate the vegetative growth and yield of two sesame cultivars in interaction with *Bacillus* spp. inoculants, as well as the physiological and biochemical behavior of plants, under water deficit conditions.

2. Results and Discussion

2.1. Gas Exchange and Fluorescence

For both cultivars, Ci (internal concentration of CO_2) increased in most treatments under water stress conditions (Figure 1a). Sesame genotypes under salt stress with average Ci below 200 μmol mol^{-1} indicate low photosynthetic activity [25]. Decreases in Ci present stomatal limitation, impairing photosynthetic performance, as the greater the stomatal opening, the greater the diffusion of carbon dioxide to the substomatic chamber, increasing Ci and consequently favoring photosynthesis. When comparing the data referring to Ci and iEC (instantaneous efficiency of carboxylation) (Figure 1e), it was possible to observe a low efficiency of carboxylation, implying that the accumulation of CO_2 present, especially in treatments under water stress, was not transferred to the substomatal chamber where it would be used in the incorporation of photoassimilates resulting from the photosynthetic process.

As expected, with the decrease in the other variables, the A (photosynthesis) (Figure 1b) was also affected in plants under water restriction, with an average reduction of 60%. Photosynthesis decreases as a function of stomatal closure and the persistence of water deficit [26]. The reduction in photosynthetic rate is correlated with several factors, such as the amount of water absorbed, the CO_2 fixed by the plant due to stomatal closure, as well as by the diffusive resistance of the stomata that limits the gaseous conduction of the leaf [27].

For E (transpiration), a reduction was observed in both cultivars in all treatments under water deficit (Figure 1c). The reduction in E comes from stomatal closure since plants under water restriction or at high temperatures tend to close their stomata so as not to lose water, as highlighted in studies with sesame plants under water deficit [28,29]. E is a crucial factor for processes such as leaf temperature regulation. E occurs in the stomata, also known as guard cells, which are microscopic pores present in the leaf epidermis that capture biotic and abiotic stimuli from the internal or external environment and respond quickly with mechanisms to close the stomata under unfavorable conditions, such as water deficit [30,31].

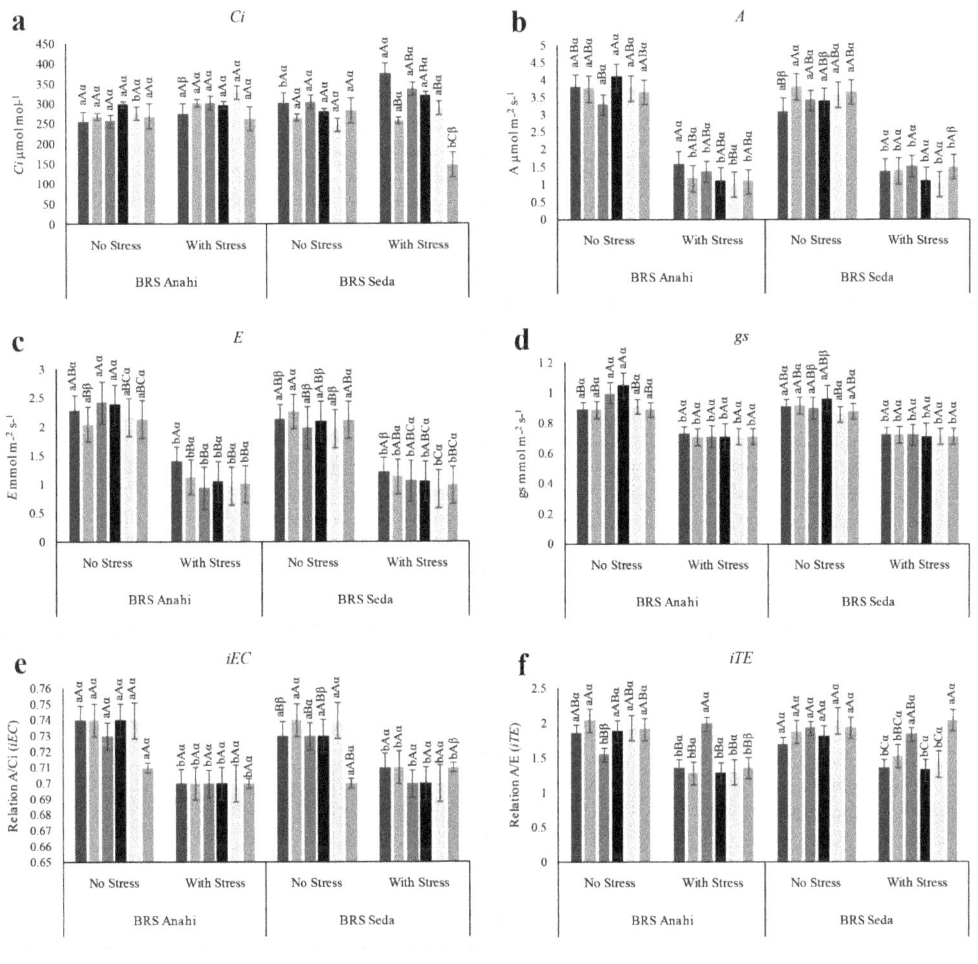

Figure 1. (**a**) Internal CO_2 concentration (Ci), (**b**) photosynthesis (A), (**c**) transpiration (E); (**d**) stomatic conductance (gs), (**e**) instantaneous efficiency of carboxylation (iEC), (**f**) instantaneous transpiration efficiency (iTE) in sesame cultivars (BRS Anahí and BRS Seda) inoculated with bacilli, under water deficit and different sources of nitrogen. WN-with nitrogen and NN-no nitrogen. Triple interactions: lowercase letters compare the water regime within each cultivar; capital letters compare treatments with inoculants within each cultivar; and Greek letters compare the cultivars.

The gs (stomatal conductance) showed no significant difference in the triple interaction in the cultivars under water deficit; however, in the irrigated condition, ESA 441 improved the dynamics of stomatal conduction in both cultivars (Figure 1d). The interaction of cultivars with the water regime showed similar behavior for both cultivars in the stress condition. The interaction between the water regime and the treatments showed a significant difference; therefore, the photosynthetic metabolism had its functioning compromised by the low absorption of CO_2 or the direct effect of the stress caused by water deficit [32].

The A/Ci ratio implies the instantaneous efficiency of carboxylation (iEC). iEC closely matches intracellular CO_2 concentration and carbon dioxide assimilation rate. Variations in the optimal temperature (between 20 °C and 30 °C) or factors considered stressful to plants, such as salinity and water deficit, can cause a restriction in the flow of CO_2 to the carboxylation site [33], thus hindering cells' metabolism in the use of substrate for plant cell biosynthesis. For iEC in the triple interaction, inoculants and nitrogen management provided greater carboxylation efficiency in the irrigated condition (Figure 1e). No statistical difference was observed in the treatments in BRS Anahí in the stressed condition; however, BRS Seda was more favored by inoculation with pant001, ESA 13, and WN in the same condition, being relatively more efficient.

ESA 402 proved to be efficient in maintaining a higher iTE standard in both cultivars, especially under water restriction (Figure 1f). Similar results were described in sesame plants under water stress and with the application of salicylic acid [34]. The same bacteria also improved the water use efficiency of sorghum under drought [7], indicating the potential of this strain for inoculating multiple crops.

When the fluorescence (Fo, Fv, Fm, and Fv/Fm) was measured, it was possible to identify positive interactions within the adopted significance levels (Figure 2). Plants once subjected to abiotic stresses (salinity, water deficit, heat stress, among others) show changes in the state of membranes, implying changes in the thylakoids' function in chloroplasts, triggering changes in the characteristics of fluorescence signals. Despite the consequences imposed by different types of stress, sesame presents good phenotypic plasticity [35].

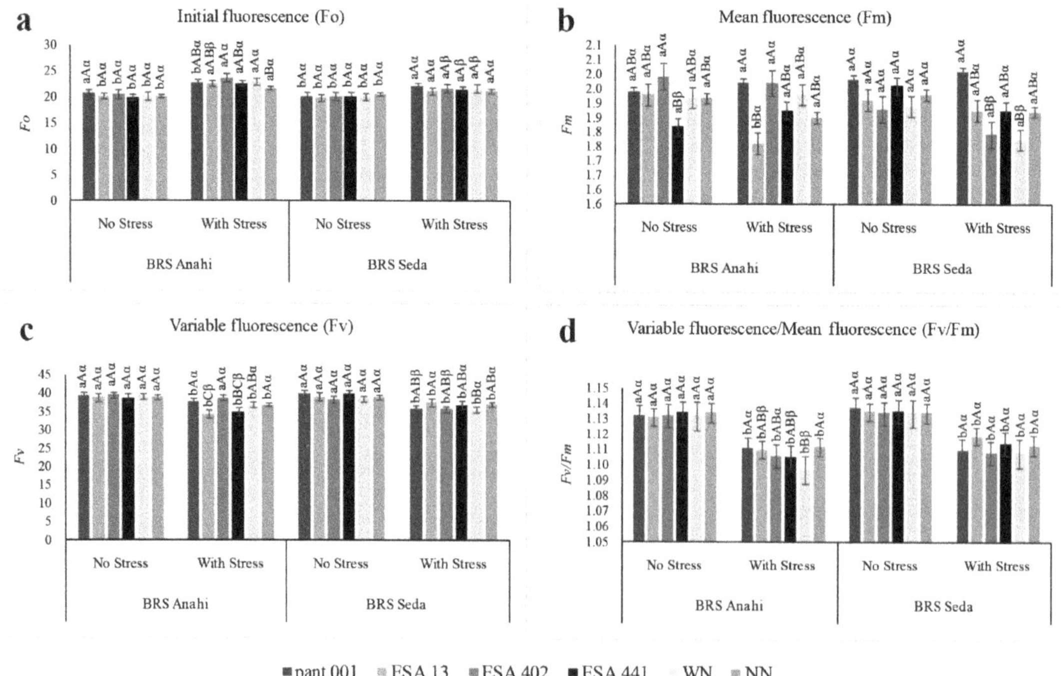

Figure 2. (**a**) Initial fluorescence (Fo), (**b**) average fluorescence (Fm), (**c**) variable fluorescence (Fv); (**d**) relationship between Fv/Fm, in sesame cultivars (BRS Anahí and BRS Seda) inoculated with bacilli, under water deficit and different nitrogen sources. WN-with nitrogen and NN-no nitrogen. Triple interactions: lowercase letters compare the water regime within each cultivar; capital letters compare treatments with inoculants within each cultivar; and Greek letters compare the cultivars.

2.2. Biomass, Growth Measures and Nitrogen

A high sensitivity to water deficit was observed in both cultivars in the first 24 h after watering was suspended, with signs of wilting, leaf curl, curving of the main stem, side branches, and the presence of trichomes. There was an accentuated floral abortion and leaf abscission in all stressed plants (Figure 3). This behavior is a strategy used by plants to save water and energy, given the critical moment to which they were subjected [36].

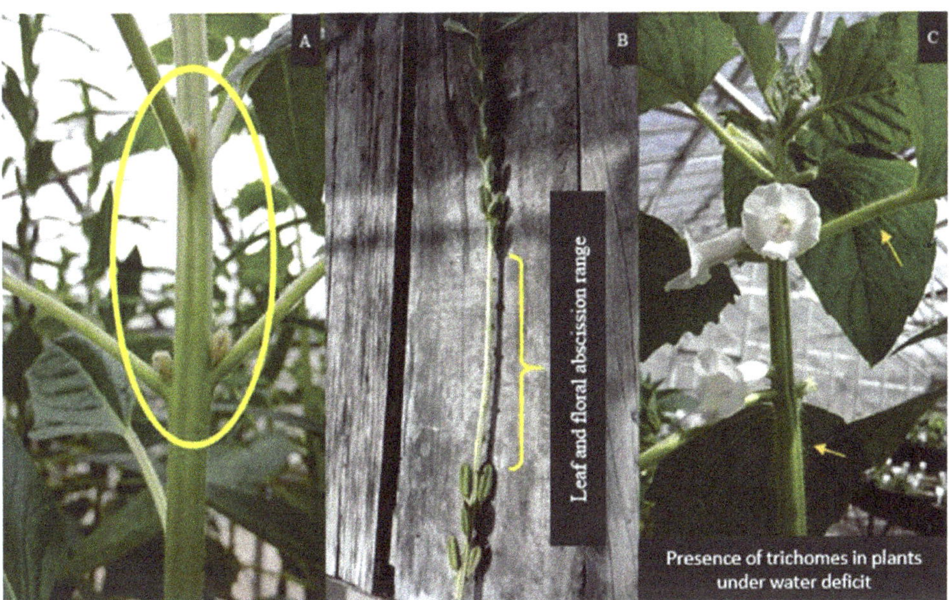

Figure 3. Observation of the events caused by water deficit in plants at BRS Anahí: (**A**) floral abortion; (**B**) leaf and floral abortion interval; and (**C**) presence of trichomes.

Sesame plants inoculated with the bacilli showed a positive interaction for plant height (Figure 4a). For the cultivar BRS Anahí there was no significant difference for the treatments in the irrigated condition; however, in the condition with water deficit pant001 and WN promoted the best averages for the variable plant height, without statistically differing from the strains ESA 13 and ESA 441. In the water deficit condition for BRS Seda, treatments with ESA 441, WN, and NN were the most significant for plant height (Figure 4a). In corn, rhizobacteria can potentially reduce fertilization with inorganic nitrogen without affecting growth parameters, proving the efficiency of *Bacillus* spp. as growth promoters when faced with a chemical nitrogen fertilizer [37]. Studies with sesame plants (BRS Seda) at different irrigation levels showed averages like those found in this work [38,39].

For the stem diameter variable, no significant difference was observed for the inoculants in the cultivar BRS Anahí, in the irrigated condition; in the water deficit condition, pant001 and nitrogen management presented the best means. The cultivar BRS Seda had a positive interaction in the irrigated condition with pant001; however, when submitted to water restriction, the strain ESA 402 was more efficient in the stem diameter, but it did not differ statically from the other treatments (Figure 4b).

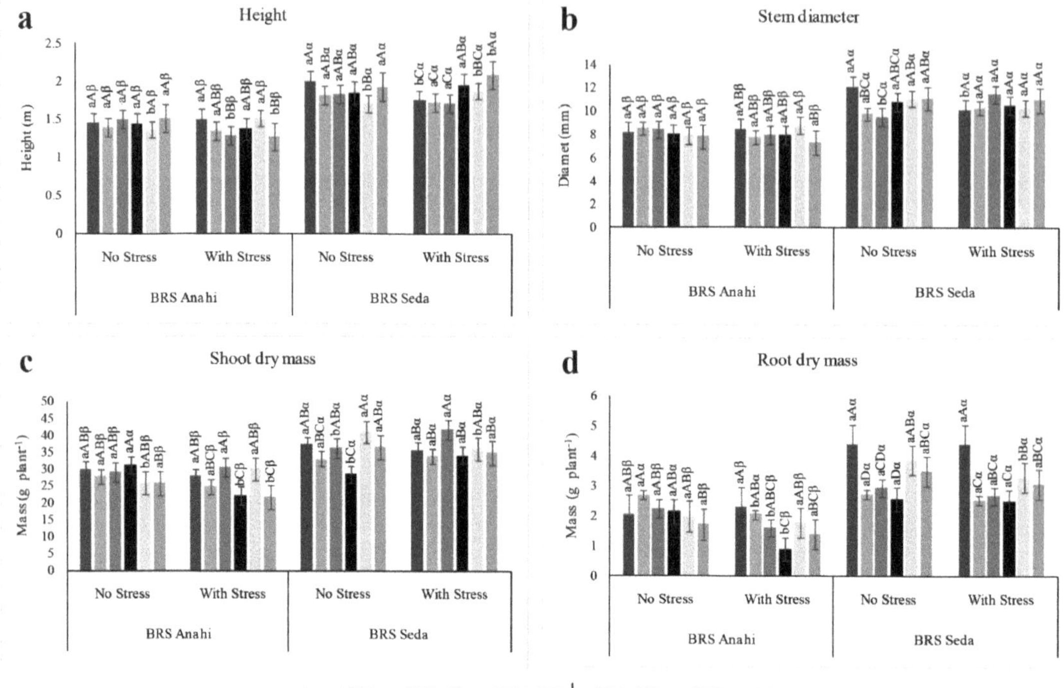

Figure 4. Growth characteristics of sesame cultivars (BRS Anahí and BRS Seda) inoculated with bacilli, under water deficit and different nitrogen sources. WN-with nitrogen and NN-no nitrogen. Plant height (**a**), stem diameter (**b**), shoot dry mass (**c**) and root dry mass (**d**). Lowercase letters compare the water regime within each cultivar; capital letters compare treatments within each cultivar; Greek letters compare cultivars.

When the shoot dry mass was evaluated for the cultivar BRS Anahí, in the irrigated condition there was a significant difference and the treatments inoculated with the bacilli showed the best means; in the water deficit condition, strain ESA 402 and WN presented the highest means, while pant001 and ESA 13 also showed positive effects when compared to NN (Figure 4c). In the cultivar BRS Seda inoculated with ESA 402, there was a significant increase in the dry mass of the shoots in the water deficit condition. Due to the characteristics of each cultivar, there was a significant difference in shoot dry mass between them (Figure 4c). Significant effects of water restriction on sesame were observed with a reduction in shoot dry mass, plant height, number of capsules per plant, and sesame productivity from the thirtieth day after planting [38]. ESA 13 already proved to be an efficient plant growth promoter for rice (*Oryza sativa*) [6], while ESA 402 showed positive effects on sorghum under full irrigation [40] and water deprivation conditions [7]. The results observed for the sesame genotypes in the present study agree with the potential of both bacilli to compose multi-crop inoculants for drylands.

In relation to dry root mass, in the irrigated condition, the cultivar BRS Anahí inoculated with ESA 13 presented the best average (Figure 4d); with water deficit, pant001 promoted the highest dry root mass, followed by ESA 13, WN, and ESA 402; with ESA 441, the dry mass of the root was reduced by more than 50% in relation to the irrigated condition. In the BRS Seda cultivar, the results for dry root mass were quite similar between irrigated and non-irrigated conditions. However, pant001 was the strain that provided the highest mean among the inoculated treatments (Figure 4d). In an experiment with peanut genotypes treated with bacterial strains and under water deficit, minimal differences were observed in relation to water restriction [41]. Despite the plants showing reductions in root growth, the bacterial strains were efficient in promoting growth and water deficit attenuation, emphasizing the importance of using inoculants to mitigate the effects of water deficit [12,41].

Based on these results, it is possible to identify a significant contribution of inoculants to the plant growth attributes of the investigated sesame cultivars, implying a beneficial action and a biosustainable alternative for the crop, with a view to reducing and/or replacing chemical fertilizers. The use of biofertilizers before, during, and after environmental stress can promote adjustments in plant defense mechanisms and increase soil water retention capacity, growth, and root performance [2,12,23,24,42].

The number of capsules per plant is a component of the final production, giving a cause-and-effect estimate: the more capsules per plant, the more seeds, which would also increase the seed mass. However, factors such as high temperature and water restriction can impair seed filling, as observed in this experiment. It is important to note that for the number of capsules per plant, strains pant001 and ESA 402 showed a positive and significant interaction for both cultivars (Figure 5a).

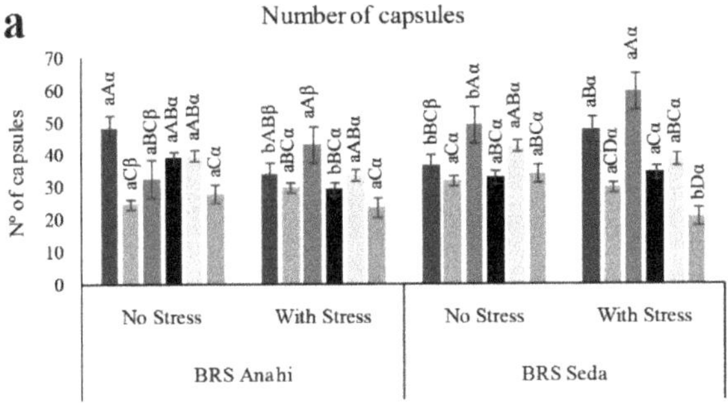

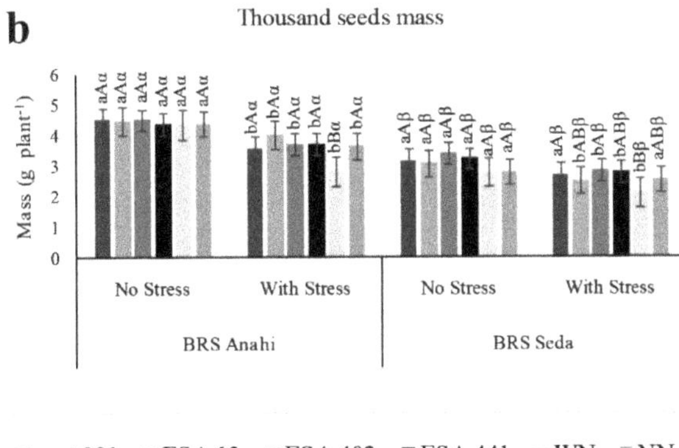

Figure 5. Number of capsules per plant (**a**) and weight of one thousand seeds (**b**) of two sesame cultivars (BRS Seda and BRS Anahí) inoculated with bacilli and under water restriction. WN-with nitrogen and NN-no nitrogen. Lowercase letters compare the water regime within each cultivar; capital letters compare treatments within each cultivar; Greek letters compare cultivars, according to Tukey's test at 5% probability.

For the mass of a thousand seeds, the inoculants promoted higher averages than the nitrogen treatment, both in the irrigated condition and in the water deficit (Figure 5b). However, in the water deficit condition, the WN treatment presented the lowest average, being more sensitive to stress in the production of viable seeds for the two cultivars.

In the double interaction analysis (water regime x treatment) for the mass of a thousand seeds, the inoculants promoted higher averages than the nitrogen treatment both in the water deficit and in the irrigated conditions (Figure 6). These results suggest that biological inoculants can mitigate the effects of water deficit and, consequently, favor production since growth variables were also favored. Nitrogen fertilization was evaluated in sesame and found to be between 2.87 g and 3.8 g for the mass of one thousand seeds [43,44]. The values mentioned above correspond to those found in the present study, in which BRS Seda presented 3.42 g with ESA 402 and BRS Anahí 4.51 g with pant001 and ESA 402, both in the irrigated condition. It is worth noting that these findings were higher than the average values described in the literature for the two cultivars used in this study when grown under irrigation conditions close to field capacity.

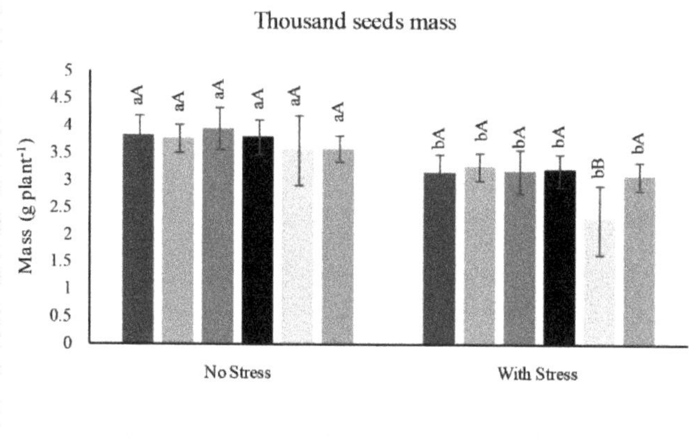

Figure 6. Interaction between water regime x treatments, with inoculants based on bacilli, with significance at $p \leq 0.05$ Tukey test for the mass of one thousand seeds of sesame cultivars (BRS Seda and BRS Anahí), under water restriction. Water regimen: no stress and with stress. WN-with nitrogen and NN-no nitrogen. Lowercase letters compare water regime; capital letters compare treatments within the water regime.

In the analysis of the nitrogen content in the leaf tissue, an increase was observed in all treatments that had the water deficit condition for the two cultivars investigated (Figure 7). This accumulation of nitrogen in leaves in the final phase of the experiment may imply a deficiency in the reallocation of this nutrient since it is required for photosynthesis and gas exchange, in addition to fruit formation, influencing the behavior of plants under water deficit [45]. Studies carried out with nitrogen application at different phenological stages of sesame plants showed differences in nitrogen partitioning and remobilization [46]. These authors highlighted the importance of using a less soluble nitrogen source to increase the efficiency of its use during the crop cycle.

For the cultivar BRS Anahí, the treatments that were inoculated with the bacilli showed no significant difference when submitted to water deficit, despite that pant001, WN, and NN revealed the highest concentrations of nitrogen in leaves, which may imply a deficit in the allocation of the macronutrient to the formation of capsules and seeds. Still, in BRS Anahí, treatments submitted to water deficit obtained higher averages of up to 50% when compared to irrigated treatments. The BRS Seda cultivar also showed an increase in nitrogen concentration in treatments under water restriction. The highest concentration of nitrogen was in the nitrogen management itself; however, it was not statistically different from the treatments with the inoculants pant001, ESA 13, and ESA 402 (Figure 7).

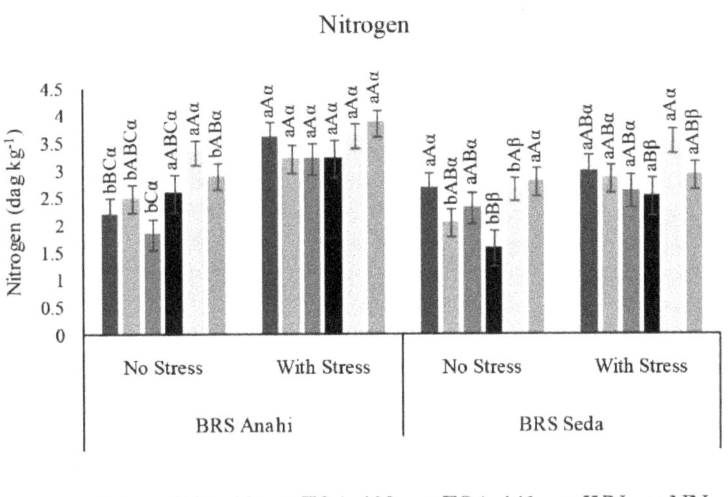

Figure 7. Accumulation of nitrogen in leaves of two sesame cultivars (BRS Seda and BRS Anahí) inoculated with bacilli, under water restriction, from crude protein ($p \leq 0.05$). WN-with nitrogen and NN-no nitrogen. Lowercase letters compare the water regime within each cultivar; capital letters compare treatments within each cultivar; Greek letters compare cultivars.

2.3. Osmoregulation, Antioxidant Enzyme Complex and Chlorophyll Content

The quantification of total free proline in sesame leaves on the 8th day of water deficit showed a significant increase in all treatments with a significance level of $p \leq 0.01$ for the triple interaction (Figure 8a). Proline plays an osmoregulatory role, binding to O_2 and free radicals produced under water stress, so its action in inducing systemic resistance in plants becomes complex, attenuating the negative effects triggered by water deficit [47]. Considering the increase in the concentration of this osmoprotective solute in sesame plants in water deficit, it is possible to affirm that, despite the limiting condition imposed, the cultivars were able to synthesize this solute in a satisfactory way, observing the highest concentrations in the BRS Anahí in inoculated treatments and nitrogen management. Its synthesis also implies that the plants were truly in a condition of stress.

Osmoregulation in plants under low water potential depends on the synthesis and accumulation of osmoprotectants or osmolytes, such as soluble proteins, sugars and sugar alcohols, quaternary ammonium compounds, and amino acids, such as proline. The synthesis and accumulation of compatible cellular solutes help plants under water deficit conditions; this process is called osmotic adjustment [48]. Osmoprotectants are made up of various inorganic ions and organic solutes that act on the cellular osmotic potential and increase water use efficiency [49]. The accumulation of proline is already recognized as an important indicator of abiotic stress in plants, favoring intracellular homeostasis [50] and proving to be capable of increasing the capacity of plants to overcome lower water potentials, since this osmolyte has a particularity of buffering under the effect of water scarcity [51,52].

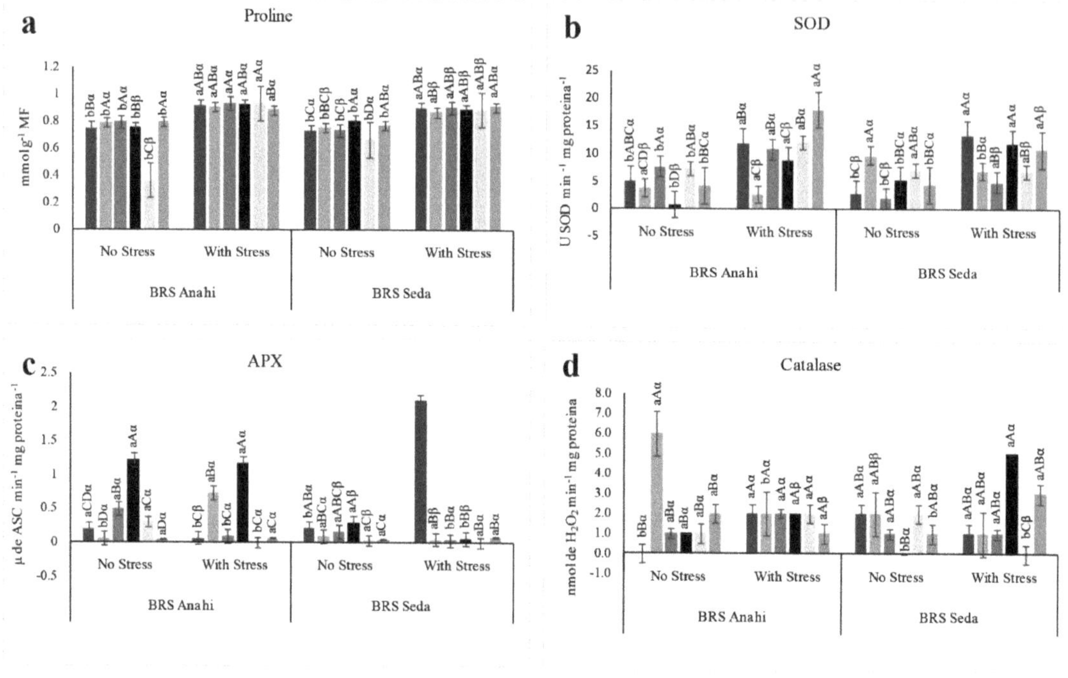

Figure 8. Concentration of proline (**a**), superoxide dismutase (SOD) (**b**), ascorbate peroxidase (APX) (**c**) and catalase (CAT) (**d**) as a function of water deficit in two sesame cultivars (BRS Seda and BRS Anahí) inoculated with bacilli. WN-with nitrogen and NN-no nitrogen.

SOD increased its activity in treatments under water stress, implying an adjustment of the plants to the imposed condition (Figure 8b). However, BRS Anahí and BRS Seda associated with ESA 13 showed a decrease in SOD enzymatic activity. The highest concentration of SOD was observed in BRS Anahí in the NN treatment. In BRS Seda, the lowest concentration of SOD was in the association of the cultivar with ESA 402 and the highest concentration was in pant001, both in the stress condition. SOD, as an antioxidant enzyme, is considered the first line of defense for the plant cell. In the presence of ROS, the enzyme acts by dismuting $O_2^{\bullet-}$ into H_2O_2, interfering with the concentration of ROS and the formation of $^{\bullet}OH$ radicals [53]. The accumulation of these free radicals damages the cellular arrangement, causing lipid peroxidation and cellular extravasation, in addition to affecting other biological molecules, including proteins and carbohydrates [54].

In the interaction between cultivars and inoculants, BRS Anahí with pant001, ESA 13, ESA 402, and WN showed higher CAT activity in the stress condition. For BRS Seda, interactions with pant001, ESA 441, and WN showed the highest CAT activities (Figure 8d). As can be seen, the performance of CAT varied; however, it is possible to state that some inoculants induced enzyme biosynthesis, helping to attenuate the oxidative effects triggered by water deficit and benefiting the cultivars. CAT is part of the antioxidant enzyme complex that defends plants that are subjected to abiotic stresses, such as water stress. Several sensitive, intermediate, and resistant sesame genotypes to water deficit presented CAT as one of the main enzymes that acts in the elimination of hydrogen peroxide (H_2O_2) generated in photorespiration and β-oxidation of fatty acids [55]. Under conditions of severe water stress, sesame inoculated with mycorrhiza (*Funneliformis mosseae* and *Rhizophagus irregularis*) presented an increase in catalase activity [56].

The highest activity of APX, under water deficit, was observed in the cultivar BRS Seda inoculated with pant001 and in BRS Anahí inoculated with ESA 13 (Figure 8c); BRS Anahí, when inoculated with ESA 441, showed increased APX activity in both water regimes. The functional activity of APX is complementary in plant defense; the biosynthetic inhibition observed in most treatments in this study may be linked to the increase in the activity of the first line of enzymatic defense, SOD (Figure 8b). Cowpea under water stress also showed a reduction in APX activity [57].

The antioxidant defense system includes a complex with several antioxidant enzymes such as SOD, CAT, and APX. When plants undergo oxidative stress due to some factor related to adverse environmental conditions, such as water or saline stress, these enzymes act in several subcellular sections to prevent damage to the plant cell [58–61]. Antioxidant enzymes act in a concatenated manner to establish maximum efficiency in plant defense under any adverse condition. Catalases are the main enzymes that convert hydrogen peroxide resulting from photorespiration into H_2O and molecular oxygen (O_2) [62]. As well as catalases, ascorbate peroxidase also acts in the primary defense, attenuating the deleterious effects of ROS in the plant cell, especially in plants under water deficit [57]. SOD enzymes catalyze the dismutation of the superoxide radical into $H_2O_2 + O_2$, CAT, and APX, which can break down $H_2O_2 \rightarrow H_2O + O_2$, antioxidants responsible for defending against free radicals, promoting detoxification caused by ROS that cause damage and compromise the plant cell functions [63].

In plants such as sesame, stress induces a complex plant response that depends on several factors, such as duration of stress, phenological phase, soil type, and genotype [64]. Sesame bears the physiological mark of stomatal resistance, which in turn provides an extension of its tolerance to drought [65] and in concomitant response to these processes are biochemical reactions that potentiate the defense system, generating a joint protection network that aims to maintain the plant's survival in the face of the limited conditions of its natural activities.

The sesame cultivars, despite the water restriction, showed positive interaction with the inoculants, increasing the concentration of chlorophyll a in the treatments with ESA 402 for BRS Anahí and ESA 13 and ESA 402 for BRS Seda (Figure 9a). In the evaluation of chlorophyll b, an increase was observed in the association of BRS Seda with ESA 13 (Figure 9b), and for total chlorophyll, BRS Seda with ESA 13 and ESA 402 was more efficient in the biosynthesis of pigments, presenting values higher than the control (Figure 9c). As for the carotenoid content, BRS Anahí inoculated with ESA 402 and BRS Seda with ESA 13, ESA 402, and ESA 441 promoted an increase superior to the control (Figure 9d). A decrease in chlorophyll a and b levels with increasing water stress was observed in sesame, with a reduction of 60% and 26%, respectively, compared to plants under optimal irrigation [56]. On the other hand, the authors highlighted an increase in carotenoids in genotypes under water deficit inoculated with mycorrhizal fungi.

The WN treatments showed higher values than the control for chlorophyll a, b, total, and carotenoids in BRS Seda and for the carotenoid variable in BRS Anahí. This behavior may be linked to the fact that nitrogen is an essential component of the chemical structure of pigments, so fertilization with this nutrient enables the synthesis of photosynthetic compounds. These pigments are directly linked to the nutritional status of plants, especially nitrogen, given that the total content of this nutrient in the leaf ratio is concentrated in chloroplasts [66]. Water stress usually results in the destruction of chloroplasts and consequently in a decrease in chlorophyll, in addition to the activity of enzymes in the Calvin cycle during the process of photosynthesis [67].

The results of this work are promising in relation to the proposed objectives and can be used as a basis to guide other field studies, knowing in advance the behavior of sesame genotypes (BRS Anahí and BRS Seda) in relation to inoculation with bacilli under water restriction. The summary of the main physiological and biochemical variables is described in Table 1.

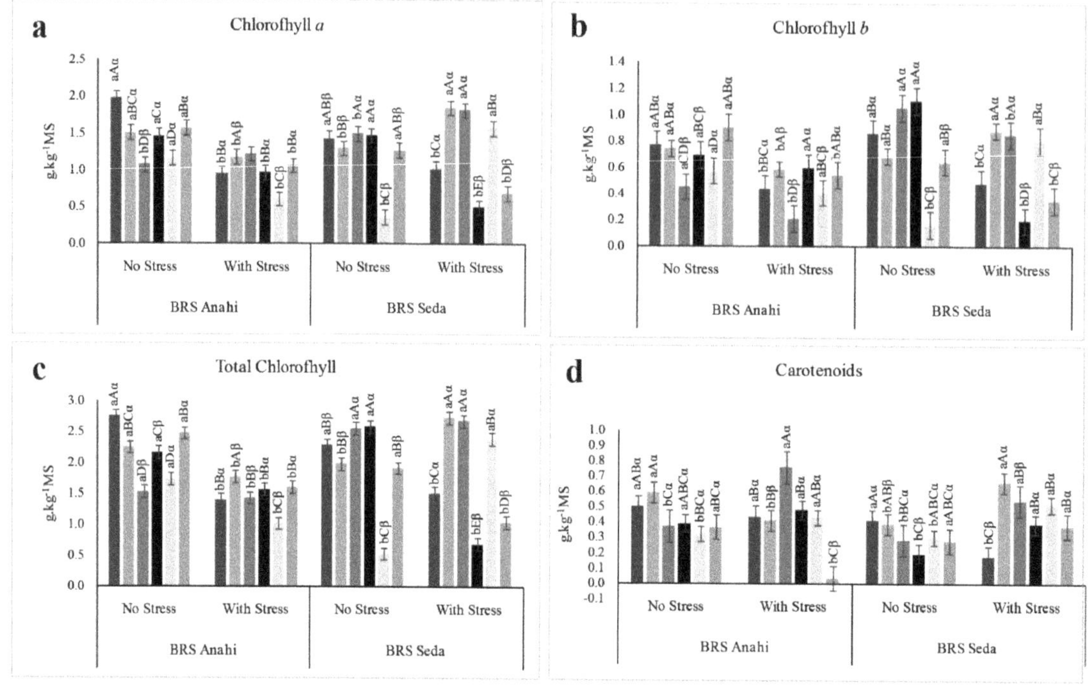

Figure 9. Photosynthetic pigments: chlorophyll a (**a**), chlorophyll b (**b**), total chlorophyll (**c**), and carotenoid (**d**) in sesame plants (BRS Seda and BRS Anahí) inoculated with bacilli, under water deficit. WN-with nitrogen and NN-no nitrogen.

Table 1. Summary of the main physiological and biochemical variables analyzed in the interaction between the two sesame genotypes (BRS Anahí and BRS Seda) and the bacilli-based inoculants, submitted to water restriction and compared to the control ($p \leq 0.01$ and $p \leq 0.05$ by the Tukey test).

	Ci	E	A	gs	Proline	SOD	APX	CAT
BRS Anahí								
pant 001	∧	∨	∨	∨	∧	∧	∨	∧
ESA 13	∧	∨	∨	∨	∧	∨	∧	∨
ESA 402	∧	∨	∨	∨	∧	∧	∨	∧
ESA 441	∧	∨	∨	∨	∧	∧	∨	∧
Nitrogen	∧	∨	∨	∨	∧	∧	∨	∧
No Nitrogen	∨	∨	∨	∨	∧	∧	∧	∨
BRS Seda								
pant 001	∧	∨	∨	∨	∧	∧	∧	∨
ESA 13	∨	∨	∨	∨	∧	∨	∨	∨
ESA 402	∧	∨	∨	∨	∧	∧	∨	∧
ESA 441	∧	∨	∨	∨	∧	∧	∨	∧
Nitrogen	∧	∨	∨	∨	∧	∨	∨	∨
No Nitrogen	∨	∨	∨	∨	∧	∧	∧	∧

Ci—internal concentration of CO_2; E—transpiration; A—photosynthesis; gs—stomatal conductance; SOD—superoxide dismutase; APX—ascorbate peroxidase; CAT—catalase; ∧—increased content; ∨—reduced content.

3. Materials and Methods

3.1. Cultivation of Bacteria and Preparation of Sesame Seeds

The *Bacillus* spp. strains ESA 13 [6], ESA 402 [7,40], and ESA 441 [68] were obtained from the "Coleção de Culturas de Micro-organismos de Interesse Agrícola da Embrapa Semiárido" (Embrapa Semiárido, Petrolina-PE, Brazil). They were streaked in LB solid medium (Luria Bertani) and incubated for 24 h at 28 °C. They were then subcultured in liquid LB medium and incubated at 28 °C, 180 rpm, for 72 h, until the exponential phase of bacterial growth (1.0×10^9 CFU mL^{-1}) [69]. The inoculant containing the strain pant001 (*Bacillus subtilis*—Panta Premium) [70] was provided by the Geoclean company, and, together with the bacteria subcultured in liquid medium, they were used directly in the inoculation of the seeds.

Sesame seeds were disinfected with pure ethanol for 15 s, 1% sodium hypochlorite for 1 min, and finally washed 10 times with sterile distilled water [69]. Then, the sesame seeds were soaked in the inoculants for 10 min and sown in pots.

3.2. Implementation and Conduction of the Experiment in a Greenhouse

The experiment was carried out in a greenhouse at Embrapa Algodão, Campina Grande-PB, Brazil (07°13' S; 53°31' W) [71]. Two sesame cultivars (BRS Seda and BRS Anahí) were grown in pots with a capacity of 20 L; the pots were filled with sandy loam soil. The soil was previously analyzed at the Laboratory of Soils and Plant Nutrition of Embrapa Algodão [72] and corrected with dolomitic limestone. The substrate, according to the treatment, was fertilized with nitrogen (ammonium sulfate, 95 kg ha^{-1}), distributed in two applications, the first at 10 days after emergence and the second at the beginning of flowering. Phosphorus (single superphosphate) and potassium (KCl) (110 kg ha^{-1} and 34 kg ha^{-1}, respectively) were applied for all treatments, based on the analysis of the soil [73].

Irrigation was performed daily, trying to maintain soil moisture close to field capacity and suspended in stressed treatments from the 30th day after emergence (beginning of flowering), for eight days, and then rehydrated.

The experimental design was completely randomized, established by a random process in a factorial scheme: 2 (cultivars) $\times$ 2 (water regime) $\times$ 6 (inoculation/fertilization treatments), totaling 24 treatments with 5 replications. The treatments were characterized as: (i) nitrogen management (with N, WN) (ammonium sulfate, 21% N); (ii) absolute control, without nitrogen (no N, NN); (iii) management with 4 inoculants based on bacilli, strains pant001, ESA 13, ESA 402, and ESA 441; and with and without irrigation.

3.3. Physiological Measures

The sesame plants were evaluated in the morning, between 9:00 a.m. and 11:00 a.m., during the water deficit period (8 days), using a portable photosynthesis analyzer (IRGA—Infra Red Gas Analyzer, model LCpro-SD), without an artificial carbon source and with an artificial light source of 1200 µmol m^{-2} m^{-1}. The following parameters were evaluated: stomatal conductance (gs) (mol m^{-2} m^{-1}); photosynthesis (A) (µmol m^{-2} m^{-1}); transpiration (E) (mmol m^{-2} m^{-1}) and internal concentration of CO_2 (Ci) (µmol mol^{-1}) [74,75]. From the obtained data, the instantaneous efficiency of carboxylation (iEC) between A and Ci (A/Ci) and the instantaneous transpiration efficiency (iTE), calculated as the photosynthesis/transpiration ratio, between A and E (A/E) [75] were estimated. To monitor stomatal closure, three assessments were performed during the water suspension period (1D; 4D; 8D—"D = Days"). When the plants reached approximately 90% stomatal closure, they were rehydrated (criterion adopted by the research team). The fluorescence of the plants was also evaluated [76], using a portable fluorometer, with which the initial fluorescence (Fo), variable fluorescence (Fv), mean fluorescence (Fm), and the relationship between Fv/Fm were estimated.

3.4. Nitrogen Content of the Shoot

At the end of the crop cycle, a sample of the aerial part of the plants was collected, stored in kraft paper bags, and placed in an oven with forced air circulation at 65 °C for 72 h, then ground in a mill. The nitrogen analysis of the aerial part of the plants was based on the sulfuric digestion method developed by Kjeldahl [77]. From the nitrogen content, the total nitrogen accumulated in the shoot was calculated by multiplying the nitrogen content by the dry mass of the shoot [78].

3.5. Antioxidant Activities and Proline Content

A sample of fresh leaves was collected on the eighth day of water restriction, immediately immersed in liquid N_2, and then stored at -80 °C. For protein extraction, the leaves (200 mg) were macerated in liquid N_2 and 3 mL of 0.1 M potassium phosphate buffer, pH 7.0, containing 100 mM EDTA, 1 mM L-ascorbic acid, and 4% polyvinylpolypyrrolidone (PVP) were added. The extracts were centrifuged at 12000 rpm for 10 min at 4 °C and the supernatants were transferred to new microtubes. Protein quantification was determined by Bradford's method [79] in a spectrophotometer at 595 nm. Superoxide dismutase (SOD) activity was determined and analyzed in a spectrophotometer at 560 nm [80]. The results were expressed in UA g MF^{-1} activity (Activity Unit g Fresh Pasta). Catalase activity (CAT) was determined, and the reading was performed in a spectrophotometer at 240 nm [81]. Ascorbate peroxidase (APX) activity was determined and analyzed in a spectrophotometer at 290 nm [82]. Free proline content was determined and analyzed in a spectrophotometer at 520 nm [83].

3.6. Chlorophyll and Carotenoid Content

Chlorophyll content of a, b, and total (a + b), and carotenoid contents were determined using the 80% acetone extraction method [84]. The entire procedure was performed in the presence of green light, thus avoiding the degradation of chlorophyll. In this methodology, 200 mg of leaves was macerated in liquid N_2 and then solubilized in 10 mL of 80% acetone. Subsequently, the solution was filtered through qualitative filter paper and read at the following absorbances: 470, 646.8, 663.2, and 710 nm.

3.7. Biomass and Growth Measures

At the end of the crop cycle, the following growth characteristics were evaluated: plant height (cm), measured from the soil surface to the apex of the main stem, using a metric tape; stem diameter, measured with a caliper; number of capsules per plant; dry mass of shoots (g) and roots (g), determined by drying the material in an oven with forced air circulation at 65 °C, for approximately 72 h, until reaching a constant mass, and weighing on a precision scale; and mass of 1000 seeds [85].

3.8. Statistical Analysis

The collected data were analyzed using the SISVAR software version 5.6 [86], submitted to an analysis of variance ($p \leq 0.05$), and the means compared by Tukey's test ($p \leq 0.05$). They were also submitted to the Shapiro–Wilk normality test to verify and correct data heterogeneity.

4. Conclusions

The cultivars BRS Anahí and BRS Seda showed semi-tolerant characteristics to water deficit. It was possible to observe a morphophysiological and biochemical adjustment, with inoculation with *Bacillus* spp. being a relevant factor for the results obtained. Under water deficit, the bacterial strains promoted positive effects on the plant height and weight of one thousand seeds variables. The interactions of BRS Anahí x ESA 13 and BRS Seda x ESA 402 promoted the greatest increases in weight of one thousand seeds with 42% and 34%, respectively. In the irrigated condition, it was observed that all inoculants promoted an increase of up to 34% for the weight of one thousand seeds in both cultivars.

In this perspective, the biofertilizer containing the strains assessed in the present study (mainly ESA 13 and ESA 402) should constitute an important agricultural input capable of mitigating the effects of the water deficit on the plants and providing positive effects on the increase in sesame production, providing greater economic viability to the crop.

Author Contributions: Conceptualization, L.M.d.L. and T.M.d.S.G.; methodology, P.I.F.-J. and T.M.d.S.G.; validation, G.M.G.d.R. and L.M.d.L.; formal analysis, G.B.P.d.L., F.d.A.S. and E.F.G.; investigation, G.M.G.d.R. and L.M.d.L.; data curation, T.M.d.S.G. and G.M.G.d.R.; writing—original draft preparation, T.M.d.S.G., P.I.F.-J., P.D.F., A.P.M. and L.M.d.L.; writing—review and editing, L.M.d.L. and A.S.d.M.; visualization, L.M.d.L. and A.S.d.M.; project administration, L.M.d.L.; funding acquisition, N.H.C.A. and L.M.d.L. All authors have read and agreed to the published version of the manuscript.

Funding: Brazilian Council for Scientific and Technological Development (CNPq) for the scholarship and the Brazilian Agricultural Research Corporation (Embrapa, SEG 20.18.01.021.00.00); Brazilian agency Coordenação de Aperfeiçoamento de Pessoal de Nível Superior—Programa de Apoio à Pós-Graduação (CAPES-PROAP, n. 8881.707827/2022-01); and Coordenação de Aperfeicoamento de Pessoal de Nível Superior (n. 001).

Data Availability Statement: All other data are presented in the paper.

Conflicts of Interest: The authors declare no conflict of interest.

References

1. Haque, M.M.; Mosharaf, M.K.; Khatun, M.; Haque, M.A.; Biswas, M.S.; Islam, M.S.; Islam, M.M.; Shozib, H.B.; Miah, M.M.U.; Molla, A.H.; et al. Biofilm producing rhizobacteria with multiple plant growth-promoting traits promote growth of tomato under water-deficit tress. *Front. Microbiol.* **2020**, *11*, 20–53. [CrossRef] [PubMed]
2. Saxena, A.K.; Kumar, M.; Chakdar, H.; Anuroopa, N.; Bagyaraj, D.J. *Bacillus* species in soil as a natural resource for plant health and nutrition. *J. Appl. Microbiol.* **2019**, *128*, 1583–1594. [CrossRef] [PubMed]
3. Glick, B.R. The enhancement of plant growth by free-living bacteria. *Can. J. Microbiol.* **1995**, *41*, 17–109. [CrossRef]
4. Kuss, A.V.; Kuss, V.V.; Lovato, T.; Flores, M.L. Fixação de nitrogênio e produção de ácido indolacético in vitro por bactérias diazotróficas endolíticas. *Pesq. Agropec. Bras.* **2007**, *42*, 1459–1465. [CrossRef]
5. Kavamura, V.N.; Santos, S.N.; Silva, J.L.; Parma, M.M.; Ávila, L.A.; Visconti, A.; Zucchi, T.D.; Taketani, R.G.; Andreote, F.D.; Melo, I.S. Screening of brazilian cacti rhizobacteria for plant growth promotion under drought. *Microbiol. Res.* **2013**, *168*, 183–191. [CrossRef]
6. Fernandes-Júnior, P.I.; Aidar, S.T.; Morgante, C.V.; Gava, C.A.T.; Zilli, J.É.; Souza, L.S.B.; Marinho, R.C.N.; Nóbrega, R.S.A.; Brasil, M.S.; Seido, S.L.; et al. The resurrection plant *Tripogon Spicatus* (Poaceae) harbors a diversity of plant growth promoting bacteria in Northeastern Brazilian Caatinga. *Rev. Bras. Ciênc. Solo* **2015**, *39*, 993–1002. [CrossRef]
7. Santana, S.R.A.; Voltolini, T.V.; Antunes, G.R.; Silva, V.M.; Simões, W.L.; Morgante, C.V.; Freitas, A.D.S.; Chaves, A.R.M.; Aidar, S.T.; Fernandes, P.I., Jr. Inoculation of plant growth-promoting bacteria attenuates the negative effects of drought on sorghum. *Arch. Microbiol.* **2020**, *202*, 1015–1024. [CrossRef]
8. Santos, A.F.J.; Morais, J.S.; Miranda, J.S.; Moreira, Z.P.M.; Feitoza, A.F.A.; Leite, J.; Fernandes, P.I., Jr. Cacti-associated rhizobacteria from Brazilian caatinga biome induce maize growth promotion and alleviate abiotic stress. *Rev. Bras. Ciênc. Agrár.* **2020**, *15*, e8221. [CrossRef]
9. Oldroyd, G.E.; Dixon, R. Biotechnological solutions to the nitrogen problem. *Curr. Opin. Biotechnol.* **2014**, *26*, 19–24. [CrossRef]
10. Bravo, G.; Vega-Celedón, P.; Gentina, J.C.; Seeger, M. Bioremediation by *Cupriavidus metallidurans* strain msr33 of mercury-polluted agricultural soil in a rotary drum bioreactor and its effects on nitrogen cycle microorganisms. *Microorganisms* **2020**, *8*, 1952. [CrossRef]
11. Rahman, M.; Miah, M.N.A.; Dudding, W. Mechanisms involved with bacilli-mediated biotic and abiotic stress tolerance in plants. In *Bacilli in Agrobiotechnology: Bacilli in Climate Resilient Agriculture and Bioprospecting*; Islam, M.T., Rahman, M., Pandey, P., Eds.; Springer: Cham, Switzerland, 2022; pp. 169–197.
12. Tsotetsi, T.; Nephali, L.; Malebe, M.; Tugizimana, F. *Bacillus* for plant growth promotion and stress resilience: What have we learned? *Plants* **2022**, *11*, 2482. [CrossRef] [PubMed]
13. Nascimento, R.C.; Cavalcanti, M.I.P.; Correia, A.J.; Escobar, I.E.C.; de Freitas, A.D.S.; Nóbrega, R.S.A.; Fernandes, P.I., Jr. Maize-associated bacteria from the brazilian semiarid region boost plant growth and grain yield. *Symbiosis* **2021**, *83*, 347–359. [CrossRef]
14. Silva, J.F.; Silva, T.R.; Escobar, I.E.C.; Fraiz, A.C.R.; Santos, J.W.M.; Nascimento, T.R.; Santos, J.M.R.; Peters, S.J.W.; Melo, R.F.; Signor, D.; et al. Screening of plant growth promotion ability among bacteria isolated from field-grown sorghum under different managements in brazilian drylands. *World J. Microbiol. Biotechnol.* **2018**, *34*, 186. [CrossRef] [PubMed]

15. Grover, M.; Bana, R.S.; Bodhankar, S. Prevalance of multifunctional *Azospirillum Formosense* strains in the rhizosphere of pearl millet across diverse edaphoclimatic regions of India. *Vegetos* **2022**, *35*, 1–11. [CrossRef]
16. Nithyapriya, S.; Lalitha, S.; Sayyed, R.Z.; Reddy, M.S.; Dailin, D.J.; Enshasy, H.A.E.; Suriani, N.L.; Herlambang, S. Production, purification, and characterization of bacillibactin siderophore of *Bacillus subtilis* and its application for improvement in plant growth and oil content in sesame. *Sustainability* **2021**, *13*, 5394. [CrossRef]
17. Varma, P.K.; Yamuna, C.; Suresh, V.; Teja, M.R.; Kumar, K.V.K. Potentiality of native *Bacillus* species in enhancing sesame seed germination and their antagonism against *Macrophomina phaseolina* under in vitro conditions. *J. Oilseeds Res.* **2017**, *34*, 98–102.
18. Ferreira, M.D.; Spricigo, P.C. Colorimetria—Princípios e aplicações na agricultura. In *Instrumentação pós-Colheita em Frutas e Hortaliças*; Ferreira, M.D., Ed.; Embrapa Instrumentação: São Carlos, Brasil, 2017; pp. 209–220.
19. Beltrão, N.E.M.; Oliveira, M.I.P. *Ecofisiologia das Culturas de Algodão, Amendoim, Gergelim, Mamona, Pinhão-Manso e Sisal*; Embrapa Informação Tecnológica: Brasília, Brazil, 2011.
20. Arriel, N.H.C.; Beltrão, N.E.M.; Firmino, P.T. Clima. O produtor pergunta a Embrapa responde. In *Gergelim*; Beltrão, N.E.M., Arriel, N.H.C., Lima, R.L.S., Eds.; Embrapa Informação Tecnológica: Brasília, Brazil, 2009; pp. 57–88.
21. Marulanda, A.; Porcel, R.; Barea, J.M.; Azcón, R. Drought tolerance and antioxidant activities in lavender plants colonized by native drought-tolerant or drought-sensitive. *Microb. Ecol.* **2007**, *54*, 543–552. [CrossRef]
22. Hashem, A.; Tabassum, B.; Allah, E.F.A. *Bacillus subtilis*: A plant-growth promoting rhizobacterium that also impacts biotic stress. *Saudi. J. Biol. Sci.* **2019**, *26*, 1291–1297. [CrossRef]
23. Patel, R.R.; Patel, D.D.; Thakor, P.; Patel, B.; Thakkar, V.R. Alleviation of salt stress in germination of *Vigna radiata* L. by two halotolerant *Bacilli* sp. isolated from saline habitats of Gujarat. *Plant Growth Regul.* **2015**, *76*, 51–60. [CrossRef]
24. Patel, R.R.; Patel, D.D.; Bhatt, J.; Thakor, P.; Triplett, L.R.; Thakkar, V.R. Induction of pre-chorismate, jasmonate and salicylate pathways by *Burkholderia* sp. RR18 in peanut seedlings. *J. Appl. Microbiol.* **2021**, *131*, 1417–1430. [CrossRef]
25. Suassuna, J.; Fernandes, P.; Brito, M.; Arriel, N.M.A.; Fernandes, J. Tolerance to salinity of sesame genotypes in different phenological stages. *Am. J. Plant Sci.* **2017**, *8*, 1904–1920. [CrossRef]
26. Lopes, M.S.; Araus, J.L.; Heerden, P.D.R.V.; Foyer, C.H. Enhancing drought tolerance in C4 crops. *J. Exp. Bot.* **2011**, *62*, 3135–3153. [CrossRef] [PubMed]
27. Dias, A.S.; Lima, G.S.L.; Gheyi, H.R.; Nobre, R.G.; Fernandes, P.D.; Silva, F.A.Z. Trocas gasosas e eficiência fotoquímica do gergelim sob estresse salino e adubação com nitrato-amônio. *Irriga* **2018**, *23*, 220–234. [CrossRef]
28. Sousa, G.G.; Viana, T.V.A.; Dias, C.N.; Silva, G.L.S.; Azevedo, B.M. Lâminas de irrigação para cultura do gergelim com biofertilizante bovino. *Magistra* **2014**, *26*, 347–356.
29. Lacerda, C.F.; Oliveira, E.V.; Neves, A.L.; Gheyi, H.R.; Bezerra, M.A.; Costa, C.A. Morphophysiological responses and mechanisms of salt tolerance in four ornamental perennial species under tropical climate. *Rev. Bras. Eng. Agric. Ambient.* **2020**, *24*, 656–663. [CrossRef]
30. Silva, A.A.R.; Lacerda, C.N.; Lima, G.S.; Soares, L.A.A.; Gheyi, H.R.; Fernandes, P.D. Morfofisiologia de Cultivares de gergelim submetidos a diferentes estratégias de uso de água salina. *Irriga* **2021**, *1*, 42–55. [CrossRef]
31. Agurla, S.; Gahir, S.; Munemasa, S.; Murata, Y.; Raghavendra, A.S. Mechanism of stomatal closure in plants exposed to drought and cold stress. *Adv. Exp. Med. Biol.* **2018**, *1081*, 215–232. [PubMed]
32. Endres, L.; Souza, J.L.S.; Teodoro, I.; Marroquim, P.M.G.; Santos, C.M.S.; Brito, J.E.D. Gas exchange alteration caused by water deficit during the bean reproductive stage. *Rev. Bras. Eng. Agric. Ambient.* **2010**, *14*, 11–16. [CrossRef]
33. Silva, E.M.; Lima, G.S.; Gheyi, H.R.; Nobre, R.G.; Sá, F.V.S.; Souza, L.P. Growth and gas exchanges in soursop under irrigation with saline water and nitrogen sources. *Rev. Bras. Eng. Agric. Ambient.* **2018**, *22*, 776–781. [CrossRef]
34. Feitosa, S.S.; Albuquerque, M.B.; Oliveira, A.P.; Pereira, W.E.; Brito Neto, J.F. Fisiologia do *Sesamum indicum* L. sob estresse hídrico e aplicação de ácido salicílico. *Irriga* **2016**, *21*, 711–723. [CrossRef]
35. Silva, E.M.; Ribeiro, R.V.; Ferreira-Silva, S.L.; Viégas, R.A.; Silveira, J.A.G. Salt stress induced damages on the photosynthesis of physic nut young plants. *Sci. Agric.* **2011**, *68*, 62–68. [CrossRef]
36. Seleiman, M.F.; Al-Suhaibani, N.; Ali, N.; Akmal, M.; Alotaibi, M.; Refay, Y.; Dindaroglu, T.; Abdul-Wajid, H.H.; Battaglia, M.L. Drought stress impacts on plants and different approaches to alleviate its adverse effects. *Plants* **2021**, *10*, 259. [CrossRef]
37. Lin, Y.; Watts, D.B.; Kloepper, J.W.; Torbert, H.A. Influence of plant growth-promoting rhizobacteria on corn growth under different fertility sources. *Commun. Soil Sci. Plant Anal.* **2018**, *49*, 1239–1255. [CrossRef]
38. Mesquita, J.B.R.; Azevedo, B.M.A.; Campelo, A.R.; Fernandes, C.N.V.; Viana, T.V.A. Crescimento e produtividade da cultura do gergelim (*Sesamum indicum* L.) sob diferentes níveis de irrigação. *Irriga* **2013**, *18*, 364–375. [CrossRef]
39. Grilo, J.A.; Azevedo, P.V. Crescimento, desenvolvimento e produtividade do gergelim 'BRS Seda' na agrovila de Canudos, em Ceará Mirim (RN). *Holos* **2013**, *2*, 19–33. [CrossRef]
40. Antunes, G.R.; Santana, S.R.A.; Escobar, I.E.C.; Brasil, M.S.; Araújo, G.G.L.; Voltolini, T.V.; Fernandes, P.I., Jr. Associative diazotrophic bacteria from forage grasses in the Brazilian semiarid region are effective plant growth promoters. *Crop Pasture Sci.* **2019**, *70*, 899–907. [CrossRef]
41. Brito, S.L.; Santos, A.B.; Barbosa, D.D.; Fernandes, P.D.; Fernandes, P.I., Jr.; Lima, L.M. *Bradyrhizobium* spp. as attenuators of water deficit stress in runner peanut genotypes based on physiological and gene expression responses. *Genet. Mol. Res.* **2019**, *18*, 1–12. [CrossRef]

42. Mitter, B.; Brader, G.; Pfaffenbichler, N.; Sessitsch, A. Next generation microbiome applications for crop production-limitations and the need of knowledge-based solutions. *Curr. Opin. Microbiol.* **2019**, *49*, 59–65. [CrossRef]
43. Jadhav, S.R.; Naiknaware, M.D.; Pawar, G.R. Effect of nitrogen, phosphorus and biofertilizers on growth, yield and quality of summer sesamum (*Sesamum indicum* L.). *Int. J. Trop. Agric.* **2015**, *33*, 475–480.
44. Campos, L.N.; Guilherme, M.F.S.; Oliveira, H.M.; Costa, V.F.; Silva, E. Avaliação biométrica de sementes de *Sesamum indicum* L. In Proceedings of the Congresso Nordestino de Biólogos, João Pessoa, PB, Brazil, 27–28 March 2014.
45. Melo, E.B.S.; Lima, L.M.; Fernandes-Junior, P.I.; Aidar, S.T.; Freire, M.A.O.; Freire, R.M.M.; Santos, R.C. Nodulation, gas exchanges and production of peanut cultivated with *Bradyrhizobium* in soils with different textures. *Comun. Sci.* **2016**, *7*, 160–166. [CrossRef]
46. Couch, A.; Jani, A.; Mulvaney, M.; Hochmuth, G.; Bennett, J.; Gloaguen, R.; Langham, R.; Rowland, D. Nitrogen accumulation, partitioning, and remobilization by diverse sesame cultivars in the humid southeastern USA. *Field Crops Res.* **2017**, *203*, 55–64. [CrossRef]
47. Dias, M.C.; Oliveira, H.; Costa, A.; Santos, C. Improving elms performance under drought stress: The retreatment with abscisic acid. *Environ. Exp. Bot.* **2014**, *100*, 64–73. [CrossRef]
48. Blum, A. Osmotic adjustment is a prime drought stress adaptive engine in support of plant production. *Plant Cell Environ.* **2017**, *40*, 4–10. [CrossRef]
49. Fang, Y.; Xiong, L. General mechanisms of drought response and their application in drought resistance improvement in plants. *Cell Mol. Life Sci.* **2015**, *72*, 673–689. [CrossRef]
50. Lin, C.C.; Hsu, Y.T.; Kao, C.H. The effect of NaCl on proline accumulation in rice leaves. *Plant Growth Regul.* **2002**, *36*, 275–285. [CrossRef]
51. Jaleel, C.A.; Manivannan, P.; Sankar, B.; Kishorekumar, A.; Gopi, R.; Somasundaram, R.; Panneerselvam, R. *Pseudomonas fluorescens* enhances biomass yield and ajmalicine production in *Catharanthus roseus* under water deficit stress. *Colloids Surf. B Biointerfaces* **2007**, *60*, 7–11. [CrossRef]
52. Ozturk, M.; Unal, B.T.; Carrapós, G.P.; Khursheed, A.; Gul, A.; Hasanuzzaman, M. Osmoregulation and its actions during the drought stress in plants. *Physiol. Plant* **2021**, *172*, 1321–1335. [CrossRef]
53. Gupta, D.K.; Plama, J.M.; Corpas, F.J. *Antioxidants and Antioxidant Enzymes in Higher Plants*; Springer International Publishing: Berlim, Germany, 2018.
54. Manivannan, P.; Jaleel, C.A.; Somasundaram, R.; Panneerselvam, R. Osmoregulation and antioxidant metabolism in drought-stressed Helianthus annuus under triadimefon drenching. *C. R. Biol.* **2008**, *331*, 418425. [CrossRef]
55. Kadkhodaie, A.; Razmjoo, J.; Zahedi, M. Peroxisase, ascorbate peroxidase e catalase activities in drought sensitive, intermediate and resistence sesame (*Sesamum indicum* L.) genotypes. *Int. J. Agron. Plant Prod.* **2013**, *4*, 3012–3021.
56. Gholinezhad, E.; Darvishzadeh, R.; Moghaddam, S.S.; Popović-Djordjević, J. Effect of mycorrhizal inoculation in reducing water stress in sesame (*Sesamum indicum* L.): The assessment of agrobiochemical traits and enzymatic antioxidant activity. *Agric. Water Manag.* **2020**, *238*, 106–234. [CrossRef]
57. Sousa, C.C.M.; Pedrosa, E.M.R.; Rolim, M.M.; Oliveira Filho, R.A.; Souza, M.A.L.M.; Pereira Filho, J.V. Crescimento e respostas enzimáticas do feijoeiro caupi sob estresse hídrico e nematoide de galhas. *Rev. Bras. Eng. Agric. Ambient.* **2015**, *19*, 113–118. [CrossRef]
58. Alscher, R.G.; Erturk, N.; Heath, L.S. Role of superoxide dismutase (SODs) in controlling oxidative stress in plants. *J. Exp. Bot.* **2002**, *53*, 1331–1341. [CrossRef] [PubMed]
59. Hussein, Y.; Amin, J.; Azeb, A.; Gahin, H. Antioxidant activities during drought stress resistance of sesame (*Sesamum indicum* L.) plant by salicylic acid and kinetin. *Res. J. Bot.* **2016**, *11*, 1–8. [CrossRef]
60. Yousefzadeh-Najafabadi, M.; Ehsanzadeh, P. Photosynthetic and antioxidative upregulation in drought-stressed sesame (*Sesamum indicum* L.) subjected to foliar applied salicylic acid. *Photosynthetica* **2017**, *55*, 611–622. [CrossRef]
61. Varamin, J.K.; Fanoodi, F.; Sinaki, J.M.; Rezvan, S.; Damavandi, A. Foliar application of chitosan and nano-magnesium fertilizers influence on seed yield, oil content, photosynthetic pigments, antioxidant enzyme activities of sesame (*Sesamum indicum* L.) under water-limited conditions. *Not. Bot. Horti Agrobot.* **2020**, *48*, 2228–2243. [CrossRef]
62. Breusegem, F.V.; Vranová, E.; Dat, J.F.; Inzé, D. The role of active oxygen species in plant signal transduction. *Plant Sci.* **2001**, *161*, 405–414. [CrossRef]
63. Carneiro, M.M.C.; Deuner, S.; Oliveira, P.V.; Teixeira, S.B.; Sousa, C.P.; Bacarin, M.A.; Moraes, D.M. Atividade antioxidante e viabilidade de sementes de girassol após estresse hídrico e salino. *Rev. Bras. Sementes* **2011**, *33*, 752–761. [CrossRef]
64. Beltrão, N.E.M.; Vieira, D.J. *O agronegócio do Gergelim No Brasil*; Embrapa Informação Tecnológica: Brasília, Brazil, 2001.
65. Perin, A.; Cruvinel, J.D.; Silva, W.J. Sesame performance as a function of NPK fertilization and soil fertility level. *Acta Sci. Agron.* **2010**, *32*, 93–98.
66. Coelho, A.P.; Leal, F.P.; Filla, V.A.; Dalri, A.B.; Faria, R.T. Estimativa da produtividade de grãos da aveia-branca cultivada sob níveis de irrigação utilizando clorofilômetro portátil. *Rev. Cient. Fac. Educ. Meio Amb.* **2018**, *9*, 662–667. [CrossRef]
67. Anaya, F.; Fghire, R.; Wahbi, S.; Loutfi, K. Influence of salicylic acid on seed germination of *Vicia faba* L. under salt stress. *J. Saudi Soc. Agric. Sci.* **2018**, *17*, 1–8. [CrossRef]
68. Silva, J.F. *Caracterização Polifásica de Bactérias Promotoras de Crescimento Vegetal Associados ao Sorgo (Sorghum Bicolor (L.) Moench) e ao milheto (Penissetum Glaucum (L.) R. Brown) Cultivados no sertão de Pernambuco*; UNIVASF: Petrolina, Brazil, 2017.
69. Vincent, J.M. *A Manual for the Practical Study of Nodule Bacteria*; Blackkwell Science Publication: Edinburgh, UK, 1970; p. 164.

70. Carvalho, M.A.C.; Sá, M.E.; Campos, D.T.S.; Machado, A.P.; Chagas, A.F., Jr. *Bacillus subtilis* UFMT-Pant001 as a plant growth promoter in soybean in a greenhouse. *Afr. J. Agric. Res.* **2023**, *19*, 161–169.
71. Silveira, C.A.P.; Bamberg, A.L.; Martinazzo, R.; Pillon, C.N.; Martins, E.S.; Piana, C.F.B.; Ferreira, L.H.G.; Pereira, I.S.; Stumpf, L. Instruções para Planejamento e Condução de Experimentos com Fertilizantes, Inoculantes, Corretivos, Biofertilizantes, Remineralizadores e Substratos para Plantas. Available online: https://www.gov.br/agricultura/pt-br/assuntos/insumos-agropecuarios/insumos-agricolas/fertilizantes/registro-estab-e-prod/registro-produtos/instrucoes-para-conducao-de-experimentos.pdf (accessed on 23 February 2023).
72. Silva, F.C. *Manual de Análises Químicas de Solos, Plantas e Fertilizantes*; Embrapa Informação Tecnológica: Brasília, Brazil, 2009.
73. Gomes, R.V.; Coutinho, J.L.B. Gergelim. In *Recomendações de Adubação Para o Estado de Pernambuco*, 2nd ed.; Cavalcanti, F.J.A., Ed.; Instituto Agronômico de Pernambuco: Recife, Brazil, 1998; p. 144.
74. Chaves, M.M.; Pereira, J.S.; Maroco, J.; Rodrigues, M.L.; Ricardo, C.P.P.; Osório, M.L.; Carvalho, I.; Faria, T.; Pinheiro, C. How plants cope with stress in the field: Photosynthesis and growth. *Ann. Bot.* **2002**, *89*, 907–916. [CrossRef]
75. Larbi, A.; Abadıa, A.; Abadıa, J.; Morales, F. Down co-regulation of light absorption, photochemistry, and carboxylation in Fe-deficient plants growing in different environments. *Photosynth. Res.* **2006**, *89*, 113–126. [CrossRef]
76. Rohácek, K. Chlorophyll fluorescence parameters: The definitions, photosynthetic meaning, and mutual relationships. *Photosynthetica* **2002**, *40*, 13–29. [CrossRef]
77. Bezerra Neto, E.; Barreto, L.P. *Análises Químicas e Bioquímicas em Plantas*; Imprensa Universitária da UFRPE: Recife, Brazil, 2011.
78. Alcantara, R.M.; Xavier, G.R.; Rumjanek, N.G.; Rocha, M.M.; Carvalho, J.S. Eficiência simbiótica de progenitores de cultivares brasileiras de feijão-caupi. *Rev. Ciênc. Agron.* **2014**, *45*, 1–9. [CrossRef]
79. Bradford, M.M. A rapid and sensitive method for the quantitation of microgram quantities of protein utilizing the principle of protein-dye binding. *Anal. Biochem.* **1976**, *7*, 248–254. [CrossRef]
80. Giannopolitis, C.N.; Ries, S.K. Superoxide desmutases occurrence in higher plants. *Plant Physiol.* **1977**, *59*, 14–309. [CrossRef]
81. Azevedo, R.A.; Alas, R.M.; Smith, R.J.; Lea, P.J. Response of antioxidant enzymes to transfer from elevated carbon dioxide to air and ozone fumigation, in the leaves and roots of wild type and a catalase deficient mutant of barley. *Physiol. Plant* **1998**, *104*, 280–292. [CrossRef]
82. Nakano, Y.; Asada, K. Hydrogen peroxide is scavenged by ascorbato-specific peroxidase in spinach chloroplasts. *Plant Cell Physiol.* **1981**, *22*, 867–880.
83. Bates, L.; Waldren, R.P.; Teare, I.D. Rapid determination of free proline for water-stress studies. *Plant Soil* **1973**, *39*, 205–207. [CrossRef]
84. Lichtenthaler, H.K. Chlorophylls and carotenoids: Pigments of photosynthetic biomembranes. *Meth. Enzymol.* **1987**, *148*, 350–382.
85. IPGRI (International Board for Plant Genetic Resources). *Descriptors for sesame (Sesamum* spp.); NBPGR: New Delhi, India, 2004; 76p.
86. Ferreira, D.F. Sisvar: A guide for its bootstrap procedures in multiple comparisons. *Ciênc. Agrotec.* **2014**, *38*, 109–112. [CrossRef]

Article

Salinity and Mulching Effects on Nutrition and Production of Grafted Sour Passion Fruit

Antônio Gustavo de Luna Souto [1], Lourival Ferreira Cavalcante [2], Edinete Nunes de Melo [2], Ítalo Herbert Lucena Cavalcante [1], Roberto Ítalo Lima da Silva [3], Geovani Soares de Lima [4,*], Hans Raj Gheyi [4], Walter Esfrain Pereira [2], Vespasiano Borges de Paiva Neto [1], Carlos Jardel Andrade de Oliveira [2] and Francisco de Oliveira Mesquita [5]

[1] Postgraduate Program in Plant Production, Federal University of the São Francisco Valley, Petrolina 56300-000, Brazil
[2] Postgraduate Program in Agronomy, Federal University of Paraiba, Areia 58397-000, Brazil
[3] Department of Plant Science, Federal University of Paraíba, Areia 58397-000, Brazil
[4] Postgraduate Program in Agricultural Engineering, Federal University of Campina Grande, Campina Grande 58429-900, Brazil
[5] Department of Soils and Mineralogy, National Institute of the Semiarid, Campina Grande 58434-700, Brazil
* Correspondence: geovani.soares@professor.ufcg.edu.br; Tel.: +55-83-99945-9864

Abstract: The Brazilian semiarid region stands out in terms of sour passion fruit production. Local climatic conditions (high air temperature and low rainfall), combined with its soil properties (rich in soluble salts), increase salinity effects on plants. This study was carried out in the experimental area "Macaquinhos" in Remígio-Paraíba (Brazil). The aim of this research was to evaluate the effect of mulching on grafted sour passion fruit under irrigation with moderately saline water. The experiment was conducted in split-plots in a $2 \times (2 \times 2)$ factorial scheme to evaluate the effects of the combination of irrigation water salinity of 0.5 dS m^{-1} (control) and 4.5 dS m^{-1} (main plot), passion fruit propagated by seed and grafted onto *Passiflora cincinnata*, with and without mulching (subplots), with four replicates and three plants per plot. The foliar Na concentration in grafted plants was 90.9% less than that of plants propagated via seeds; however, it did not affect fruit production. Plastic mulching, by reducing the absorption of toxic salts and promoting greater absorption of nutrients, contributed to greater production of sour passion fruit. Under irrigation with moderately saline water, the plastic film in the soil and seed propagation promote higher production of sour passion fruit.

Keywords: *Passiflora edulis* f. flavicarpa Degener; abiotic stress; rootstock; plastic film; mineral composition; yield fruit

Citation: de Luna Souto, A.G.; Cavalcante, L.F.; de Melo, E.N.; Cavalcante, Í.H.L.; da Silva, R.Í.L.; de Lima, G.S.; Gheyi, H.R.; Pereira, W.E.; de Paiva Neto, V.B.; de Oliveira, C.J.A.; et al. Salinity and Mulching Effects on Nutrition and Production of Grafted Sour Passion Fruit. *Plants* **2023**, *12*, 1035. https://doi.org/10.3390/plants12051035

Academic Editor: Mingjun Li

Received: 26 January 2023
Revised: 16 February 2023
Accepted: 21 February 2023
Published: 24 February 2023

1. Introduction

In arid and semi-arid regions, soil salinity and irrigation management have a direct relationship and affect plants as a function of soluble salt concentrations and compositions of water sources [1]. Soil salinity is affected by irrigation with saline water from dams or artesian wells and saline wastewater (brine) discharged by desalination plants and process industries such as oil and gas, textile, leather, food, dairy, agriculture, and pharmaceutical industries [2–4]. Under high salt concentrations, crop yields may be severely affected by water deficit due to low soil-solution osmotic potential (osmotic effect) and by nutritional imbalance, which may be induced by salinity associated with excessive absorption of toxic ions (Na$^+$ and Cl$^-$) or nutrient availability, transport, or partition within the plant [5–7].

In Brazil, high saline levels in the soil or irrigation water are more common in semi-arid regions of the northeast regions due to low rainfall and high air temperatures [8,9]. This region produces about 71.2% of the Brazilian sour passion fruit (*Passiflora edulis* f. flavicarpa Degener) [10]. The water sources available often have moderate to high concentrations of soluble salts, which contribute to soil degradation, nutritional imbalance, and yields below 10.0 t ha^{-1} [10,11].

According to the threshold salinity of the crops, most varieties of sour passion fruit cultivated behave as salt-sensitive species, with significant reductions in their yields from irrigation water salinity, leading to electrical conductivity of irrigation water (ECiw) = 1.3 dS m^{-1} [11,12]. In addition, the greater or lesser sensitivity of plants to salt stress varies depending on differences in climate, soil, and cultural management factors in each growing region [13]. Recent studies have shown that wild species of *Passiflora* ssp. have greater tolerance to salinity than the sour passion fruit [14,15]. Therefore, they can be used as the rootstock of commercial species for cultivation in saline areas [9,16]. The need for information on the tolerance and mineral nutrition of plants in saline zones, and therefore, on the impact of salinity on fruit production, has a direct economic impact [17].

The excess of toxic elements in cells, such as sodium (Na$^+$) and chloride (Cl$^-$) ions, increases oxidative stress by increasing the production of reactive oxygen species (ROS), which causes damage to proteins, lipids, and nucleic acids [18]. Over time, some species have developed tolerance mechanisms to acclimate to saline environments, such as exclusion, compartmentalization of toxic ions, and preference for absorption of essential elements by plants, called ionic homeostasis [19]. Salt-tolerant rootstocks reduce leaf concentrations of Na$^+$ and Cl$^-$ in melons (*Cucumis melo*) [20] and citrus fruits—*Citrus macrophylla* and *Citrus reticulata* [21], reducing their absorption by roots [7,22]. Such tolerant species also maintain essential elements, such as potassium, calcium, and magnesium, at adequate levels in leaves [23,24]. These results are crucial, since the decreasing order of nutritional demand of sour passion fruit is N > K > Ca > S > P > Mg, as reported by [25].

Another alternative to mitigate salinity effects on plants is plastic mulching (PM) on the soil surface. The technique is often used in agriculture of semi-arid regions to promote an adequate soil microclimate [26,27], favoring water-and nutrient-use efficiencies [28–30]. PM benefits are undeniable for arid and semi-arid areas affected by salinity problems, where the evaporative demand is high and soil and water naturally have high levels of soluble salts [31,32] which migrate by the capillary rise from deeper layers to the surface. Therefore, some studies have shown that PM reduces salinity within the root-zone, increasing fruit yields of species irrigated with saline water, as observed by [31] for grapevines (*Vitis* sp.) and by [33] for raspberries (*Rubus idaues*).

Grafting has been used to induce abiotic stress tolerance in several fruit species [21–24]. However, such a propagation method has progressed little for sour passion fruit, despite the salt-tolerant wild species [9–11]. Plastic mulching has recently been used in the exploitation of fruit species. According to [30], studies still lack progress in different edaphoclimatic conditions. There are gaps to be filled regarding the production benefits in several fruit species that are mainly irrigated with saline water. This study hypothesizes that the use of plastic mulching and the grafting technique with wild species of Passiflora, respectively, reduce the accumulation of salts in the root zone of the soil and increase the selectivity of absorption of essential elements to toxic ions (Na$^+$ and Cl$^-$), influencing the nutrition and productivity of sour passion fruit. Therefore, this study aimed to evaluate saline water and plastic mulching effects on the nutritional status and fruit production of sour passion fruit grafted on *Passiflora cincinnata*.

2. Results

2.1. Macronutrients

Regarding leaf concentrations of macronutrients, sour passion fruit plants responded differently to sources of variation (Table 1). While Ca responded to the interaction water salinity (WS) × propagation (Pg) × plastic mulching (PM), P levels were affected by the interaction of Pg × PM. Leaf K concentrations responded to interactions of WS × Pg, WS × PM, and Pg × PM. Leaf concentrations of Mg were influenced by the interactions of WS × Pg and Pg × PM, while leaf S concentrations were influenced by the interaction of WS × PM. Finally, N responded to PM application.

Table 1. Variance analysis summary and mean concentrations of nitrogen (N), phosphorus (P), potassium (K), calcium (Ca), magnesium (Mg), and sulfur (S) in sour passion fruit leaves as a function of water salinity (WS), plant propagation (Pg), and plastic mulching (PM).

Source of Variation	N	P	K	Ca	Mg	S
	\multicolumn{6}{c}{g kg^{-1}}					
Water salinity (WS)						
Low salinity (0.5 dS m^{-1})	38.6 a	2.1 a	9.2 a	14.0 a	3.8 a	2.7 a
Moderately saline (4.5 dS m^{-1})	37.7 a	2.1 a	10.9 a	14.6 a	3.5 b	2.9 a
Propagation (Pg)						
Seed (SP)	37.8 a	2.2 a	8.7 b	12.6 b	3.1 b	2.7 a
Grafting (GP)	38.8 a	2.0 b	11.4 a	16.1 a	4.2 a	2.9 a
Plastic mulching (PM)						
Without	37.3 b	2.0 b	10.7 a	15.5 a	3.8 b	3.0 a
With	39.3 a	2.2 a	9.4 b	13.2 b	3.4 a	2.6 b
Analysis of variance mean squares						
WS × Pg	7.04 [ns]	0.18 [ns]	5.04 **	73.5 **	2.0 *	1.04 **
WS × PM	1.04 [ns]	0.03 [ns]	5.04 **	6.0 [ns]	0.04 [ns]	0.37 [ns]
Pg × PM	12.04 [ns]	0.12 **	22.04 **	88.2 **	2.0 *	0.04 [ns]
WS × Pg × PM	5.04 [ns]	0.33 [ns]	0.37 [ns]	8.2 *	1.04 [ns]	0.04 [ns]
Mean	38.3	2.1	10.0	14.3	3.6	2.8
CV1 (%)	5.6	6.75	31.7	9.9	0.1	12.7
CV2 (%)	4.2	4.6	6.4	7.9	13.8	11.9

CV = Coefficient of variation; ns, * and ** = not significant, significant at 0.05 and 0.01 probability level by the F-test, respectively; (a and b) means with equal letters do not differ from each other by the 'Tukey' test for water salinity, propagation and plastic mulching, respectively.

Soil plastic mulching enhanced N concentrations in leaves from 37.3 to 39.3 g kg^{-1} (Figure 1A), representing an increase of 5.36%. Figure 1B indicates no difference in P leaf concentrations between sour passion fruits irrigated with low salinity and moderately saline water. However, in plants irrigated with low salinity water, SP plants showed a P concentration in leaves that was 19.4% higher than the GP seedlings. A higher P concentration was also verified in grafted plants grown in mulched soil (Figure 1C), with a P concentration 19.3% higher than non-grafted plants.

Irrigation water salinity did not affect the leaf concentration of K in either SP or GP plants (Figure 1D). However, when plants were irrigated with moderately saline water, K leaf concentrations were 50.4% higher in GP than in SP plants. Figure 1F shows that leaf K concentration in GP plants grown in non-mulched soil was higher than in mulched soil. Under non-mulched treatment, the lack of protection against water loss promoted a leaf K concentration that was 57% higher in grafted plants than in SP. Leaf concentrations of K did not differ between passion fruits grown in mulched and non-mulched soil (Figure 1E). However, an irrigation water salinity of 4.5 dS m^{-1} reduced leaf K concentration by 24.4% in plants grown in non-mulched soil but did not affect plants in mulched soil.

For SP plants, Ca concentrations in the leaf did not differ between plants grown in mulched and non-mulched soil, regardless of the irrigation water salinity level (Figure 2A). However, for GP plants, the highest leaf Ca concentrations were observed in plants grown in non-mulched soil, especially for plants irrigated with moderately saline water. These GP plants showed leaf Ca concentrations that were 66.4% higher than SP plants irrigated with low salinity water and 114.4% higher than SP irrigated with moderately saline water.

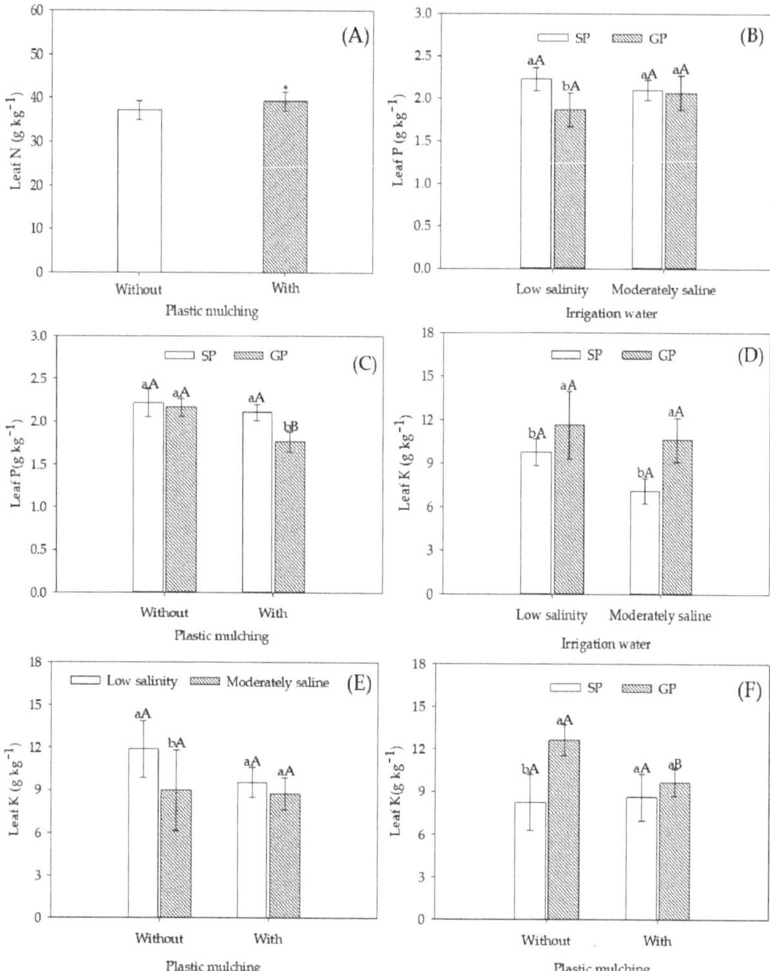

Figure 1. Concentration of macronutrients in leaves of sour passion fruit by seed-propagated and grafted propagated irrigated with low salinity and moderately saline waters with and without plastic mulching. (**A**) N concentration of sour passion fruit in mulched soil; (**B**) P concentration of seed-propagated (SP) and grafted propagated (GP) on *P. cincinnata* irrigated with low and moderately saline water; (**C**) K concentration of seed-propagated (SP) and grafted propagated (GP) on *P. cincinnata* in soil without and with plastic mulching; (**D**) K concentration of seed-propagated (SP) and grafted propagated (GP) on *P. cincinnata* irrigated with low salinity and moderately saline water; (**E**) K concentration in leaves of sour passion fruit irrigated with low salinity and moderately saline water and in soil without and with mulching plastic; (**F**) K concentration of seed-propagated (SP) and grafted propagated (GP) on *P. cincinnata* fruit in soil without and with mulching plastic. Vertical bars represent the standard error of the mean ($n = 4$). Bars with an asterisk (*) differ from each other for soil with and without plastic mulching by the F-test ($p > 0.05$) (**A**). Bars with the same lower-case letter are similar for soil with and without plastic mulching (**C,E,F**) or for low salinity and moderately saline irrigation water (**B,D**) by the F-test ($p > 0.05$). Bars with the same uppercase letter are similar for seed propagation and grafting (**B–D,F**) or low salinity and moderately saline irrigation water (**E**) by the F-test ($p > 0.05$).

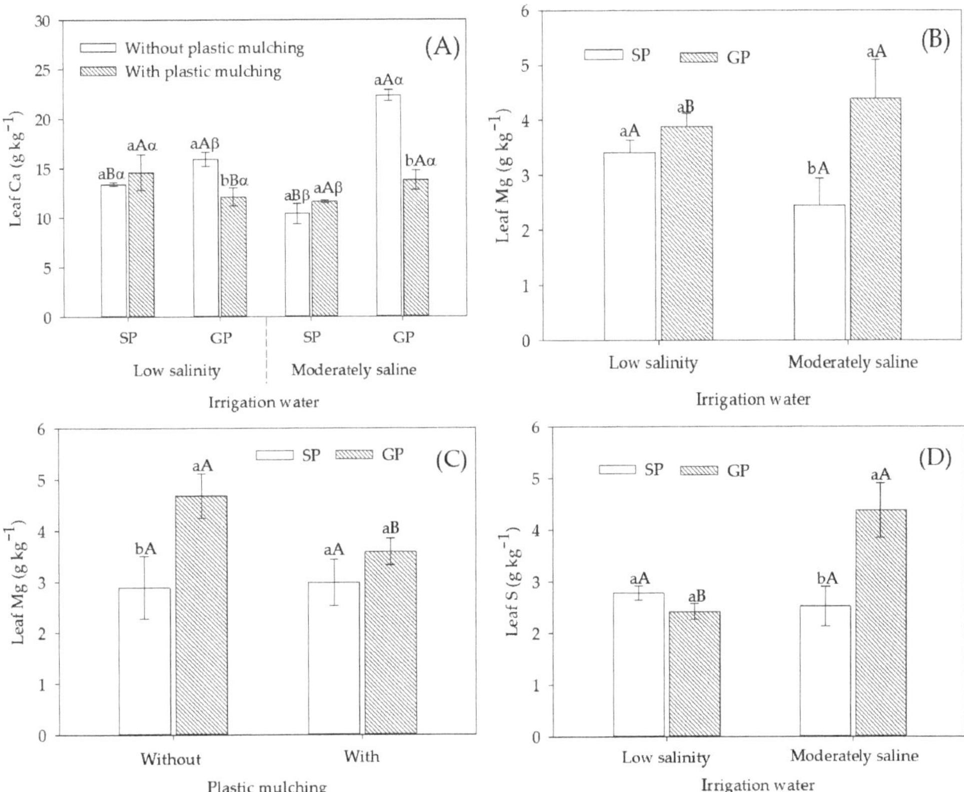

Figure 2. Concentration of macronutrients in leaves of sour passion fruit by seed-propagated and grafted propagated irrigated with low salinity and moderately saline waters with and without plastic mulching. (**A**) Ca concentration in sour passion fruit seed-propagated (SP) and grafted-propagated (GP) on *Passiflora cincinnata* irrigated with low salinity and moderately saline water and in soil without and with mulching plastic; (**B**) Mg concentration in sour passion fruit seed-propagated (SP) and grafted-propagated (GP) on *Passiflora cincinnata* irrigated with low salinity and moderately saline water; (**C**) Mg concentration in sour passion fruit seed-propagated (SP) and grafted-propagated (GP) on *Passiflora cincinnata* in soil without and with plastic mulching, and (**D**) S concentration in sour passion fruit seed-propagated (SP) and grafted-propagated (GP) on *Passiflora cincinnata* irrigated with low salinity and moderately saline water. Vertical bars represent the standard error of the mean (n = 4). Bars with the same lower-case letter are similar for soil without and with plastic mulching (**A**) or for seed propagation and grafting (**B–D**) by the F-test ($p > 0.05$). Bars with the same uppercase letter are similar for seed propagation and grafting (**A**) or low salinity and moderately saline irrigation water (**B,D**) or soil without and with plastic mulching (**D**) by the F-test ($p > 0.05$). Bars with the same Greek letter are similar for low salinity and moderately saline irrigation water (**A**) by the F-test ($p > 0.05$).

Moderately saline water irrigation in GP plants increased the leaf Mg concentration (Figure 2B). Sour passion fruit plants grafted on *P. cincinnata* increased leaf Mg concentration by 78.8% when compared to SP plants irrigated with 4.5 dS m^{-1} water. On the other hand, plastic mulching caused no significant effect on leaf Mg concentration in SP. However, GP plants grown in non-mulched soil had a higher nutrient concentration than those that were grown in in mulched soil (Figure 2C). When comparing propagation forms under both soil mulching conditions, GP had a leaf Mg concentration that was 62.5% higher than SP. When irrigated with moderately saline water, the sour passion fruit grafted propagated showed a higher S concentration than the plants seed-propagated (Figure 2D). Furthermore, under irrigation with moderately saline water, the sour passion fruit showed an increase in S concentration of 80.1% compared to irrigation with low salinity water.

2.2. Micronutrients and Sodium

Leaf concentrations of Cu, Fe, Mn, Zn, and Na were influenced by the interaction of WS $\times$ Pg $\times$ PM (Table 2). In addition, the interaction of Pg $\times$ PM affected leaf B concentrations, while Cl responded to the interaction of WS $\times$ PM.

Table 2. Variance analysis summary and mean concentrations of copper (Cu), iron (Fe), manganese (Mn), zinc (Zn), boron (B), chlorine (Cl), and sodium (Na) in leaves of sour passion fruit plants as a function of water salinity (WS), plant propagation method (Pg), and plastic mulching (PM).

SV	Cu	Fe	Mn	Zn	B	Cl	Na
	mg kg^{-1}						
Water salinity (WS)							
Low salinity (0.5 dS m^{-1})	6.5 a	184.6 a	32.8 a	49.8 a	34.8 b	13.8 b	1151 b
Moderately saline (4.5 dS m^{-1})	3.8 b	202.8 a	33.8 a	47.7 a	30.7 a	21.3 a	6327 a
Propagation (Pg)							
Seed (SP)	5.8 a	192.9 a	33.2 a	51.3 a	36.3 a	19.4 a	6671 a
Grafting (GP)	4.5 b	194.5 a	33.4 a	46.1 b	29.1 b	15.8 b	807 b
Plastic mulching (PM)							
Without	5.7 a	231.1 a	33.9 a	56.9 a	33.6 a	22.4 a	4750 a
With	4.6 b	156.2 b	32.7 a	40.6 b	31.8 a	12.8 b	2728 b
Analysis of variance mean squares							
WS $\times$ Pg	32.7 **	27,405 **	651 **	3504 **	0.37 ns	2.7 ns	1302 $\times$ 106 **
WS $\times$ PM	8.2 **	3675 ns	273 *	80 *	84.4 ns	216 **	167 $\times$ 106 **
Pg $\times$ PM	20.2 **	6633 *	376 *	522 **	247 **	20.2 ns	130 $\times$ 106 **
WS $\times$ Pg $\times$ PM	8.2 **	22,632 **	360 *	1872 **	22 ns	8.2 ns	156 $\times$ 106 **
Mean	5.2	193.7	33.3	48.8	32.7	17.6	3739.4
CV1 (%)	3.95	23.73	21.77	4.83	14.07	2.01	1.97
CV2 (%)	13.69	15.66	20.01	6.20	14.03	15.55	2.74

CV = Coefficient of variation; ns, *, and ** = non-significant, significant at 0.05, and significant at 0.01 probability level by the F-test, respectively; (a and b) means with equal letters do not differ from each other by the 'Tukey' test for water salinity, propagation and plastic mulching, respectively.

The highest leaf Cu concentration was observed in SP sour passion fruit irrigated with low salinity water and grown without mulching, with an increase of 178.32% compared to GP plants (Figure 3A). However, under irrigation with 4.5 dS m^{-1} water and in mulched soil, the leaf Cu concentration was 48.4% higher in GP plants than in SP plants.

Under irrigation with low salinity water, leaf Fe and Zn concentrations were higher in seed-propagated plants in non-mulched soil; concentrations were 61% and 100.2% superior to those of grafted-propagated plants, respectively (Figure 3B,D). However, no significant ($p > 0.05$) difference was observed for Mn concentration (Figure 3C). The opposite behavior was observed in plants under moderately saline water irrigation, but with grafted plants: the leaf Fe, Mn, and Zn concentrations were higher than those of seed-propagated plants grown in soil without mulching, with increases of 46.6, 108.1, and 134.8%, respectively.

Leaf B concentrations did not differ significantly between irrigation with low salinity and moderately saline water (Figure 4A). However, SP plants showed higher foliar B concentration than GP plants, with 24.4% and 25.7% increments in plants irrigated with 0.5 and 4.5 dS m^{-1} water, respectively.

Sour passion fruit irrigated with moderately saline water had higher leaf Cl concentration, but plastic mulching considerably reduced its concentrations in leaf tissues (Figure 4B). Mulching reduced leaf Na concentration in the sour passion fruit, regardless of the irrigation water salinity (Figure 4C). Moreover, GP plants under salt stress had leaf Na concentrations similar to those of plants irrigated with water at 0.5 dS m^{-1}. Such findings are significant compared to SP plants under moderately saline water irrigation (14,982.2 mg kg^{-1}), in which the Na concentration was 996.7% higher than that in GP (1366.1 mg kg^{-1}).

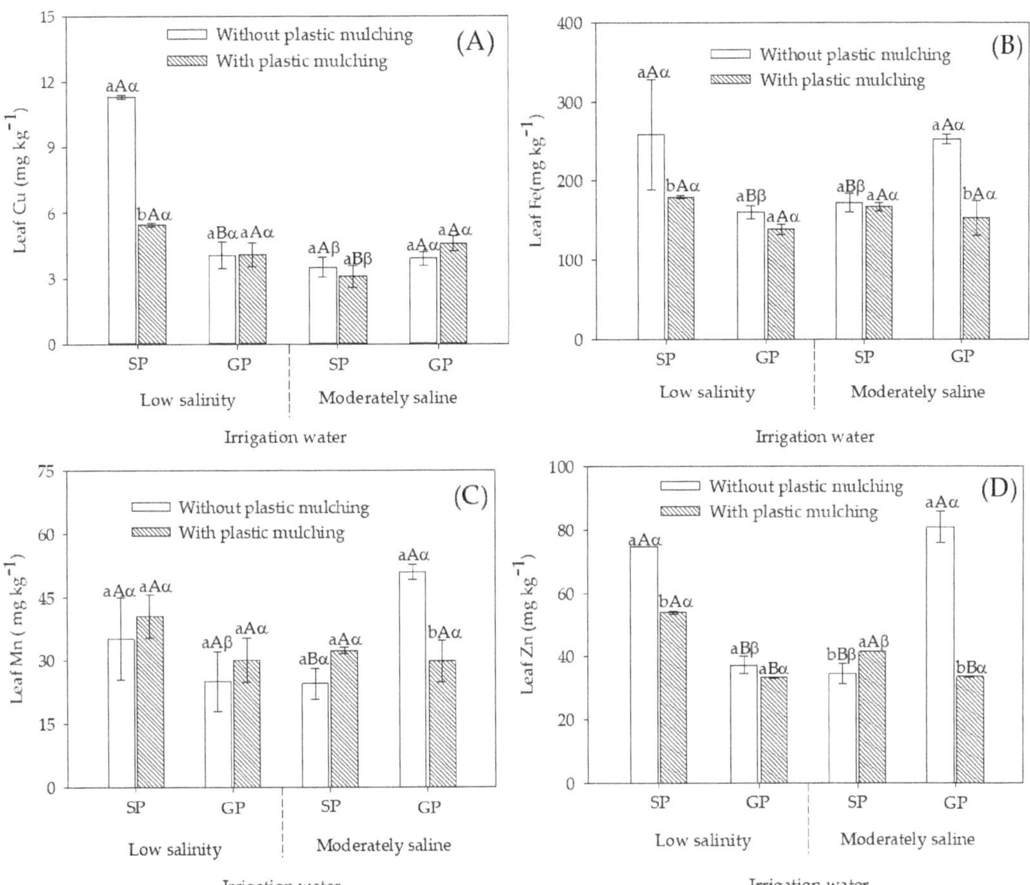

Figure 3. Concentration of micronutrients in leaves of sour passion fruit by seed-propagated and grafted propagated irrigated with low salinity and moderately saline waters with and without plastic mulching. (**A**) Cu, (**B**) Fe, (**C**) Mn, (**D**) Zn concentration in sour passion fruit seed-propagated (SP) and grafted propagated (GP) on *Passiflora cincinnata* irrigated with low salinity and moderately saline water in soil with and without plastic mulching. Vertical bars represent the standard error of the mean (n = 4). Bars with the same lower-case letter are similar for soil without and with plastic mulching by the F-test ($p > 0.05$). Bars with the same uppercase letter are similar for seed propagation and grafting by the F-test ($p > 0.05$). Bars with the same Greek letter are similar for irrigation with low salinity and moderately saline water by the F-test ($p > 0.05$).

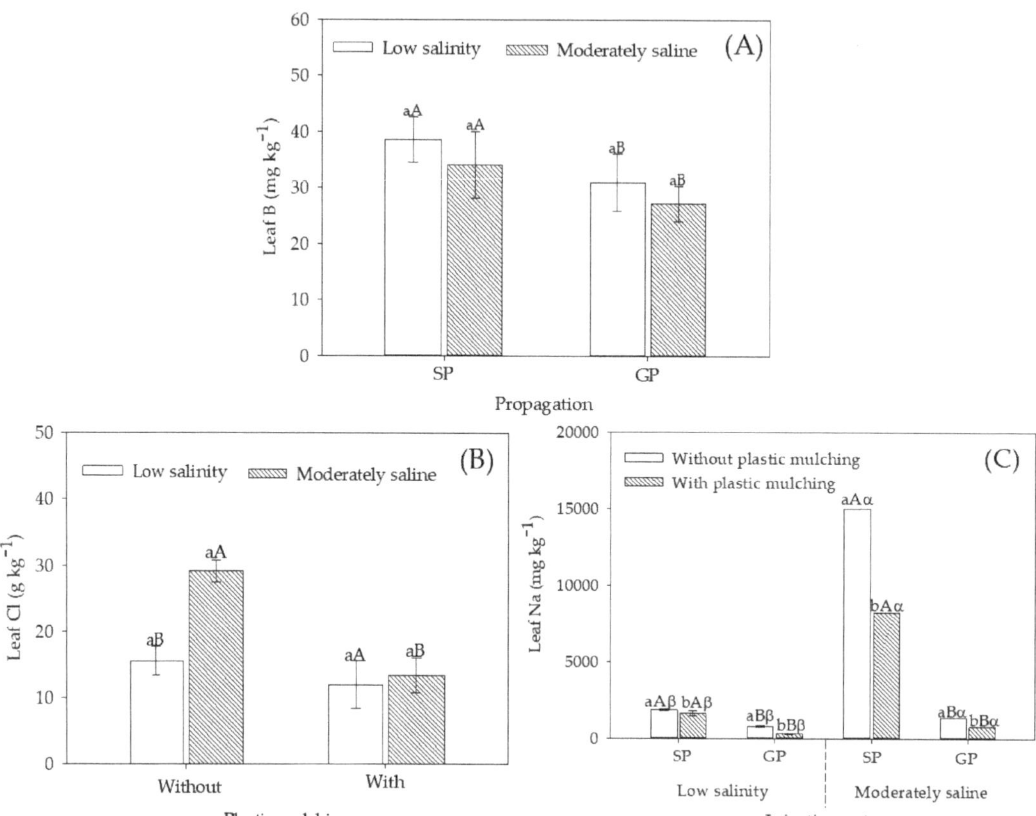

Figure 4. Concentration of micronutrients (boron and chlorine) and sodium in leaves of sour passion fruit by seed-propagated and grafted propagated, irrigated with low salinity and moderately saline waters, with and without plastic mulching. (**A**) B concentration in sour passion fruit seed-propagated (SP) and grafted propagated (GP) on *Passiflora cincinnata* irrigated with low salinity and moderately saline water; (**B**) Cl concentration in sour passion fruit irrigated with low salinity and moderately saline water in soil with and without plastic mulching; (**C**) Na concentration in sour passion fruit seed-propagated (SP) and grafted propagated (GP) on *Passiflora cincinnata* irrigated with low salinity and moderately saline water in soil with and without plastic mulching. Vertical bars represent the standard error of the mean (n = 4). Bars with the same lower-case letter are similar for soil with and without plastic mulching (**A**) or low salinity and moderately saline irrigation water (**B,C**) by the F-test ($p > 0.05$). Bars with the same uppercase letter are similar for soil with and without plastic mulching (**B**) or seed propagation and grafting (**A,C**) by the F-test ($p > 0.05$). Bars with the same Greek letter are similar for irrigation with low saline and moderate saline water (**A**) by the F-test ($p > 0.05$).

2.3. Production of Fruits per Plant

The fruit yield per plant was affected by the interactions of WS × PM (F = 70.62; $p = 0.0001$) and Pg × PM (F = 37.96; $p = 0.0001$). The salinity of the irrigation water did not affect the production of sour passion fruit (Figure 5A). In addition, the use of plastic cover in the soil increased fruit production from 11.26 to 15.03 kg per plant (low salinity water) and from 8.65 to 16.93 kg per plant (moderately saline water). However, SP sour passion fruit showed higher production than the GP ones, mainly in plants grown in mulched soil, with an increase of 259.5% (Figure 5B). Soil protection with mulching increased production per plant by 57.1% in SP and 78.7% in GP plants.

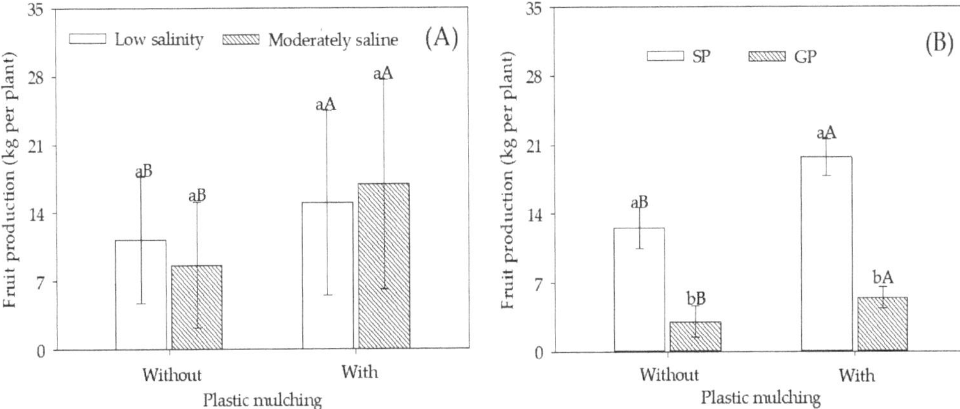

Figure 5. Fruit production of sour passion fruit seed-propagated and grafted propagated irrigated with low salinity and moderately saline waters and with and without plastic mulching. (**A**) Fruit production of sour passion fruit irrigated with low salinity and moderately saline water in soil with and without plastic mulching; (**B**) Fruit production of sour passion fruit seed-propagated (SP) and grafted propagated (GP) on *Passiflora cincinnata* in soil with and without plastic mulching. Vertical bars represent the standard error of the mean (n = 4). Bars with the same lower-case letter are similar for soil without and with plastic mulching (**A,B**) and bars with the same uppercase letter are similar for low salinity and moderately saline irrigation water (**A**) or seed propagation and grafting (**B**) by the F-test ($p > 0.05$).

3. Discussion

Irrigation with moderately saline water had no significant effect on the leaf N concentration in passion fruit; this agrees with the results presented by [34] for the same crop under irrigation with the same type of water. Increases in leaf N concentration in yellow passion fruit grown in soil under plastic mulching (Figure 1A) can be attributed to decreases in water losses by evaporation and N losses by leaching. These reductions are due to improvements in thermal amplitude and soil moisture, increasing N absorption and nutrient-use efficiency by plants [28–30]. Nevertheless, sour passion fruit plants had an adequate N concentration in both treatments, within the adequate range of 36.0 to 46.0 g kg^{-1} [35].

Figure 1B,C show that only grafted sour passion fruits under irrigation with low salinity water and in mulched soil had leaf P concentrations outside the recommended range of 2.0–3.0 g kg^{-1} [35]. Zucarelli et al. [36] verified the same trend in the purple passion fruit grafted on *Passiflora cincinnata*, which had leaf P concentrations lower than non-grafted plants. Moreover, fertigation with potassium sulfate can reduce P absorption due to ionic antagonism between $H_2PO_4^-$ and SO_4^{2-} ions [6].

The use of *Passiflora cincinnata* as rootstock for sour passion fruit increased tolerance or adaptability to salinity and efficiency in K acquisition (Figure 1D) regardless of the mulching condition (Figure 1F), maintaining sufficient leaf K concentrations. The higher K absorption capacity of plants grafted on *P. cincinnata* tends to restrict the absorption and transport of toxic ions (Na^+ and Cl^-) of the irrigation water, as reported by [24] in grafted and non-grafted pomegranate (*Punica granatum* L.) under irrigation with 7.0 dS m^{-1} water.

The benefits of mulching on the soil by reducing heat and increasing humidity enhanced K absorption and accumulation in sour passion fruit leaves (Figure 1E); this impacted soil microbiota, which in turn increased K availability in plants through decomposition and cycling of nutrients in the soil [29,37]. Despite the increases, sour passion fruit plants were deficient in K, since the sufficiency range is between 24.0 and 32.0 g kg^{-1} [35].

Leaf Ca, Mg, and S concentrations were higher in GP than in SP plants, mainly under moderately saline water irrigation (Figure 2). In several crops, tolerant species have been used as rootstocks for salt sensitive commercial species, such as tomatoes—*Solanum lycopersicum* [5], melon—*Cucumis melo* [20,23], pumpkins—*Cucurbita ficifolia, Cucurbita moschata* L. landraces [6,7], and pomegranate [24].

As rootstock, *P. cincinnata* provided salt tolerance in sour passion fruit by selective absorption of nutrients and reduction in absorption and transport of Na^+ and Cl^- ions, in addition to accumulation and compartmentalization of toxic ions in root cells [7–24]. Under saline conditions, sour passion fruit grafted on *P. cincinnata* was properly supplied with Ca, Mg, and S, according to their crop sufficiency ranges of 17–28 g kg^{-1}, 2.1 g kg^{-1}, and 4.4 g kg^{-1}, respectively [35].

In the present study, the employment of *P. cincinnata* as rootstock increased absorption and leaf concentrations of micronutrients (Cu, Fe, Mn, and Zn) in sour passion fruit under salt stress (Figure 3). In cucumbers irrigated with 5.7 dS m^{-1} water, grafting raised both leaf concentrations of micronutrients and crop yield [6]. Micronutrients are involved in many metabolic and cellular functions essential to plant growth, such as energy metabolism, synthesis of primary and secondary metabolites, hormonal balance, and signal transduction [38].

Despite the higher leaf B concentrations in SP sour passion fruit, in both irrigation water salinities (Figure 4A), it was not enough according to the nutritional requirements of the plant (39 to 47 mg kg^{-1}), as reported by [35]. López-Gómez et al. [22] described similar results for grafted loquat (*Eriobotrya japonica* Lindl.) under salt stress and fertilized with B. These authors reported that leaf B concentrations increase in grafted plants, reducing lipid peroxidation by salt stress and improving cell membrane protection.

Plastic mulching minimizes soil water losses through evaporation [39]. Under such a situation, sour passion fruit plants had lower leaf Cl^- concentrations (Figure 4B). Therefore, higher moisture in the soil irrigated with moderately saline water reduced soluble salts, such as chloride, in the topsoil layer, wherein absorbing roots are significantly concentrated [32,40]. The application of plastic mulching is important in conditions where the water has high concentrations of Cl^- ion. In this case, the absorption of this ion is accompanied by a decrease in the concentration of $N\text{-}NO_3^-$ in the aerial parts of the plants [17].

Based on leaf concentrations of Na (Figure 4C) and other nutrients (Figures 1–3) in both SP and GP plants, as rootstock, *P. cincinnata* acts as a filter of ions mobilized to tillers. Generally, species native to saline environments have saline stress tolerance genes that can be transmitted to commercial species to obtain more tolerant hybrids [41]. Ferreira et al. [42] observed that the genes involved in sodium transport (SOS1 and SOS3) were upregulated in sour passion fruit under a water salinity of 12 dS m^{-1}. Another factor that also contributes to selectivity in salt absorption is the membrane transporters that regulate ionic homeostasis in cells, especially Na^+/H^+ and K^+/H^+, transporters of sucrose and amino acids [43].

Lima et al. [41] attributed the reduction of up to 50% of the foliar concentration of Na^+ in *Passiflora mucronata* Lam compared to *P. edulis*, both irrigated with saline water (150 mM NaCl), to the possible presence of these genes in the wild species. Fanny irrigated tomato (*Lycopersicon esculentum* Mill) cv Pwith 60 mM NaCl water. The use of rootstock AR-9704 reduced foliar sodium concentration by 29.16% compared to non-grafted plants [5]. The use of grafting on citrus Cleopatra Mandarin (*Citrus reticulata* Blanco) on Alemow (*Citrus macrophylla*) irrigated with saline water reduced the presence of Na in the aerial parts by 63% in comparison with non-grafted plants [21]. Its ability to accumulate Na and excrete salts by roots, as already verified in other Passiflora species [11,42], acts as a retention mechanism and prevents damage to plant shoots [23,24], reducing Na concentrations in the leaf tissue of the scion.

For plants irrigated with moderately saline water and grown in mulched soil (Figure 5A), fruit production overcame the maximum of 10.76 kg $plant^{-1}$, observed by [8], for sour passion fruit irrigated with saline water and fertilized with bovine biofertilizer. Soil plastic mulching promoted a higher increase in fruit production of plants irrigated with moderately saline water (+95.7%) than irrigated with low salinity water (+33.7%). This finding highlights the benefits of plastic mulching by maintaining irrigated water volume, suitable edaphic microclimate, and reducing toxic salts within the soil layer below the root zone [31,44].

The root system absorbs water and nutrients from the soil and is the organ that is most affected under limiting conditions, such as low water availability or high levels of toxic ions [27,45]. Thus, plastic mulching in soil promoted favorable conditions for the absorption of water and nutrients and increased the production of sour passion fruit.

Fruit production in SP plants was always higher than in GP plants (Figure 5B). These results are similar to those reported by [46], who evaluated the productive capacity of sour passion fruit propagated by cutting and grafting on sweet passion fruit (*Passiflora alata*) and passion fruit Giberti (*Passiflora gibertii*). The authors verified that grafted plants were less vigorous during vegetative growth, forming lighter fruits [14]. According to [47], higher fruit productions in non-grafted plants are due to an increased average mass of harvested fruits than in grafted plants, as observed in our study (SP = 240.7 g and GP = 204.6 g—results not presented). Despite the greater accumulation of nutrients and reduction of foliar Na^+ and Cl^- in the grafted plants, this was not reflected in fruit production and is due to the loss of vigor of the grafted plants observed in the field over time. This demonstrates that propagation by grafting, a technique recently used in passion fruit, still requires more investigations related to the grafting material and the most appropriate technique.

4. Materials and Methods

4.1. Characterization of the Experimental Area

The experiment was carried out in an experimental area located at 'Macaquinhos Farm', in the municipality of Remígio (7°00′1.95″ S, 35°47′55″ W and 562-m above sea level), Paraíba State, Brazil, between September 2019 and February 2021. According to Köppen's classification, the local climate is classified as As' type, which means that it is tropical hot and humid and has a dry season in winter [48]. The average air temperature, relative air humidity, and rainfall during the experimental period were 26.3 °C, 57.6%, and 375.8 mm, respectively (Figure 6).

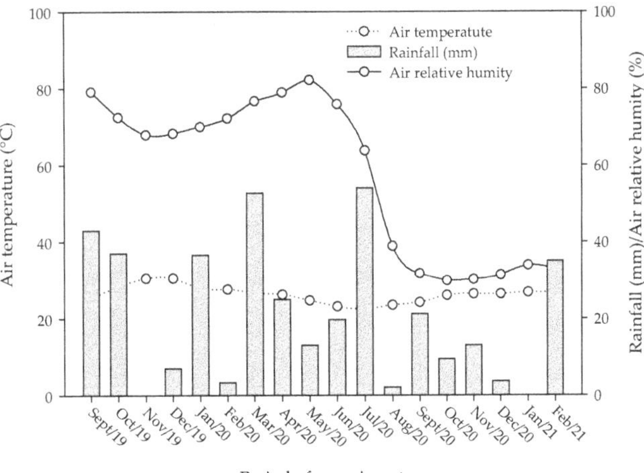

Figure 6. Meteorological data—temperature, relative humidity of the air, and rainfall collected at the experimental site during the study period.

The soil of the experimental area (0–0.40 m) was classified, according to the criteria of the [49], as arenic *Psamment*. Before the installation of the experiment, soil samples were collected in the area, mixed. Then, a composite sample was analyzed for chemical (fertility and salinity) and physical analyses according to [50], as presented in Table 3.

Table 3. Chemical (fertility and salinity) and physical properties of the soil (0–0.40 m depth) of the experimental area before the installation of experiment.

Soil Fertility		Soil Salinity		Soil Physical Properties	
pH	6.00	pHsp (H_2O)	6.16	Sand (g kg^{-1})	831.5
P (mg dm^{-3})	16.63	EC (dS m^{-1})	0.22	Silt (g kg^{-1})	100.0
K^+ (cmol$_c$ dm^{-3})	0.08	SO_4^{2-} (mmol$_c$ L^{-1})	3.91	Clay (g kg^{-1})	68.5
Ca^{2+} (cmol$_c$ dm^{-3})	1.09	Ca^{2+} (mmol$_c$ L^{-1})	5.12	DW (g (kg^{-1})	0.00
Mg^{2+} (cmol$_c$ dm^{-3})	1.12	Mg^{2+} (mmol$_c$ L^{-1})	15.25	FD (kg dm^{-3})	1000
Na^+ (cmol$_c$ dm^{-3})	0.05	K^+ (mmol$_c$ L^{-1})	0.89	SD (g cm^{-3})	1.53
SB (cmol$_c$ dm^{-3})	2.34	Na^+ (mmol$_c$ L^{-1})	5.70	PD (g cm^{-3})	2.61
$H^+ + Al^{3+}$ (cmol$_c$ dm^{-3})	1.24	CO_3^{2-} (mmol$_c$ L^{-1})		TP (m^3 m^{-3})	0.42
Al^{3+} (cmol$_c$ dm^{-3})	0	Cl^- (mmol$_c$ L^{-1})	15.00	H0.01 MPa (g kg^{-1})	65
CEC (cmol$_c$ dm^{-3})	3.58	SAR (mmol L^{-1})^{0,5}	0.28	H0.03 MPa (g kg^{-1})	49
V (%)	65.36	ESP (%)	1.39	H1.50 MPa (g kg^{-1})	28
OM (g kg^{-1})	13.58	Classification	Non saline non sodic	Textural class	Loamy sand

SB—Sum of bases (K^+ + Ca^{2+} + Mg^{2+}+ N a$^+$); CEC—Cation exchange capacity (K^+ + Ca^{2+} + Mg^{2+} + Na^++ [H^++Al^{3+}]); V—Base saturation ([SB/CEC] × 100); OM—Organic matter; EC—Electric conductivity in 1:2 soil water suspension; SAR—Sodium adsorption ratio; ESP—Exchangeable sodium percentage; AD—Dispersed clay; FD—Flocculation degree; SD—Soil density; PD—Particle density; TP—Total porosity; U0.01MPa—Soil moisture at field capacity; U0.03MPa—Soil moisture at 80% field capacity; U1.5MPa—Soil moisture at permanent wilting point.

4.2. Experimental Design and Plant Material Used

The experimental design was in randomized blocks and split plots in a $2 \times (2 \times 2)$ factorial scheme. The main plots were represented by low salinity (0.5 dS m^{-1}) and moderately saline (4.5 dS m^{-1}) irrigation water. The subplots were sour passion fruit propagated by seeds (SP) and grafted on wild passion fruit (GP) grown in plastic-mulched and bare soil (without mulch) conditions (Figure 7), with four replicates and three plants per plot.

Figure 7. Experimental design of sour passion fruit propagated by seeds (SP) and grafted on *P. cincinnata* (GP) irrigated with low salinity (0.5 dS m^{-1}) and moderately saline (4.5 dS m^{-1}) irrigation water and grown in plastic-mulched and bare soil.

Seed-propagated seedlings of the sour passion fruit accession 'Guinezinho' (SP) and seedlings grafted on wild passion fruit (*Passiflora cincinnata*) (GP) were evaluated in the experiment. The choice of passion fruits 'Guinezinho' and *Passiflora cincinnata* is due to the proven tolerance of plant materials to biotic stress and saline environments, respectively, compared to commercial varieties and wild species [9,51]. Seeds of non-grafted seedlings were collected in an orchard near the experimental area from fruits at the physiological maturation stage [52]. The scion variety was obtained from tertiary branches at the vegetative stage of plants in an orchard near the experimental area (Figure 1). Rootstock variety was obtained from seeds collected from fruits of plants grown in the municipality of Cerro Corá, in Rio Grande do Norte (6°2′45″ S, 36°20′45″ W), Brazil (Figure 7). The grafting technique employed was the full cleft, on the rootstocks, 90 days after sowing (DAS).

4.3. Experiment Installation and Performance

Holes were dug and measured $0.40 \times 0.40 \times 0.40$ m (64 dm^3), separating soil from the 0–0.20 and 0.20–0.40 m depth layers. To the 0–0.20 m soil layer, 20 L well-decomposed cattle manure (Table 4) and 50 g FTE-BR12 fertilizer (S = 3.9%, B = 1.8%, Cu = 0.85%, Mn = 2.0%, and Zn = 9.0%) [53] were added for fertilization, as well as 120 g dolomitic limestone (CaO = 47%, MgO = 3.4%, and RPTN = 82%) to raise soil base saturation to 70%. It was then immediately returned to the hole.

Table 4. Chemical characterization of the bovine manure used in the experiment.

Macronutrients		Micronutrients	
Organic carbon (g kg^{-1})	159.1	Boron (mg kg^{-1})	58.0
Nitrogen (g kg^{-1})	8.3	Copper (mg kg^{-1})	941.0
Carbon: nitrogen ratio	19.17	Iron (mg kg^{-1})	250.0
Phosphorus (g kg^{-1})	19.2	Manganese (mg kg^{-1})	8.0
Potassium (g kg^{-1})	10.4	Zinc (mg kg^{-1})	21.3
Calcium (g kg^{-1})	8.2	Sodium (mg kg^{-1})	79.0
Magnesium (g kg^{-1})	5.0	Hydrogen potential (H$_2$O)	8.81
Sulfur (g kg^{-1})	1.8		

Passion fruit vines were trained on the espalier system, using smooth wire #12 fixed on 2.30-m high stakes buried 0.30 m into the ground and spaced 3 m apart. At the end of the line, the stake diameter was 0.20 m. This was to withstand the tension of the training system and plants. The planting spacing was 3 m between plants and 2 m between rows, totaling 1667 ha^{-1} [54]. Seedlings were transplanted when they reached from 0.25 to 0.30 m in height and had four fully expanded leaf pairs.

Low salinity water (0.5 dS m^{-1}) was collected from a surface dam near the experimental area, and moderately saline water (4.5 dS m^{-1}) was obtained by dissolving non-iodinated NaCl (94% purity) into the low salinity water (Table 5). Electrical conductivity was measured using a portable Instrutherm model CD-850 conductivity meter. Over the first 30 days after transplanting (DAT), plants were irrigated with low salinity water (0.5 dS m^{-1}) to allow root system establishment.

Table 5. Chemical characteristics of surface dam water used for irrigation with low salinity water (0.5 dS m^{-1}) and to prepare moderately saline water (4.5 dS m^{-1}).

EC	pH	K$^+$	Ca^{2+}	Mg^{2+}	Na$^+$	Cl$^-$	CO$_3$$^{2-}$	SO$_4$$^{2-}$	SAR	Classification
dS m^{-1}					mmol$_c$ L^{-1}				(mmol L^{-1})$^{1/2}$	
0.5	6.10	0.28	0.65	0.27	1.88	1.87	0.00	0.51	2.77	C1S1

EC = electrical conductivity at 25 °C; C1S1 = Low risk of salinization and sodification of the soil, according to [55].

Afterwards, plants were irrigated according to each treatment to replace evapotranspiration losses. The crop evapotranspiration (ETc) was estimated as the product of potential evapotranspiration (ET_0) and crop coefficient (**kc**), according to methods described by [56,57]:

$$ETc = ET_0 \times kc \tag{1}$$

Crop coefficients adopted were 0.69 for the vegetative stage, 0.82 for flowering, and 1.09 for fruiting [58]. A Class-A tank was installed near the experimental area, and its evaporation (**ETa**) was used to determine ET_0 by multiplying ET_a by a correction coefficient (0.75), as suggested by [59]:

$$ETa = ET_a \times 0.75 \tag{2}$$

For irrigation, a drip system was used. The system was installed before the seedling transplanting, and the soil was covered with plastic mulch. Four pressure auto-compensating drippers were used for each plant (two facing east and two facing west at 0.20 and 0.40 m apart from the plant stem, respectively). The system was set to a flow rate and service pressure of 4 L h^{-1} and 0.2 MPa, respectively.

The soil was covered with a 320-μ-thick white plastic film to protect the soil (mulching) under the three plants in treated plots. The plastic film dimensions were 2.0 m wide and 12 m long, and it was fixed at a distance of 2 m between rows, covering a surface of 24.0 m^2. At the points where the seedlings were transplanted, 0.40-m diameter holes were dug. Then, the unprotected area was covered with a plastic sheet to prevent evaporation.

Nitrogen (N), phosphorus (P), and potassium (K) topdressings were performed through fertigation using a Venturi injector [55]. N and K were supplied every 15 days at a ratio of 1:1 as urea (45% N) and potassium sulfate (50% K_2O and 45% S). Phosphorus was supplied monthly as mono-ammonium phosphate—MAP (50% P_2O_5 and 10% N). Micronutrients (boron [B], copper [Cu], iron [Fe], manganese [Mn], molybdenum [Mo], and zinc [Zn]) were applied via foliar fertilization following recommendations of [60].

4.4. Traits Analyzed

4.4.1. Plant Nutritional Status

At the full flowering stage (120 DAT), in the treatments of each block (four blocks), eight intact and healthy leaves were sampled from the middle part of sour passion fruit plants: four to the east and four to the west from the third or fourth leaf pairs. According to the recommendation of [61], for the sour passion fruit, leaf sampling is carried out at the time of full bloom, as this is the phase with the highest nutritional demand for the crop, and its purpose is to guide possible corrections in fertilization. The samples were analyzed for nutritional status in terms of macro and micronutrients, as well as sodium per dry matter weight [62]. The determination of the nutritional status of the plants was carried out as follows: nitrogen (N) by the Kjeldahl method (wet digestion); phosphorus (P) by molybdenum blue spectrometry; potassium (K) and sodium (Na) by atomic emission spectroscopy; calcium (Ca), magnesium (Mg), sulfur (S), copper (Cu), and iron (Fe) using an atomic absorption spectrophotometer at wavelengths of 422.7, 285.2, 400.0, 3274.7, and 508.0 nm, respectively; boron (B) by UV–vis spectrophotometry at a wavelength of 460.0 nm; manganese (Mn) zinc (Zn) by flame-acetylene atomic absorption spectrometry; and, chloride (Cl$^-$) by the volumetric method of Mohr [63].

4.4.2. Fruit Production per Plant

Fruits were harvested daily as their peels turned predominantly yellow, which occurred 60 days after anthesis [52]. The harvested fruits were counted and weighed on an electronic scale to calculate production per plant (kg per plant).

4.5. Statistical Analysis

Data were subjected to analysis of variance (ANOVA) by the F-test at a 0.05 probability level, after performing a test for normality and data homogeneity using the Shapiro–Wilk test. The means referring to the sources of variation and the interactions were compared by the Tukey test ($p > 0.05$). Data were analyzed using the statistical software SISVAR 5.6 [64].

5. Conclusions

Our results point out that, as rootstock, *Passiflora cincinnata* can alleviate harmful effects of water salinity on sour passion fruit plants increasing absorption of nutrients (K, Ca, Mg, S, Fe, Mn, and Zn) and restricting sodium absorption or transport to the scion variety, but without positive effects on fruit production. Mulching with plastic film, by reducing the presence of toxic salts close to the root zone, promoted greater absorption of elements such as N and Mg and reduced Na and Cl, contributing to greater production of sour passion fruit. Sour passion fruit propagated by seeds and grafted accumulate foliar macronutrients in the following order: N > Ca > K > Mg > S > P; and micronutrients and sodium: Na > Fe > Zn > Mn > B > Cl > Cu (seeds) and Na > Fe > Zn > B > Mn > Cl > Cu (grafted). The use of plastic mulch film in sour passion fruit irrigated with moderately saline water reduced leaf Na^+ and Cl^- concentrations and increased production per plant compared to bare soil. The results in fruit production suggest that plastic mulch attenuates the effects of salts and increases the production capacity of sour passion fruit plants, with an emphasis on seed-propagated plants. Even though *Passiflora cincinnata* rootstock increased absorption of nutrients and decreased sodium concentrations in leaf tissue, it was not reflected in high fruit yields due to loss of production vigor. For future studies, we suggest that studies related to the biochemical and molecular activity of sour passion fruit grafted on *Passiflora cincinnata* be investigated to elucidate possible tolerance mechanisms present in wild species and how they are transferred to commercial cultivars.

Author Contributions: Conceptualization, A.G.d.L.S., L.F.C. and R.Í.L.d.S.; methodology, A.G.d.L.S., L.F.C. and E.N.d.M.; software, F.d.O.M.; validation, A.G.d.L.S., C.J.A.d.O. and R.Í.L.d.S.; formal analysis, A.G.d.L.S., F.d.O.M. and Í.H.L.C.; investigation, A.G.d.L.S., L.F.C. and E.N.d.M.; resources, A.G.d.L.S., L.F.C. and W.E.P.; data collection, A.G.d.L.S., W.E.P. and H.R.G.; writing—original draft preparation, A.G.d.L.S., Í.H.L.C. and V.B.d.P.N.; writing—review and editing, Í.H.L.C., G.S.d.L. and H.R.G.; visualization, Í.H.L.C., A.G.d.L.S. and W.E.P.; supervision, L.F.C., Í.H.L.C. and W.E.P.; project administration, A.G.d.L.S. and L.F.C.; funding acquisition, A.G.d.L.S. and L.F.C. All authors have read and agreed to the published version of the manuscript.

Funding: CNPq (National Council for Scientific and Technological Development, Brazil, Process: 160146/2019-4), CAPES (Coordination for the Improvement of Higher Education Personnel, Brazil, Financial code-001) and UFPB (Universidade Federal da Paraíba, Areia –PB, Brazil).

Acknowledgments: The authors thank the National Council for Scientific and Technological Development (CNPq) for the financial support and research grant to the first author.

Conflicts of Interest: The authors declare no conflict of interest.

References

1. Corwin, D.L. Climate change impacts on soil salinity in agricultural areas. *Eur. J. Soil Sci.* **2020**, *22*, 842–862. [CrossRef]
2. Panagopoulos, A.; Haralambous, K.J. Minimal liquid discharge (MLD) and zero liquid discharge (ZLD) strategies for wastewater management and resource recovery—Analysis, challenges and prospects. *J. Environ. Chem. Eng.* **2020**, *8*, 104418. [CrossRef]
3. Panagopoulos, A. Brine management (saline water & wastewater effluents): Sustainable utilization and resource recovery strategy through minimal and zero liquid discharge (MLD & ZLD) desalination systems. *Chem. Eng. Process.* **2022**, *176*, 108944. [CrossRef]
4. Panagopoulos, A. Techno-economic assessment of zero liquid discharge (ZLD) systems for sustainable treatment, minimization and valorization of seawater brine. *J. Environ. Manag.* **2022**, *306*, 114488. [CrossRef]
5. Fernández-García, N.; Martínez, V.; Carvajal, M. Effect of salinity on growth, mineral composition, and water relations of grafted tomato plants. *J. Plant Nutr. Soil Sci.* **2004**, *167*, 616–622. [CrossRef]
6. Huang, Y.; Bie, Z.; He, S.; Hua, B.; Zhen, A.; Liu, Z. Improving cucumber tolerance to major nutrients induced salinity by grafting onto *Cucurbita ficifolia. Environ. Exp. Bot.* **2010**, *69*, 32–38. [CrossRef]
7. Usanmaz, S.; Abak, K. Plant growth and yield of cucumber plants grafted on different commercial and local rootstocks grown under salinity stress. *Saudi J. Biol. Sci.* **2019**, *26*, 1134–1139. [CrossRef]
8. Dias, T.J.; Cavalcante, L.F.; Leon, M.J.; Santos, G.P.; Albuquerque, R.P.F. Produção do maracujazeiro e resistência mecânica do solo com biofertilizante sob irrigação com águas salinas. *Rev. Ciênc. Agron.* **2011**, *42*, 644–651. [CrossRef]
9. Moura, R.S.; Soares, T.L.; Lima, L.K.S.; Gheyi, H.R.; Dias, E.A.; Jesus, O.N.; Coelho Filho, M.A. Effects of salinity on growth, physiological and anatomical traits of *Passiflora* species propagated from seeds and cuttings. *Rev. Bras. Bot.* **2020**, *44*, 17–32. [CrossRef]

10. Instituto Brasileiro de Geografia e Estatística [IBGE]. *Brazilian Production of Passion Fruit*; IBGE: Rio de Janeiro, Brazil, 2021.
11. Hurtado-Salazar, A.; Silva, D.F.P.; Ocampo, J.; Ceballos-Aguirre, N.; Bruckner, C.H. Salinity tolerance of *Passiflora tarminiana* Coppens & Barney. *Rev. Colomb. Cienc. Hortic.* **2018**, *12*, 11–19. [CrossRef]
12. Soares, F.A.L.; Gheyi, H.R.; Viana, S.B.A.; Uyeda, C.A.; Fernandes, P.D. Water salinity and initial development of yellow passion fruit. *Sci. Agric.* **2002**, *59*, 491–497. [CrossRef]
13. Lima, G.S.; Silva, J.B.; Pinheiro, F.W.A.; Soares, L.A.A.; Gheyi, H.R. Potassium does not attenuate salt stress in yellow passion fruit under irrigation management strategies. *Rev. Caatinga* **2020**, *33*, 1082–1091. [CrossRef]
14. Souto, A.G.L.; Cavalcante, L.F.; Melo, E.N.; Cavalcante, I.H.L.; Oliveira, C.J.A.; Silva, R.I.L.; Mesquita, E.F.; Mendonça, R.M.M. Gas exchange and yield of grafted yellow passion fruit under salt stress and plastic mulching. *Rev. Bras. Eng. Agríc. Ambient.* **2022**, *26*, 823–830. [CrossRef]
15. Souza, G.L.F.; Nascimento, A.P.J.; Silva, J.A.S.; Bezerra, F.T.C.; Silva, R.Í.L.; Cavalcante, L.F.; Mendonça, R.M.N. Growth of wild passion fruit (*Passiflora foetida* L.) rootstock under irrigation water salinity. *Rev. Bras. Eng. Agríc. Ambient.* **2023**, *27*, 114–120. [CrossRef]
16. Faleiro, F.G.; Junqueira, N.T.V.; Junghans, T.G.; Jesus, O.N.; Miranda, D.; Otoni, W.C. Advances in passion fruit (*Passiflora* spp.) propagation. *Rev. Bras. Frutic.* **2019**, *41*, e155. [CrossRef]
17. Zuazo, V.H.D.; Raya, A.M.; Ruiz, J.A.; Tarifa, D.F. Impact of salinity on macro- and micronutrient uptake in mango (*Mangifera indica* L. cv. Osteen) with different rootstocks. *Span. J. Agric. Res.* **2004**, *2*, 121–133. [CrossRef]
18. Amin, I.; Rasool, S.; Mir, M.A.; Wani, W.; Masoodi, K.Z.; Ahmad, P. Ion homeostasis for salinity tolerance in plants: A molecular approach. *Physiol. Plant.* **2021**, *171*, 578–594. [CrossRef]
19. Basu, S.; Kumar, A.; Benazir, I.; Kumar, G. Reassessing the role of ion homeostasis for improving salinity tolerance in crop plants. *Physiol. Plant.* **2021**, *171*, 502–519. [CrossRef]
20. Colla, G.; Roupahel, Y.; Cardarelli, M. Effect of salinity on yield, fruit quality, leaf gas exchange, and mineral composition of grafted watermelon plants. *HortScience* **2006**, *41*, 622–627. [CrossRef]
21. Bleda, F.J.; Madrid, R.; Garcia-Torres, A.L.; Garcia-Lidón, A.; Porras, I. Chlorophyll fluorescence and mineral nutrition in citrus leaves under salinity stress. *J. Plant Nutr.* **2011**, *34*, 1579–1592. [CrossRef]
22. López-Gómez, E.; San Juan, M.A.; Diaz-Vivancos, P.; Beneyto, J.M.; García-Legaz, M.F.; Hernández, J.A. Effect of rootstocks grafting and boron on the antioxidant systems and salinity tolerance of loquat plants (*Eriobotrya japonica* Lindl.). *Environ. Exp. Bot.* **2007**, *60*, 151–158. [CrossRef]
23. Yetisir, H.; Uygur, V. Responses of grafted watermelon onto different gourd species to salinity stress. *J. Plant Nutr.* **2010**, *33*, 313–327. [CrossRef]
24. Karimi, H.; Hassanpour, N. Effects of salinity, rootstock and position of sampling on macro nutrient concentration of pomegranate cv. Gabri. *J. Plant Nutr.* **2017**, *40*, 2269–2278. [CrossRef]
25. Cavalcante, L.F.; Cavalcante, I.H.L.; Rodolfo Júnior, F.; Beckmann-Cavalcante, M.Z. Leaf-macronutrient status and fruit yield of biofertilized yellow passion fruit plants. *J. Plant Nutr.* **2012**, *35*, 176–191. [CrossRef]
26. Shiukhy, S.; Raeini-Sarjaz, M.; Chalavi, V. Colored plastic mulch microclimates affect strawberry fruit yield and quality. *Int. J. Biometeorol.* **2015**, *59*, 1061–1066. [CrossRef]
27. Zhang, S.; Zhang, G.; Xia, Z.; Wu, M.; Bai, J. Optimizing plastic mulching improves the growth and increases grain yield and water use efficiency of spring maize in dryland of the Loess Plateau in China. *Agric. Water Manag.* **2022**, *271*, 107729. [CrossRef]
28. Qin, W.; Hu, C.; Oenema, O. Soil mulching significantly enhances yields and water and nitrogen use efficiencies of maize and wheat: A meta-analysis. *Sci. Rep.* **2015**, *5*, 16210. [CrossRef]
29. Amare, G.; Desta, B. Coloured plastic mulches: Impact on soil properties and crop productivity. *Chem. Biol. Technol. Agric.* **2021**, *8*, 4. [CrossRef]
30. Zhang, H.; Miles, C.; Gerdeman, B.; LaHue, D.G.; DeVetter, L. Plastic mulch use in perennial fruit cropping systems—A review. *Sci. Hortic.* **2021**, *281*, 109975. [CrossRef]
31. Aragüés, R.; Medina, E.T.; Clavería, I. Effectiveness of inorganic and organic mulching for soil salinity and sodicity control in a grapevine orchard drip-irrigated with moderately saline waters. *Span. J. Agric. Res.* **2014**, *12*, 501–508. [CrossRef]
32. Zhao, Y.; Li, Y.; Pang, H.; Li, Y. Buried straw layer plus plastic mulching reduces soil salinity and increases sunflower yield in saline soils. *Soil Tillage Res.* **2016**, *155*, 363–370. [CrossRef]
33. Zhang, H.; Miles, C.; Ghimire, S.; Benedict, C.; Zasada, I.; DeVetter, L. Polyethylene and biodegradable plastic mulches improve growth, yield, and weed management in floricane red raspberry. *Sci. Hortic.* **2019**, *250*, 371–379. [CrossRef]
34. Souza, J.T.A.; Nunes, J.C.; Nunes, J.A.S.; Pereira, W.E.; Freire, J.L.O. Effects of water salinity and organomineral fertilization on leaf composition and production in *Passiflora edulis*. *Rev. Bras. Eng. Agríc. Ambient.* **2018**, *22*, 535–540. [CrossRef]
35. Cantarutti, R.B.; Barros, N.F.; Martinez, H.E.P.; Novais, R.F. Soil fertility assessment and fertilizer recommendations. In *Fertilidade do Solo*; Novais, R.F., Alvarez, V.H., Barros, N.F., Fontes, R.L.F., Cantarutti, R.B., Neves, J.C.L., Eds.; Sociedade Brasileira de Ciência do Solo: Viçosa, Brazil, 2007; pp. 769–850.
36. Zucarelli, V.; Ono, E.O.; Boaro, C.F.; Brambilla, L.P. Desenvolvimento inicial de maracujazeiros (*Passiflora edulis* f. flavicarpa, *P. edulis f. edulis* e *P. alata*) enxertados sobre *Passiflora cincinnata*. *Semin. Cienc. Agrar.* **2014**, *35*, 2325–2339. [CrossRef]

37. Bandopadhyay, S.; González, J.E.L.; Henderson, K.B.; Anunciado, M.B.; Hayes, D.G.; DeBruyn, J.M. Soil microbial communities associated with biodegradable plastic mulch films. *Front. Microbiol.* **2020**, *11*, 2840. [CrossRef] [PubMed]
38. Hänsch, R.; Mendel, R.R. Physiological functions of mineral micronutrients (Cu, Zn, Mn, Fe, Ni, Mo, B, Cl). *Curr. Opin. Plant Biol.* **2009**, *12*, 259–266. [CrossRef]
39. Qin, S.; Li, S.; Yang, K.; Hu, K. Can plastic mulch save water at night in irrigated croplands? *J. Hydrol.* **2018**, *564*, 667–681. [CrossRef]
40. Sintim, H.Y.; Bandopadhyay, S.; English, M.E.; Bary, A.I.; DeBruyn, J.M.; Schaeffer, S.M.; Miles, C.A.; Reganold, J.P.; Flury, M. Impacts of biodegradable plastic mulches on soil health. *Agric. Ecosyst. Environ.* **2019**, *273*, 36–49. [CrossRef]
41. Lima, L.K.S.; Jesus, O.N.; Soares, T.L.; Santos, I.S.; Oliveira, E.J.; Coelho Filho, M.A. Growth, physiological, anatomical and nutritional responses of two phenotypically distinct passion fruit species (*Passiflora* L.) and their hybrid under saline conditions. *Sci. Hortic.* **2020**, *263*, 109037. [CrossRef]
42. Ferreira, J.F.S.; Liu, X.; Suddarth, S.R.P.; Nguyen, C.; Sandhu, D. NaCl accumulation, shoot biomass, antioxidant capacity, and gene expression of *Passiflora edulis* f. Flavicarpa Deg. in response to irrigation waters of moderate to high salinity. *Agriculture* **2022**, *12*, 1856. [CrossRef]
43. Gürel, F.; Öztürk, Z.N.; Uçarli, C.; Roselline, D. Barley genes as tools to confer abiotic stress tolerance in crops. *Front. Plant Sci.* **2016**, *7*, 1137. [CrossRef] [PubMed]
44. Danierhan, S.; Shalamu, A.; Tumaerbai, H.; Guan, D. Effects of emitter discharge rates on soil salinity distribution and cotton (*Gossypium hirsutum* L.) yield under drip irrigation with plastic mulch in an arid region of Northwest China. *J. Arid. Land* **2013**, *5*, 51–59. [CrossRef]
45. Bezborodov, G.A.; Shadmanov, D.K.; Mirhashimov, R.T.; Yuldashev, T.; Qureshi, A.S. Mulching and water quality effects on soil salinity and sodicity dynamics and cotton productivity in Central Asia. *Agric. Ecosyst. Environ.* **2010**, *138*, 95–102. [CrossRef]
46. Cavichioli, J.C.; Corrêa, L.S.; Boliani, A.C.; Santos, P.C. Desenvolvimento e produtividade do maracujazeiro-amarelo enxertado em três porta-enxertos. *Rev. Bras. Frutic.* **2011**, *33*, 558–566. [CrossRef]
47. Melo, N.J.A.; Negreiros, A.M.P.; Sarmento, J.D.A.; Morais, P.L.D.; Sales Junior, R. Physical-chemical characterization of yellow passion fruit produced in different cultivation systems. *Emir. J. Food Agric.* **2020**, *32*, 897–908. [CrossRef]
48. Alvares, C.A.; Stape, J.L.; Sentelhas, P.C.; Gonçalves, J.L.M.; Sparovek, G. Köppen's climate classification map for Brazil. *Meteorol. Z.* **2013**, *22*, 711–728. [CrossRef] [PubMed]
49. US Soil Survey Staff. *Keys to Soil Taxonomy*; United States Department of Agriculture and Natural Resources Conservation Service: Lincoln, NE, USA, 2014; 332p.
50. Silva, F.C. *Analysis Manual Soil Chemistry, Plants and Fertilizers*, 2nd ed.; Embrapa: Brasília, Brazil, 2009.
51. Moura, R.S.; Gheyi, H.R.; Silva, E.M.; Dias, E.A.; Cruz, C.S.; Coelho Filho, M.A. Salt stress on physiology, biometry and fruit quality of grafted *Passiflora edulis*. *Biosci. J.* **2020**, *36*, 731–742. [CrossRef]
52. Vianna-Silva, T.; Lima, R.V.; Azevedo, I.G.; Rosa, R.C.C.; Souza, M.S.; Oliveira, J.G. Determinação da maturidade fisiológica de frutos de maracujazeiro-amarelo colhidos na Região Norte do Estado do Rio de Janeiro, Brasil. *Rev. Bras Frutic.* **2010**, *32*, 55–66. [CrossRef]
53. Borges, A.L.; Souza, L.D. *Lime and Fertilization Recommendations for Passion Fruit*; Cruz das Almas, Brazil, 2010.
54. Morais, R.R.; Macêdo, J.P.S.; Cavalcante, L.F.; Lobo, J.T.; Souto, A.G.L.; Mesquita, E.F. Arranjo espacial e poda na produção e qualidade química de maracujá irrigado com água salina. *Irriga* **2020**, *25*, 549–561. [CrossRef]
55. Richards, L.A. *Diagnosis and Improvement of Saline Alkali Soils, Agriculture*; Handbook 60; US Department of Agriculture: Washington, DC, USA, 1954.
56. Valipour, M. Calibration of mass transfer-based models to predict reference crop evapotranspiration. *Appl. Water Sci.* **2017**, *7*, 628–635. [CrossRef]
57. Anwar, S.A.; Mamadou, O.; Diallo, I.; Sylla, M.B. On the influence of vegetation cover changes and vegetation-runoff systems on the simulated summer potential evapotranspiration of tropical Africa using RegCM4. *Earth Syst. Environ.* **2021**, *5*, 883–897. [CrossRef]
58. Freire, J.L.O.; Cavalcante, L.F.; Rebequi, A.M.; Dias, T.J.; Souto, A.G.L. Necessidade hídrica do maracujazeiro amarelo cultivado sob estresse salino, biofertilização e cobertura do solo. *Rev. Caatinga* **2011**, *24*, 82–91.
59. Cunha, P.C.R.; Nascimento, J.L.; Silveira, P.M.; Alves Júnior, J. Eficiência de métodos para o cálculo de coeficientes do tanque classe A na estimativa da evapotranspiração de referência. *Pesq. Agropec. Trop.* **2013**, *43*, 114–122. [CrossRef]
60. Borges, A.L.; Coelho, E.F. Fertigation in tropical fruit trees. In *Portuguese with English Summary*, 2nd ed.; Embrapa: Cruz das Almas, Brazil, 2009.
61. Costa, A.F.S.; Costa, A.N.; Ventura, J.A.; Fanton, C.J.; Lima, I.M.; Caetano, L.C.S.; Santana, E.N. *Technical Recommendations for the Passion Fruit Cultivation*; Incaper: Vitória, Brazil, 2008.
62. Malavolta, E.; Vitti, G.C.; Oliveira, S.A. *Evaluation of the Nutritional State of Plants: Principles and Applications*, 2nd ed.; Potafos: Piracicaba, Brazil, 1997.

63. Gaines, T.P.; Parker, M.B.; Gascho, G.J. Automated determination of chlorides in soil and plant tissue by sodium nitrate extraction. *Agron. J.* **1984**, *76*, 371–374. [CrossRef]

64. Ferreira, D.F. Sisvar: A computer analysis system to fixed effects split plots type designs. *Rev. Bras. Biom.* **2019**, *37*, 529–535. [CrossRef]

Article

Influence of Foliar Application of Hydrogen Peroxide on Gas Exchange, Photochemical Efficiency, and Growth of Soursop under Salt Stress

Jessica Dayanne Capitulino [1], Geovani Soares de Lima [1,*], Carlos Alberto Vieira de Azevedo [1], André Alisson Rodrigues da Silva [1], Thiago Filipe de Lima Arruda [1], Lauriane Almeida dos Anjos Soares [2], Hans Raj Gheyi [1], Pedro Dantas Fernandes [1], Maria Sallydelândia Sobral de Farias [1], Francisco de Assis da Silva [1] and Mirandy dos Santos Dias [1]

1   Academic Unit of Agricultural Engineering, Federal University of Campina Grande, Campina Grande 58430-380, PB, Brazil
2   Academic Unit of Agrarian Sciences, Federal University of Campina Grande, Pombal 58840-000, PB, Brazil
*   Correspondence: geovani.soares@professor.ufcg.edu.br; Tel.: +55-83-99945-9864

Citation: Capitulino, J.D.; Lima, G.S.d.; Azevedo, C.A.V.d.; Silva, A.A.R.d.; Arruda, T.F.d.L.; Soares, L.A.d.A.; Gheyi, H.R.; Dantas Fernandes, P.; Sobral de Farias, M.S.; Silva, F.d.A.d.; et al. Influence of Foliar Application of Hydrogen Peroxide on Gas Exchange, Photochemical Efficiency, and Growth of Soursop under Salt Stress. *Plants* **2023**, *12*, 599. https://doi.org/10.3390/plants12030599

Academic Editor: Dayong Zhang

Received: 30 December 2022
Revised: 17 January 2023
Accepted: 27 January 2023
Published: 29 January 2023

Abstract: Hydrogen peroxide at low concentrations has been used as a salt stress attenuator because it induces a positive response in the antioxidant system of plants. This study aimed to assess the gas exchange, quantum yield, and development of soursop plants cv. Morada Nova grown with saline water irrigation and foliar hydrogen peroxide application. The experiment was carried out under greenhouse conditions using a randomized block design in a 4×4 factorial scheme corresponding to four levels of electrical conductivity of irrigation water, ECw (0.8, 1.6, 2.4, and 3.2 dS m^{-1}), and four doses of hydrogen peroxide, H_2O_2 (0, 10, 20, and 30 µM), with three replicates. The use of irrigation water with electrical conductivity above 0.8 dS m^{-1} inhibited stomatal conductance, internal CO_2 concentration, transpiration, maximum fluorescence, crown height, and vegetative vigor index of the Morada Nova cultivar of soursop. Compared to untreated plants, the hydrogen peroxide concentration of 30 µM resulted in greater stomatal conductance. Water salinity of 0.8 dS m^{-1} with hydrogen peroxide concentrations of 16 and 13 µM resulted in the highest variable fluorescence and quantum efficiency of photosystem II, respectively, of soursop plants cv. Morada Nova at 210 days after transplantation.

Keywords: *Annona muricata* L.; water salinity; attenuator; abiotic stress; reactive oxygen species

1. Introduction

The soursop (*Annona muricata* L.), a fruit tree native to Central America and the Peruvian Valleys, is notable for its commercialization potential in the domestic market with significant economic importance and export prospects because of the high acceptance of its fruit and pulp, primarily for numerous food and pharmaceutical applications [1,2]. Despite the production potential of soursop for the Northeast region, one of the limitations to its production system is the occurrence of low rainfall and high evapotranspiration in most months of the year [3]. In addition, the water sources commonly used in irrigation have high concentrations of salts [4,5].

High salt concentration in water and/or soil inhibits plant growth due to restrictions of water absorption (osmotic effect) and changes in metabolism, as well as ionic imbalance (specific ion effect), which affects photosynthetic pigments, harms cellular components, and causes lipid peroxidation of the membrane [6,7].

In soursop, salt stress induces loss of photosynthetic activity due to stomatal and non-stomatal limitations [8,9]. It compromises the quantum efficiency of photosystem II (PSII), demonstrating that PSII reaction centers have experienced photoinhibitory damage, which results in the formation of reactive oxygen species (ROS) [10] capable of inducing oxidative damage to proteins and other biological components [11].

Considering the socioeconomic importance of soursop in Brazilian agribusiness, it is essential to study strategies that enable the use of saline water in its cultivation in semi-arid regions. Among the alternatives used to reduce the effects of salt stress on plants, the application of hydrogen peroxide (H_2O_2) on leaves stands out [12]. Hydrogen peroxide is a reactive oxygen species (ROS) that, when applied in low concentrations, acts in acclimatization and/or signaling to salt stress due to metabolic alterations that are responsible for increasing its tolerance to stress, thus enabling its use of water with high concentrations of salts [13–15]. It must be considered that the beneficial effect of the application of H_2O_2 depends on several factors, including the concentration, plant species analyzed, development stage, and application method [16].

In recent years, studies have reported that foliar application with hydrogen peroxide can attenuate the deleterious effects caused by salt stress in several crops, for example, cotton [12], passion fruit [13], mini-watermelon [17], orange [18], tomato [19], maize [20], wheat [16], and rice [21]. However, information about its use in fruit plants is incipient, especially in the soursop crop.

This study is based on the hypothesis that the soursop crop suffers smaller losses from salt stress under the application of hydrogen peroxide through the regulation of physiological processes, which contributes to an increase in photosynthetic and antioxidant activity, avoiding lipid peroxidation caused by ROS and an increase in the rate of CO_2 assimilation and stomatal regulation, thus leading to an improvement in the growth of the soursop plant. In this context, this study aimed to evaluate the gas exchange, quantum yield, and growth of soursop cv. Morada Nova irrigated with saline water under the foliar application of hydrogen peroxide.

2. Results

2.1. Leaf Gas Exchange

The salinity of irrigation water had a significant effect on the soursop plants' stomatal conductance (gs), internal CO_2 concentration (Ci), and transpiration € (Table 1). H_2O_2 concentrations, as a single factor, influenced stomatal conductance (gs). The CO_2 assimilation rate (A), instantaneous water use efficiency ($WUEi$), and instantaneous carboxylation efficiency (CEi) were all significantly affected by the interaction between water salinity levels and H_2O_2 concentrations.

Table 1. Analysis of variance (F-test) summary of stomatal conductance (gs), transpirati€(E), CO_2 assimilation rate (A), internal CO_2 concentration (Ci), instantaneous carboxylation efficiency (CEi), and instantaneous water use efficiency ($WUEi$) of soursop plants cv. Morada Nova irrigated with saline water and subjected to foliar application of hydrogen peroxide at 210 days after transplantation.

Source of Variation	DF	Mean Squares					
		gs	E	A	Ci	CEi	$WUEi$
Salinity levels (SL)	4	0.000547 **	0.1781 **	0.880 ns	87.729 **	0.00009 ns	0.3614 ns
Linear regression	1	0.000240 *	0.111 **	-	122.90 **	-	-
Quadratic regression	1	0.00010 ns	0.1220 ns	-	124.32 ns	-	-
Hydrogen peroxide (H_2O_2)	4	0.001836 **	0.0206 ns	2.650 *	3594.42 ns	0.000083 **	3.1369 **
Linear regression	1	0.004160 **	-	0.0811 ns	-	0.00009 ns	5.693 **
Quadratic regression	1	0.000300 ns	-	4.656 **	-	0.00030 **	0.0965 ns

Table 1. *Cont.*

Source of Variation	DF	Mean Squares					
		gs	E	A	Ci	CEi	$WUEi$
Interaction (SL $\times$ H$_2$O$_2$)	16	0.001775 [ns]	0.3500 [ns]	9.450 **	2442.85 [ns]	0.000186 **	1.074 **
Blocks	3	0.00015 [ns]	0.0687 [ns]	0.1692 [ns]	23.003 [ns]	0.000053 [ns]	0.2020 [ns]
Residual	30	0.000149	0.0347	0.334	105.27	0.000060	0.112
CV (%)		15.65	10.33	12.39	3.97	10.26	13.91

[ns], * and ** are, respectively, not significant and significant at $p \leq 0.05$ and $p \leq 0.01$. CV: coefficient of variation; DF: degrees of freedom.

The stomatal conductance of soursop plants cv. Morada Nova was linearly reduced by the increase in the electrical conductivity of irrigation water (Figure 1A), with a decrease of 18.08% per unit increment in ECw. When comparing the gs of plants subjected to ECw of 3.2 dS m^{-1} to that of plants that received the lowest salinity level (0.8 dS m^{-1}), there was a reduction of 50.74%.

Regarding the effects of H$_2$O$_2$ concentrations on the stomatal conductance of soursop plants (Figure 1B), there was a linear increase corresponding to 5.25% per unit increment in H$_2$O$_2$ concentration. In relative terms, the foliar application of 30 µM resulted in an increase of 0.0870 H$_2$O m^{-2} s^{-1} compared to plants grown under 0 µM of hydrogen peroxide.

Regarding the effects of the electrical conductivity of irrigation water on the internal CO$_2$ concentration (Figure 1C) and transpiration (Figure 1D), there were linear reductions of 6.72 and 9.47%, respectively, per unit increase in ECw. When comparing the Ci and E of plants grown under water salinity of 3.2 dS m^{-1} to the values of those that received ECw of 0.8 dS m^{-1}, there were reductions of 17.04 and 24.59%, respectively, at 210 days after transplantation.

For the interaction between the ECw levels and H$_2$O$_2$ concentrations (Figure 1E), the maximum estimated value of the CO$_2$ assimilation rate (3.20 µmol CO$_2$ m^{-2} s^{-1}) was obtained in plants irrigated with an ECw of 0.80 dS m^{-1} and in the absence of H$_2$O$_2$ application. On the other hand, the minimum value of A (0.05 µmol CO$_2$ m^{-2} s^{-1}) was reached in plants subjected to ECw of 3.2 dS m^{-1} and the foliar application of hydrogen peroxide at a concentration of 30 µM.

For instantaneous carboxylation efficiency (*CEi*) (Figure 2A) and instantaneous water use efficiency (*WUEi*) (Figure 2B), irrigation with water of electrical conductivity of 0.8 dS m^{-1} in the absence of foliar application of hydrogen peroxide (0 µM) resulted in higher estimated values of CEi (0.0278 (µmol m^{-2} s^{-1}) (µmol mol^{-1})$^{-1}$) and *WUEi* (2.67 (µmol CO$_2$ m^{-2} s^{-1}) (mmol H$_2$O m^{-2} s^{-1})$^{-1}$). On the other hand, irrigation with an ECw of 3.2 dS m^{-1} and the foliar application of H$_2$O$_2$ at concentrations of 27 and 30 µM, respectively, led to the estimated minimum values of 0.008 ((µmol m^{-2} s^{-1}) (µmol mol^{-1})$^{-1}$) in *CEi* and 1.46 ((µmol m^{-2} s^{-1}) (mmol H$_2$O m^{-2} s^{-1})$^{-1}$) in *WUEi*.

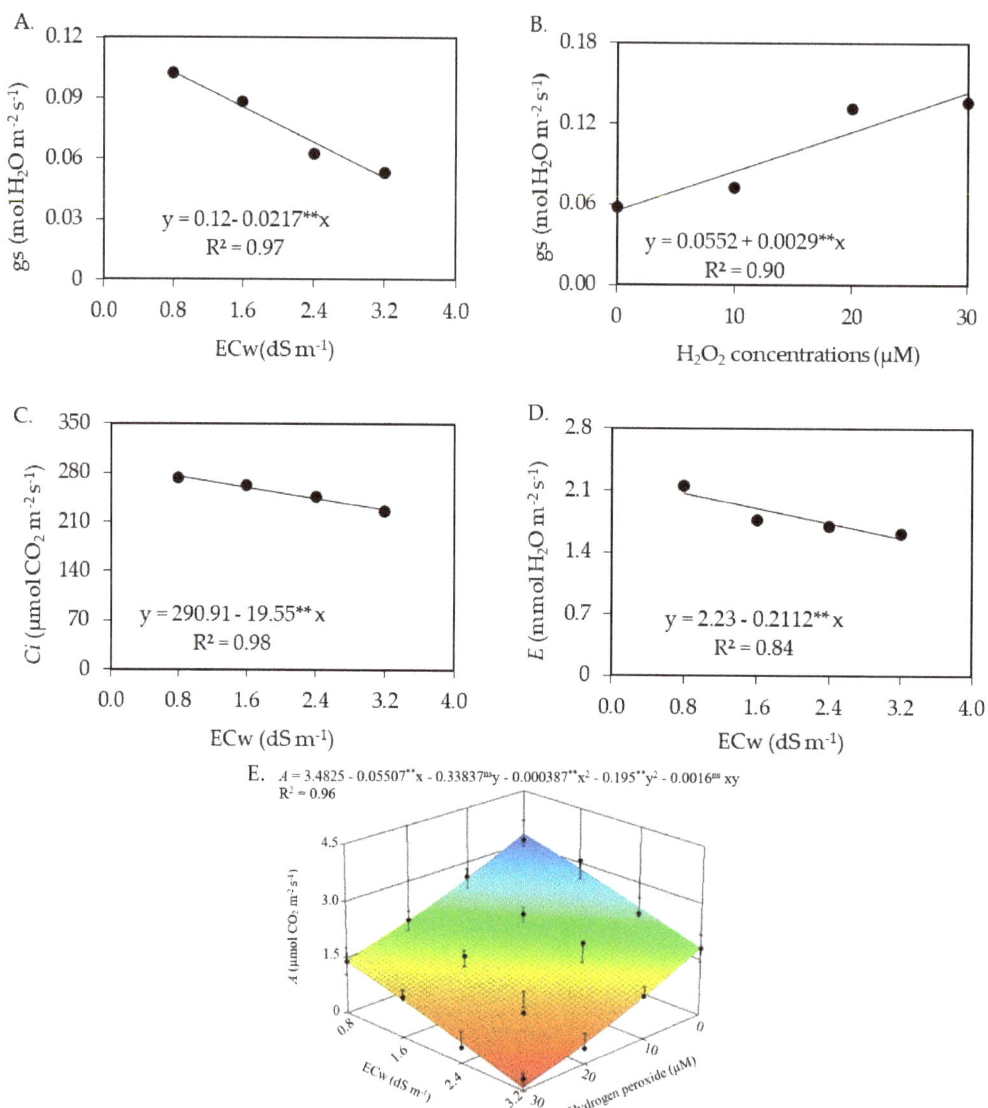

Figure 1. Stomatal conductance (*gs*) of soursop plants cv. Morada Nova as a function of the salinity of irrigation water (ECw) (**A**) and concentrations of hydrogen peroxide (H_2O_2) (**B**), internal CO_2 concentration (*Ci*) (**C**), and transpiration (*E*) (**D**) as a function of ECw and CO_2 assimilation rate (*A*) (**E**) as a function of the interaction between ECw levels and H_2O_2 at 210 days after transplantation. ** is, significant at $p \leq 0.01$ by the F-test. x and y correspond to ECw and hydrogen peroxide (H_2O_2) concentrations, respectively.

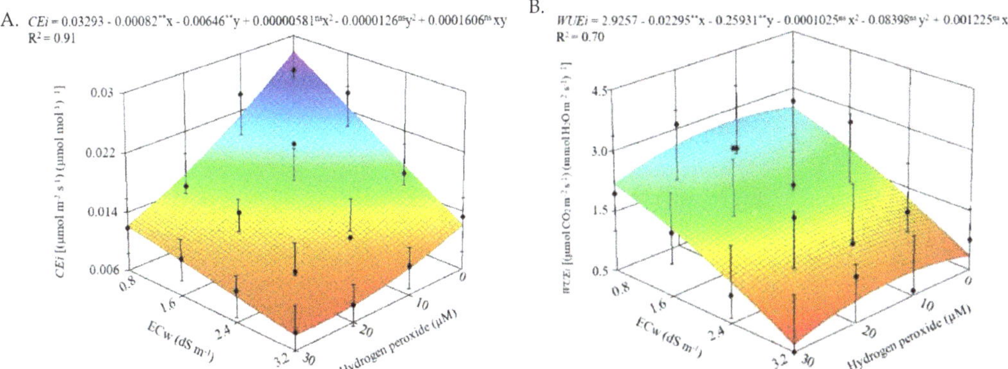

Figure 2. Instantaneous carboxylation efficiency (*CEi*) (**A**) and instantaneous water use efficiency (*WUEi*) (**B**) of soursop plants cv. Morada Nova as a function of the interaction between the irrigation water salinity levels (ECw) and concentrations of hydrogen peroxide (H_2O_2) at 210 days after transplantation. ns and ** are, respectively, not significant and significant at $p \leq 0.01$ by the F-test. x and y correspond to ECw and hydrogen peroxide (H_2O_2) concentrations, respectively.

2.2. Quantum Yield

The initial fluorescence, variable fluorescence, and quantum efficiency of photosystem II of soursop plants at 210 DAT were significantly affected by the interaction between water salinity levels and hydrogen peroxide concentrations (Table 2). The salinity levels of the irrigation water significantly influenced the maximum fluorescence of the soursop. Hydrogen peroxide concentrations, as a single factor, influenced the initial and variable fluorescence and the quantum efficiency of photosystem II of soursop plants.

Table 2. Analysis of variance (F-test) summary of initial fluorescence (F_0), maximum fluorescence (*Fm*), variable fluorescence (*Fv*), and quantum efficiency of photosystem II (*Fv/Fm*) of soursop plants cv. Morada Nova irrigated with saline water and subjected to foliar application of hydrogen peroxide at 210 days after transplantation.

Source of Variation	DF	Mean Squares			
		F_0	*Fm*	*Fv*	*Fv/Fm*
Salinity levels (SL)	4	7842.51 **	75277.36 **	139489.6 **	0.002239 **
Linear regression	1	19634.8 **	190976.0 **	415168.01 **	0.006202 **
Quadratic regression	1	2257.7 ns	32870.5 ns	3088.02 ns	0.000033 ns
Hydrogen peroxide (H_2O_2)	4	658.99 **	1771.91 ns	4881.95 *	0.000928 **
Linear regression	1	436.32 *	-	608.01 ns	0.000375 ns
Quadratic regression	1	21.60 ns	-	13736.3 **	0.002408 **
Interaction (SL $\times$ H_2O_2)	16	362.87 **	7207.74 ns	5538.60 **	0.000163 *
Blocks	3	140.15 ns	16949.8 ns	1220.47 ns	0.000065 ns
Residual	30	82.83	3186.54	1279.87	0.000049
CV (%)		1.36	2.42	2.08	0.97

ns, * and ** are, respectively, not significant and significant at $p \leq 0.05$ and ≤ 0.01. DF: degrees of freedom; CV: coefficient of variation.

The interaction between water salinity levels and hydrogen peroxide concentrations significantly influenced the initial fluorescence of soursop (Figure 3A), with the highest estimated value (674.52) attained in plants irrigated with ECw of 0.8 dS m^{-1} without the foliar application of H_2O_2. Plants under irrigation with water of 3.2 dS m^{-1} and foliar application of 30 µM of H_2O_2 reached the estimated minimum value of 552.10.

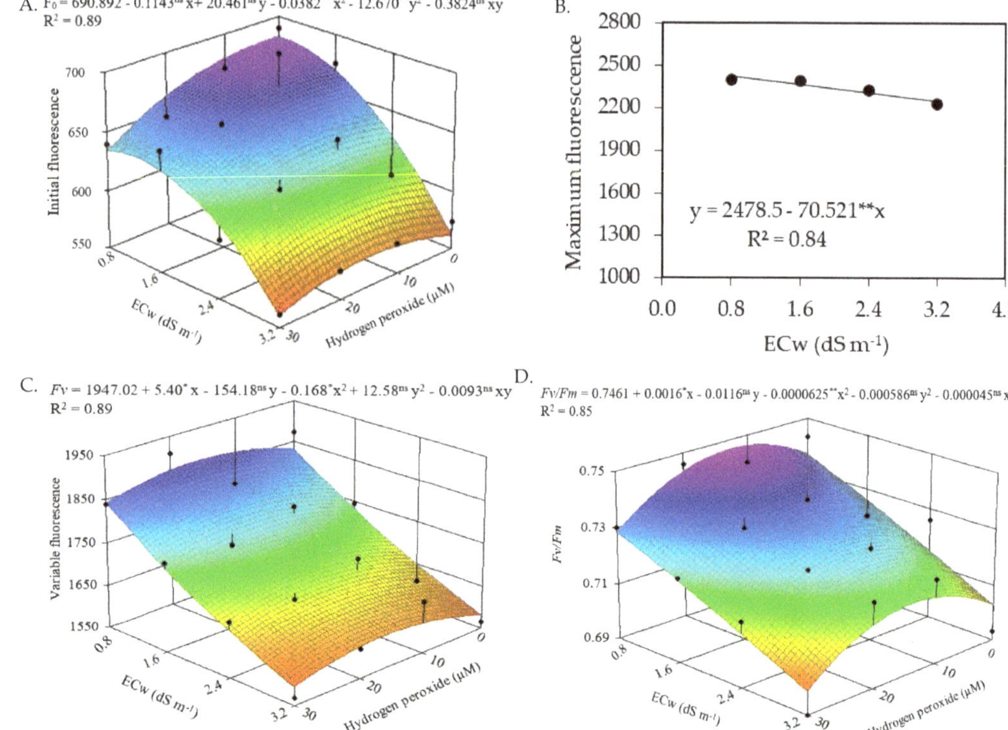

Figure 3. Initial fluorescence (*F0*) (**A**), variable fluorescence (*Fv*) (**C**), and quantum efficiency of photosystem II (*Fv/Fm*) (**D**) of soursop plants cv. Morada Nova as a function of the interaction between the salinity levels of irrigation water (ECw) and hydrogen peroxide concentrations, and maximum fluorescence (*Fm*) (**B**) as a function of the irrigation water salinity (ECw) at 210 days after transplantation. ns, * and ** are, respectively, not significant and significant at $p \leq 0.05$ and ≤ 0.01 by the F-test. x and y correspond to ECw and hydrogen peroxide (H_2O_2) concentrations, respectively.

The maximum fluorescence of soursop plants was linearly reduced by the increase in the electrical conductivity of irrigation water (Figure 3B), with a decrease of 2.85% per unit increment in ECw. When comparing the Fm of plants irrigated with an ECw of 3.2 dS m^{-1} to that of plants under irrigation with the lowest level of water salinity (0.8 dS m^{-1}), a decrease of 6.99% was verified.

Irrigation with water of 0.8 dS m^{-1} associated with the application of hydrogen peroxide concentrations of 16 and 13 µM, respectively, promoted the highest estimated values of variable fluorescence (1875) and quantum efficiency of photosystem II (0.746). However, the combination between the highest salinity of irrigation water (3.2 dS m^{-1}) and the highest concentration of H_2O_2 (30 µM) resulted in the lowest estimated values of *Fv* (1592.37) and *Fv/Fm* (0.690).

2.3. Morphological Parameters

The interaction between irrigation water salinity levels and hydrogen peroxide concentrations significantly affected stem diameter (SD) and crown diameter (D_{crown}) (Table 3). The single factor of irrigation water salinity influenced the crown height (H_{crown}) and the vegetative vigor index (VVI) of soursop plants 210 days after transplantation. There was a significant effect of hydrogen peroxide concentrations only on the V_{crown} of soursop plants.

Table 3. Analysis of variance (F-test) summary of stem diameter (SD), crown height (H_{crown}), crown diameter (D_{crown}), crown volume (V_{crown}), and vegetative vigor index (VVI) of soursop plants cv. Morada Nova irrigated with saline water and subjected to foliar application of hydrogen peroxide at 210 days after transplantation.

Source of Variation	DF	Mean Squares				
		SD	H_{crown}	D_{crown}	V_{crown}	VVI
Salinity level (SL)	4	0.00002 ns	0.0250 *	0.0748 ns	0.00262 *	0.0007 *
Linear regression	1	-	0.0297 *	-	0.00662 **	0.00001 *
Quadratic regression	1	-	0.0285 ns	-	0.00063 ns	0.00002 ns
Hydrogen peroxide (H_2O_2)	4	0.000021 ns	0.0129 ns	0.0344 ns	0.01284 **	0.00006 ns
Linear regression	1	-	-	-	0.01460 **	-
Quadratic regression	1	-	-	-	0.0107 ns	-
Interaction (SL × H_2O_2)	16	0.000021 **	0.0109 ns	0.1119 **	0.0067 ns	0.00005 ns
Blocks	3	0.000030 ns	0.0520 ns	0.0230 ns	0.00129 ns	0.000075 ns
Residual	30	0.000023	0.078	0.0201	0.0045	0.00004
CV (%)		7.29	7.28	15.81	15.39	1.63

ns, * and ** are, respectively, not significant and significant at $p \leq 0.05$ and ≤ 0.01. CV: coefficient of variation; DF: degrees of freedom.

The salinity of irrigation water negatively affected the crown height (Figure 4A) and vegetative vigor index (Figure 4B) of soursop plants at 210 days after transplantation. Regression equations (Figure 4A,B) showed linear reductions of 3.32 and 4.23% per unit increment in ECw in the H_{crown} and VVI of soursop plants, respectively. When comparing the H_{crown} and VVI of plants irrigated with water of an electrical conductivity of 3.2 dS m^{-1} to the values of those subjected to water salinity of 0.8 dS m^{-1}, there were reductions of 8.20 and 10.51%, respectively.

The interaction between water salinity levels and hydrogen peroxide concentrations significantly affected the crown volume (Figure 4C), crown diameter (Figure 4D), and stem diameter (Figure 4E) of soursop plants. Irrigation with water of 0.8 dS m^{-1} in the absence of the foliar application of H_2O_2 promoted the maximum values of 0.125 m^3 and 1.47 m, respectively, for V_{crown} and D_{crown}. However, water salinity levels of 0.8 and 3.2 dS m^{-1} under the foliar application of 30 μM of H_2O_2 contributed to the minimum values of V_{crown} (0.1238 m^3) and D_{crown} (0.762 m), respectively, in soursop plants cv. Morada Nova.

Regarding stem diameter (Figure 4E), it was verified that plants receiving water of 0.8 dS m^{-1} and exposed to a 30 μM H_2O_2 concentration achieved a stem diameter (SD) of 21.11 mm. On the other hand, water salinity of 3.2 dS m^{-1} associated with the foliar application of 11 μM of H_2O_2 resulted in a lower value of the stem diameter (17.96 mm).

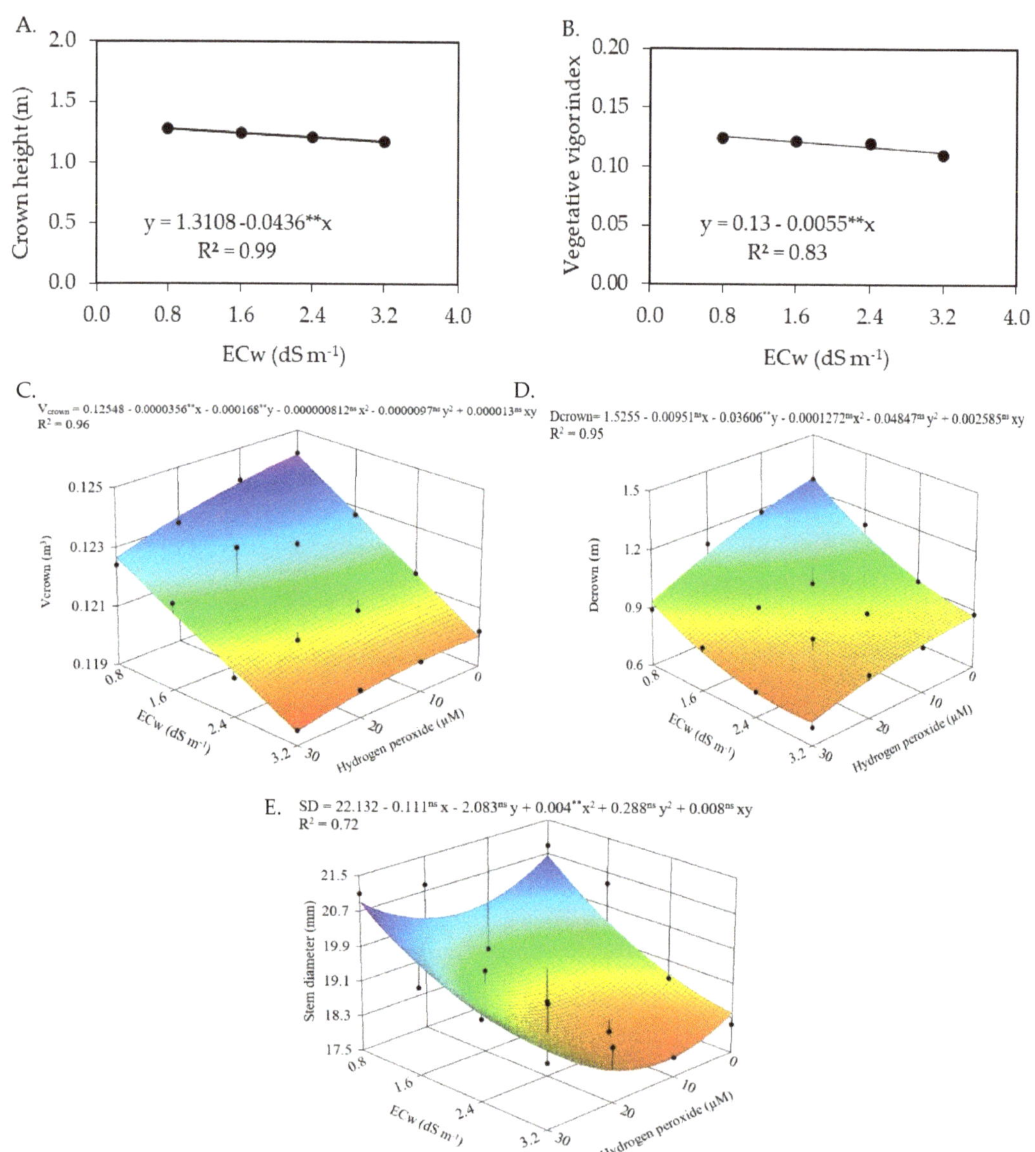

Figure 4. Crown height (Hcrown) (**A**) and vegetative vigor index (VVI) (**B**) of soursop plants cv. Morada Nova as a function of the irrigation water salinity (ECw), and crown volume (Vcrown) (**C**), crown diameter (Dcrown) (**D**), and stem diameter (SD) (**E**), as a function of the interaction between the salinity of irrigation water (ECw) and hydrogen peroxide concentrations at 210 days after transplantation. ns and ** are, respectively, not significant and significant at $p \leq 0.01$ by the F-test. x and y correspond to ECw and hydrogen peroxide (H_2O_2) concentrations, respectively.

3. Discussion

The excess of salts present in irrigation water induces salt stress in plants, which negatively affects their metabolism and limits their growth and development, standing out as a limiting factor for irrigated agriculture, especially in semi-arid regions [22]. In the present study, it was verified that the increase in the electrical conductivity of the irrigation water reduced the gas exchanges, the quantum yield, and the growth of the soursop plant, but the deleterious effects caused by the saline stress were partially mitigated by the foliar application of hydrogen peroxide.

The reduction of gs (Figure 1A) with increasing water salinity is a way for plants to minimize water losses in the form of vapor to the atmosphere and maintain turgor pressure inside their cells, in addition to reducing the absorption of salts [23,24]. Reductions in stomatal conductance due to salt stress were also observed in studies with other fruit plants such as acerola [24], guava [25], and custard apple [26]. The beneficial effect of H_2O_2 observed on the plants' stomatal conductance (Figure 1B) may have occurred due to the defense mechanisms of the plant, inducing the system of antioxidant enzymes, thus minimizing the harmful effects of salinity [27,28]. Silva et al. [15], in their study evaluating the induction of salt stress tolerance (ECw ranging from 0.6 to 3.0 dS m^{-1}) in soursop seedlings using hydrogen peroxide (from 0 to 20 µM), found that the application of hydrogen peroxide at a concentration of 20 µM promoted greater stomatal conductance compared to the control treatment (0 µM H_2O_2) for all salinity levels at 110 days after transplantation.

Increasing the electrical conductivity of irrigation water also reduced the transpiration of soursop plants (Figure 1D). A decrease in E is a strategy of plants to reduce water loss through transpiration, constituting a mechanism of tolerance to salt stress [29].

The decrease in A (Figure 1E) may be related to the lower concentrations of CO_2 found in the substomatal chamber due to the partial closure of the stomata and possible metabolic restrictions to the Calvin cycle and, consequently, a diminution in the synthesis of sugars in the photosynthetic process and in the substrate for RuBisCo [30,31]. The stress-induced decrease in CO_2 assimilation rate in plants can be caused by stomatal and/or non-stomatal factors, leading to changes in the metabolic processes of photosynthesis and affecting the activities of a number of enzymes in the stroma involved in CO_2 reduction [32,33].

Salt stress in plants causes significant losses in the functioning of the photosystem due to the degradation of the proteins involved in the photosynthetic activity. This may explain the reductions in *CEi* (Figure 2A) and *WUEi* (Figure 2B) in soursop plants cv. Morada Nova with the increase in the electrical conductivity of irrigation water [34]. Silva et al. [35] also observed a decrease in the carboxylation efficiency in passion fruit under salt stress (electrical conductivity of water varying from 0.7 to 2.8 dS m^{-1}) and application of hydrogen peroxide (0, 25, 50, and 75 µM) at 60 days after transplantation.

In the present study, it was verified that the foliar application of hydrogen peroxide at a concentration of 30 µM intensified the effects of salt stress on the initial fluorescence (Figure 3A). In this case, the concentration of 30 µM may have induced oxidative damage to the cell membrane and possibly had a negative influence on the initial fluorescence of soursop plants cv. Morada Nova at 210 DAT. At high concentrations, H_2O_2 causes damage to plants, presumably because of alterations in their metabolism, mainly as a consequence of oxidative stress, which limits photosynthetic activities [36].

The restriction of maximum fluorescence (Figure 3B) by salt stress indicates a slowdown in photosynthetic activity aimed at mitigating the toxic effects of salinity [37]. In a study conducted by Silva et al. [38] evaluating the fluorescence of chlorophyll *a* in soursop plants under saline stress (ECw ranging from 0.8 to 4.0 dS m^{-1}), a reduction of 3.31% in maximum fluorescence was also observed by an increase in the electrical conductivity of the irrigation water; the authors attributed this fact to the low efficiency in the photoreduction of quinones and in the flow of electrons between the photosystems, which results in the low activity of photosystem II in the thylakoid membrane, directly influencing the flow of electrons between the photosystems.

The variable fluorescence (Figure 3C) and quantum efficiency of photosystem II (Figure 3D) benefited from the application of hydrogen peroxide at estimated concentrations of 16 and 13 µM, respectively. Thus, it can be concluded that the use of hydrogen peroxide at low concentrations contributed to greater efficiency in the photoreduction of quinone A and the flow of electrons between the photosystems, promoting the adequate activity of PSII in the membrane of thylakoids, directly influencing the flow of electrons between the photosystems, which indicates that PS II was not damaged because, when the photosynthetic apparatus is intact, the values of *Fv/Fm* vary between 0.75 and 0.85 [39]. This result is similar to that reported by Veloso et al. [40], who evaluated the photochemical efficiency and growth of soursop rootstocks subjected to salt stress (ECw ranging from 0.6 to 3.0 dS m^{-1}) and hydrogen peroxide (0 and 20 μM) and found that applications of hydrogen peroxide at the concentration of 20 μM minimized the negative effects of salinity on the initial fluorescence and favored the variable fluorescence and quantum efficiency of photosystem II at 120 days after sowing.

The results obtained in the present study indicate that the growth of soursop plants is negatively affected by the increase in the salinity of the irrigation water. Inhibition of plant growth may be a consequence of the effect caused by excess salts in the root zone, which imposes water limitations, negatively affecting cell elongation and expansion. In addition, the partial closure of stomata compromises photosynthesis, resulting in lower growth [3,41].

On the other hand, the foliar application of hydrogen peroxide at an estimated concentration of 11 μM promoted an increase in growth in stem diameter. Veloso et al. [42], while evaluating the physiological changes and growth of soursop cultivated under saline waters and H$_2$O$_2$ in the post-grafting stage, reported that the exogenous application of H$_2$O$_2$ at 20 μM reduced the harmful effect of water salinity on the stem diameter of the rootstock and scion of soursop plants irrigated with water of 1.6 dS m^{-1} at 150 days after transplantation.

In general, the results of this research reveal that the salt stress caused by the irrigation water up to 3.2 dS m^{-1} negatively affected the gas exchanges, the photochemical efficiency, and the growth of the soursop plants under the conditions of a protected environment. These alterations may be related to osmotic and ionic effects, particularly of Na$^+$ and Cl$^-$, which interfere with the metabolic processes, causing damage to the membrane, nutritional imbalance, changes in the levels of growth regulators, and decreases in the synthesis of chlorophyll [43,44]. However, the foliar application of hydrogen peroxide in low concentrations can reduce the harmful effects of the salinity of irrigation water on the soursop plant. This fact may be related to enzymatic antioxidant defense mechanisms (catalase and peroxidase) in plants, reducing the negative effect of reactive oxygen species [45,46]. Additionally, hydrogen peroxide can improve the absorption of water and nutrients, including elements such as N, P, and K that are essential for plant growth and development [47].

On the other hand, at higher concentrations of hydrogen peroxide, an effect was observed on the analyzed variables. It is important to note that the beneficial effect of hydrogen peroxide depends on several variables, including the concentration of the solution; i.e., at higher concentrations, H$_2$O$_2$ can exert hazardous effects on plants [12,48]. Hydrogen peroxide is the most stable reactive oxygen species in cells and, at high concentrations, can quickly spread across the subcellular membrane, resulting in oxidative damage to the plasma membrane [16]. Additionally, at high concentrations, hydrogen peroxide can react with O$_2$ and possibly become responsible for the dissociation of the pigment-protein complex of the internal antenna of the PS II light-gathering system within the photosynthetic apparatus, leading to enzymatic inactivation, pigment discoloration, and lipid peroxidation [45,49].

4. Materials and Methods

4.1. Location of the Experiment

The experiment was carried out between April and November 2020 under the conditions of the greenhouse that belongs to the Academic Unit of Agricultural Engineering of the Federal University of Campina Grande, situated in the municipality of Campina Grande, Paraíba, Brazil, at the geographic coordinates of 07°15′18″ S, 35°52′28″ W with a mean altitude of 550 m. Figure 5 shows the temperature (maximum and minimum) and average relative air humidity data for the experimental site.

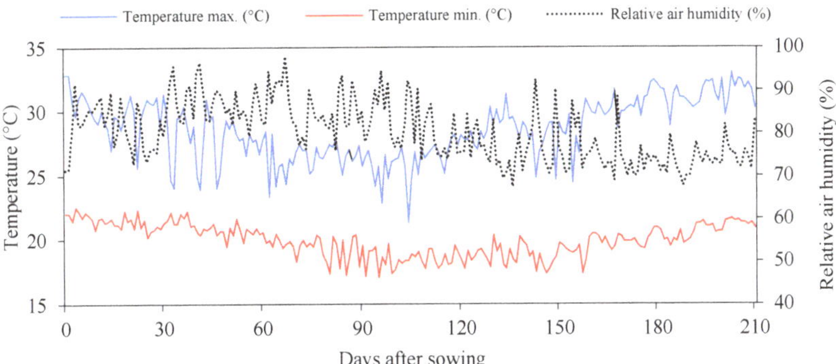

Figure 5. Observed daily temperature (maximum and minimum) and average relative humidity of air in the internal area of the greenhouse during the experimental period.

4.2. Treatments and Experimental Design

The experimental design was randomized in a 4 × 4 factorial arrangement. The treatments consisted of the combination of two factors: four levels of electrical conductivity of irrigation water (ECw; 0.8, 1.6, 2.4, and 3.2 dS m^{-1}) associated with four concentrations of hydrogen peroxide (H$_2$O$_2$; 0, 10, 20, and 30 µM), with three replicates and one plant per plot, totaling forty-eight experimental units.

The salinity levels of the water were established based on a study conducted by [15], who observed that water up to 2.0 dS m^{-1} can be used to produce soursop seedlings with an acceptable average reduction (up to 10%) in growth. Concentrations of hydrogen peroxide were based on the results of an assay conducted by [9], who verified that the use of hydrogen peroxide at the concentration of 20 µM attenuated the harmful effects of irrigation water salinity on the initial growth and gas exchange of soursop cv. Morada Nova.

4.3. Description of the Experiment

Plastic recipients adapted as drainage lysimeters with 200 L capacity were used to grow the plants, and each lysimeter was drilled at the base to allow the drainage of excess water and connected to a transparent drain of 16 mm diameter. The end of the drain inside the lysimeter was wrapped with a nonwoven geotextile (Bidim OP 30) to prevent clogging by soil material. A container was placed below each drain to collect drained water and determine water consumption by the plants (Figure 6).

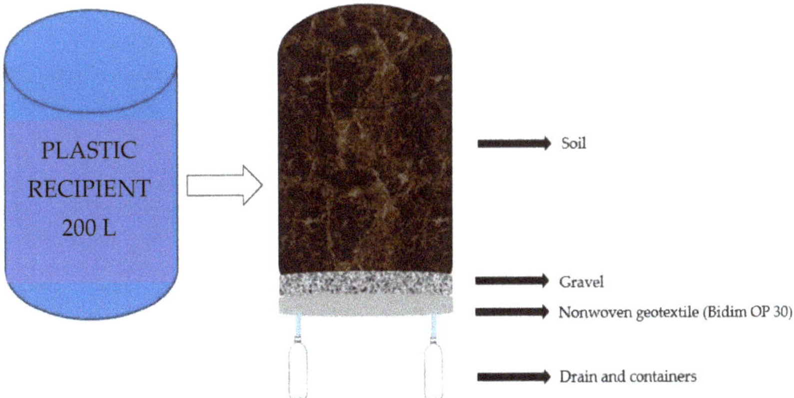

Figure 6. Illustration of filling drainage lysimeters.

The pots were filled with a soil classified as *Entisol* with sandy loam texture from the rural area (0–0.30 m layer) of the municipality of Riachão de Bacamarte, PB, Brazil, whose chemical and physical characteristics (Table 4) were obtained according to the methodologies recommended by [50].

Table 4. Chemical and physical characteristics of the soil (0–0.30 m depth) used in the experiment.

				Chemical characteristics				
pH_{H2O}	OM	P	K^+	Na^+	Ca^{2+}	Mg^{2+}	Al^{3+}	H^+
1:2.5	$g\,dm^{-3}$	$\frac{mg}{dm^{-3}}$			$cmol_c\,kg^{-1}$			
6.5	8.1	79	0.24	0.51	14.9	5.4	0	0.9

	Chemical characteristics				Physical characteristics			
EC_{se}	CEC	SAR_{se}	ESP	Particle-size fraction ($g\,kg^{-1}$)			Moisture ($dag\,kg^{-1}$)	
$dS\,m^{-1}$	$cmol_c\,kg^{-1}$	$(mmol\,L^{-1})^{0.5}$	%	Sand	Silt	Clay	$33.42\,kPa^1$	$1519.5\,kPa^2$
2.15	16.54	0.16	3.08	572.7	100.7	326.6	25.91	12.96

pH: hydrogen potential, OM: Organic matter, Walkley–Black Wet Digestion; Ca^{2+} and Mg^{2+} extracted with 1 M potassium chloride at pH 7.0; Na^+ and K^+ extracted with 1 M ammonium acetate at pH 7.0; Al^{3+} + H^+ extracted with 0.5 M calcium acetate at pH 7.0; ECse: electrical conductivity of saturated paste extract; CEC: cation exchange capacity; SARse: sodium adsorption ratio of saturated paste extract; ESP: exchangeable sodium percentage; superscripts 1 and 2 correspond to field capacity and permanent wilting point, respectively.

The saline waters were obtained by the addition of sodium chloride, calcium chloride, and magnesium chloride salts in the equivalent proportion of 7:2:1, a predominant ratio found in the principal sources of water in northeastern Brazil [51], following the relationship between electrical conductivity and the salt concentration [52], according to Equation (1):

$$Q = 640 \times ECw \tag{1}$$

where:

Q = Quantity of salts to be applied ($mg\,L^{-1}$);
ECw = Electrical conductivity of water ($dS\,m^{-1}$).

After transplanting the seedlings to the lysimeters, irrigation was performed manually and applied daily to each container at 5 p.m., with the volume corresponding to that obtained by the water balance, according to Equation (2):

$$VI = \frac{(Va - Vd)}{(1 - LF)} \tag{2}$$

where:

VI = Volume of water to be applied in the irrigation event (mL);
Va = Volume applied in the previous irrigation event (mL);
Vd = Volume drained (mL);
LF = Leaching fraction of 0.10.

The soursop seedlings cv. Morada Nova were acquired from a commercial nursery accredited in the Registry of Seeds and Seedlings, located in the District of São Gonçalo, Sousa-PB, species registered under No. 23458 in the national registry of cultivars (RNC) of the Ministry of Agriculture, Livestock and Supply of Brazil, and were produced in polyethylene bags with dimensions of 10 × 20 cm. Mineral fertilization was performed according to the recommendations of [53], applying 40 g of N, 60 g of K_2O, and 40 g of P_2O_5 per plant per year. Urea, potassium chloride, and monoammonium phosphate (MAP) were used as sources of nitrogen, potassium, and phosphorus, respectively.

The fertilizer doses were split into 24 portions and applied every 15 days. Micronutrients were applied from 60 days after transplantation (DAT) and continued at fortnightly intervals with Dripsol micro solution (2.5 g L^{-1}) with the following composition: N (15%),

P_2O_5 (15%), K_2O (15%), Ca (1%), Mg (1.4%), S (2.7%), Zn (0.5%), B (0.05%), Fe (0.5%), Mn (0.05%), Cu (0.5%), and Mo (0.02%), by spraying on the adaxial and abaxial sides of the leaves.

The different concentrations of hydrogen peroxide were obtained by dilution in distilled water, followed by calibration in a spectrophotometer at an absorbance wavelength of 240 nm. Foliar applications started at 30 DAT of the seedlings to the lysimeters and were performed at 30-day intervals, spraying the abaxial and adaxial sides of the leaves to obtain complete wetting using a backpack sprayer between the hours of 17:00 and 18:00. On average, 330 mL of H_2O_2 solution was applied per plant in each application. The air drift between treatments was controlled by a plastic tarpaulin curtain which involved the entire plant as the hydrogen peroxide solution was applied (Figure 7).

Figure 7. Application of hydrogen peroxide on the abaxial and adaxial sides of soursop leaves.

Formative pruning was carried out as the plant reached 60 cm height when the apical meristem bud was cut. Of the shoots that emerged, three well-distributed and equidistant branches were selected, and these branches, in turn, were pruned when they reached 40 cm in length [54]. During the experimental period, the emergence of pests and diseases was monitored by observing their incidence, and they were eradicated by chemical control using recommended insecticides/pesticides.

4.4. Variables Analyzed

Gas exchange, chlorophyll *a* fluorescence, and the growth of soursop cv. Morada Nova were evaluated at 210 DAT. Gas exchange was evaluated by CO_2 assimilation rate (*A*) (μmol CO_2 m^{-2} s^{-1}), transpiration (*E*) (mmol H_2O m^{-2} s^{-1}), stomatal conductance (*gs*) (mol H_2O m^{-2} s^{-1}), and internal CO_2 concentration (*Ci*) (μmol CO_2 m^{-2} s^{-1}) in the leaves of the middle third of the plants using a portable IRGA (infrared gas analyser, LCpro-SD model, ADC BioScientific, UK). The ratios *A/gs* and *A/Ci* were utilized to obtain *WUEi*, the water use efficiency ((μmol CO_2 m^{-2} s^{-1}) (mmol H_2O m^{-2} s^{-1})$^{-1}$), and *CEi*, the carboxylation efficiency ((μmol m^{-2} s^{-1}) (μmol mol^{-1})$^{-1}$), respectively. Observations were taken between 07:00 and 10:00 a.m. on the third fully expanded leaf counted from the apical bud under natural conditions of air temperature, CO_2 concentration, and employing an artificial source of radiation established through the photosynthetic response curve to light and determination of the point of photosynthetic saturation by light [55].

Chlorophyll *a* fluorescence measurements were made on the same leaves utilizing a pulse-modulated fluorometer, OS5p model from Opti Science. Initial fluorescence (F_0), maximum fluorescence (*Fm*), variable fluorescence (*Fv*), and the quantum efficiency of photosystem II (*Fv/Fm*) were measured; this protocol was performed after adaptation of the leaves to the dark for 30 minutes between 06:00 and 09:00 a.m., utilizing a clip of the device to ensure that all the primary acceptors were fully oxidized.

The growth of soursop was evaluated by measuring crown height (H_{crown}), stem diameter (SD), and crown diameter (D_{crown}), which was considered the average crown diameter in the row direction (RD) and in the inter-row direction (IRD). Crown volume (V_{crown}) and the vegetative vigor index (VVI) were determined by Equations (3) and (4), respectively, following the methodology of [56]:

$$V_{crown} = \frac{\pi \times H \times RD \times IRD}{6} \tag{3}$$

$$VVI = \frac{[H + D_{crown} + (SD + 10)]}{100} \tag{4}$$

where:

V_{crown}—Crown volume (m^3);
VVI—Vegetative vigor index;
H—Crown height (m);
RD—Crown diameter in the row direction (m);
IRD—Crown diameter in the inter-row direction (m);
SD—Stem diameter (m).

4.5. Statistical Analysis

The collected data were subjected to the distribution normality test (Shapiro–Wilk test) at a 0.05 probability level. Subsequently, analysis of variance was performed at a 0.05 and 0.01 probability level, and in the cases of significance, a regression analysis was performed using the statistical program SISVAR-ESAL [57]. The choice of model was based on the coefficient of determination. In the case of the significance of the interaction between factors, TableCurve 3D software was used to create the response surfaces.

5. Conclusions

Physiological indices of soursop are negatively affected by increases in the electrical conductivity of irrigation water above 0.8 dS m^{-1}. However, the foliar application of hydrogen peroxide between concentrations of 10 and 30 μM mitigates the harmful effects of salinity on the gas exchange, quantum yield, and growth in stem diameter of soursop cv. Morada Nova at 210 days after transplantation. These results reinforce the hypothesis that the foliar application of hydrogen peroxide in adequate concentrations can act as a signaling molecule, influential in mitigating salt stress in soursop plants, which can potentialize the use of brackish water in irrigated agriculture, mainly in regions with a scarcity of fresh water. However, further studies are necessary to understand how hydrogen peroxide acts in salt stress signaling through biochemical analysis. In addition, it is fundamental to perform research under field conditions to prove the beneficial effects of hydrogen peroxide in the attenuation of salt stress in soursop plants.

Author Contributions: Conceptualization, J.D.C. and G.S.d.L.; methodology, A.A.R.d.S., J.D.C., F.d.A.d.S. and M.d.S.D.; software, A.A.R.d.S. and T.F.d.L.A.; validation, J.D.C., G.S.d.L. and L.A.d.A.S.; formal analysis, M.S.S.d.F. and L.A.d.A.S.; investigation, G.S.d.L. and J.D.C.; resources, G.S.d.L. and L.A.d.A.S.; data collection, A.A.R.d.S.; writing—original draft preparation, J.D.C. and G.S.d.L.; writing—review and editing, A.A.R.d.S., G.S.d.L., C.A.V.d.A., P.D.F. and H.R.G.; supervision, H.R.G.; project administration, G.S.d.L., C.A.V.d.A. and L.A.d.A.S. All authors have read and agreed to the published version of the manuscript.

Funding: CNPq (National Council for Scientific and Technological Development) (Proc. 306245/2017-5 and 430525/2018-4), CAPES (Coordination for the Improvement of Higher Education Personnel), and UFCG (Universidade Federal de Campina Grande).

Data Availability Statement: Not applicable.

Acknowledgments: The authors thank the Graduate Program in Agricultural Engineering of the Federal University of Campina Grande, the National Council for Scientific and Technological Development (CNPq), and the Coordination for the Improvement of Higher Education Personnel (CAPES) for the financial support in carrying out this research.

Conflicts of Interest: The authors declare no conflict of interest.

References

1. Cavalcante, L.F.; Carvalho, S.D.; Lima, E.D.; Feitosa Filho, J.C.; Silva, D.A. Produção e qualidade da graviola sob irrigação e cobertura do solo com resíduo de sisal. *Magistra* **2016**, *28*, 91–101.
2. Bento, E.B.; Monteiro, Á.B.; Lemos, I.C.S.; Brito Junior, F.E.; Oliveira, D.R.; Menezes, I.R.A.; Kerntopf, M.R. Estudio etnofarmacológico comparativo en la región del Araripe de la *Annona muricata* L. (Graviola). *Rev. Cubana Plant. Med.* **2016**, *21*, 9–19.
3. Da Silva, A.A.R.; de Lima, G.S.; de Azevedo, C.A.V.; Veloso, L.L.D.S.A.; Souza, L.D.P.; de Fátima, R.T.; Silva, F.D.A.D.; Gheyi, H.R. Exogenous application of salicylic acid on the mitigation of salt stress in *Capsicum annuum* L. *Ciênc. Rural.* **2023**, *53*, e20210447. [CrossRef]
4. Silva, A.A.R.; Veloso, L.L.S.A.; Lima, G.S.; Azevedo, C.A.V.; Gheyi, H.R.; Fernandes, P.D. Hydrogen peroxide in the acclimation of yellow passion fruit seedlings to salt stress. *Rev. Bras. Eng. Agríc. Ambient.* **2021**, *25*, 116–123. [CrossRef]
5. De Lima, G.S.; Pinheiro, F.W.A.; Gheyi, H.R.; Soares, L.A.D.A.; Sousa, P.F.D.N.; Fernandes, P.D. Saline water irrigation strategies and potassium fertilization on physiology and fruit production of yellow passion fruit. *Rev. Bras. Eng. Agric. Ambient.* **2022**, *26*, 180–189. [CrossRef]
6. De Lima, G.S.; dos Santos, J.B.; Soares, L.A.D.A.; Gheyi, H.R.; Nobre, R.G.; Pereira, R.F. Irrigação com águas salinas e aplicação de prolina foliar em cultivo de pimentão 'All Big'. *Comun. Sci.* **2016**, *7*, 513–522. [CrossRef]
7. Pinheiro, F.W.A.; de Lima, G.S.; Gheyi, H.R.; Soares, L.A.D.A.; de Oliveira, S.G.; da Silva, F.A. Gas exchange and yellow passion fruit production under irrigation strategies using brackish water and potassium. *Rev. Cien. Agron.* **2022**, *53*, e20217816. [CrossRef]
8. Da Silva, E.M.; De Lima, G.S.; Gheyi, H.R.; Nobre, R.G.; Sá, F.V.D.S.; Souza, L.D.P. Growth and gas exchanges in soursop under irrigation with saline water and nitrogen sources. *Rev. Bras. Eng. Agríc. Ambient.* **2018**, *22*, 776–781. [CrossRef]
9. Veloso, L.L.D.S.A.; da Silva, A.A.R.; de Lima, G.S.; de Azevedo, C.A.V.; Gheyi, H.R.; Moreira, R.C.L. Growth and gas exchange of soursop under salt stress and hydrogen peroxide application. *Rev. Bras. Eng. Agríc. Ambient.* **2022**, *26*, 119–125. [CrossRef]
10. Veloso, L.L.D.S.A.; Nobre, R.G.; Souza, L.D.P.; Gheyi, H.R.; Cavalcante, I.T.S.; Araujo, E.B.G.; Da Silva, W.L. Formation of soursop seedlings irrigated using waters with different salinity levels and nitrogen fertilization. *Biosci. J.* **2018**, *34*, 151–160. [CrossRef]
11. Dos Santos, A.N.; Silva, E.F.S.; Silva, G.S.; Bezerra, R.R.; Pedrosa, M.R.P. Antioxidant response of cowpea co-inoculated with plant growth-promoting bacteria under salt stress. *Braz. J. Micro.* **2018**, *49*, 513–521. [CrossRef]
12. Veloso, L.L.D.S.A.; de Azevedo, C.A.V.; Nobre, R.G.; de Lima, G.S.; Capitulino, J.D.; Silva, F.D.A.D. H_2O_2 alleviates salt stress effects on photochemical efficiency and photosynthetic pigments of cotton genotypes. *Rev. Bras. Eng. Agríc. Ambient.* **2023**, *27*, 34–41. [CrossRef]
13. Andrade, E.M.G.; De Lima, G.S.; De Lima, V.L.A.; Da Silva, S.S.; Gheyi, H.R.; Da Silva, A.A.R. Gas exchanges and growth of passion fruit under saline water irrigation and H_2O_2 application. *Rev. Bras. Eng. Agríc. Ambient.* **2019**, *23*, 945–951. [CrossRef]
14. Bagheri, M.; Gholami, M.; Baninasab, B. Hydrogen peroxide-induced salt tolerance in relation to antioxidant systems in pistachio seedlings. *Sci. Hortic.* **2019**, *243*, 207–213. [CrossRef]
15. Da Silva, A.A.R.; De Lima, G.S.; De Azevedo, C.A.V.; Veloso, L.L.D.S.A.; Capitulino, J.D.; Gheyi, H.R. Induction of tolerance to salt stress in soursop seedlings using hydrogen peroxide. *Comun. Sci.* **2019**, *10*, 484–490. [CrossRef]
16. Liu, L.; Huang, L.; Lin, X.; Sun, C.O. Hydrogen peroxide alleviates salinity-induced damage by increasing proline buildup in wheat seedlings. *Plant Cell Rep.* **2020**, *39*, 567–575. [CrossRef]
17. Da Silva, A.A.R.; Sousa, P.F.D.N.; de Lima, G.S.; Soares, L.A.D.A.; Gheyi, H.R.; de Azevedo, C.A.V. Hydrogen peroxide reduces the effect of salt stress on growth and postharvest quality of hydroponic mini watermelon. *Water Air Soil Pollut.* **2022**, *233*, 198. [CrossRef]
18. Tanou, G.; Job, C.; Rajjou, L.; Arc, E.; Belghazi, M.; Diamantidis, G.; Molassiotis, A.; Job, D. Proteomics reveals the overlapping roles of hydrogen peroxide and nitric oxide in the acclimation of citrus plants to salinity. *Plant J.* **2009**, *60*, 795–804. [CrossRef]
19. Hajivar, B.; Zare-Bavani, M.R. Alleviation of salinity stress by hydrogen peroxide and nitric oxide in tomato plants. *Adv. Hortic. Sci.* **2019**, *33*, 409–416. [CrossRef]
20. Gondim, F.A.; Miranda, R.S.; Gomes-Filho, E.; Prisco, J.T. Enhanced salt tolerance in maize plants induced by H_2O_2 leaf spraying is associated with improved gas exchange rather than with non-enzymatic antioxidant system. *Theor. Exp. Plant Physiol.* **2013**, *25*, 251–260. [CrossRef]

21. Wang, X.; Hou, C.; Liu, J.; He, W.; Nan, W.; Gong, H.; Bi, Y. Hydrogen peroxide is involved in the regulation of rice (*Oryza sativa* L.) tolerance to salt stress. *Acta Physiol. Plant.* **2013**, *35*, 891–900. [CrossRef]
22. Mendonça, A.J.T.; da Silva, A.A.R.; de Lima, G.S.; Soares, L.A.D.A.; Oliveira, V.K.N.; Gheyi, H.R.; de Lacerda, C.F.; de Azevedo, C.A.V.; de Lima, V.L.A.; Fernandes, P.D. Salicylic acid modulates okra tolerance to salt stress in hydroponic system. *Agriculture* **2022**, *12*, 1687. [CrossRef]
23. De Lima, G.S.; De Andrade, J.N.F.; De Medeiros, M.N.V.; Soares, L.A.D.A.; Gheyi, H.R.; Nobre, R.G.; Fernandes, P.D.; De Lacerda, C.N. Gas exchange, growth, and quality of passion fruit seedlings cultivated with saline water. *Semin. Ciências Agrárias* **2021**, *42*, 137–154. [CrossRef]
24. Dias, A.S.; De Lima, G.S.; Sá, F.V.D.S.; Gheyi, H.R.; Soares, L.A.D.A.; Fernandes, P.D. Gas exchanges and photochemical efficiency of West Indian cherry cultivated with saline water and potassium fertilization. *Rev. Bras. Eng. Agric. Ambient.* **2018**, *22*, 628–633. [CrossRef]
25. Abrar, M.M.; Sohail, M.; Saqib, M.; Akhtar, J.; Abbas, G.; Wahab, H.A.; Xu, M. Interactive salinity and water stress severely reduced the growth, stress tolerance, and physiological responses of guava (*Psidium guajava* L.). *Sci. Rep.* **2022**, *12*, 18952. [CrossRef]
26. Sá, F.V.D.S.; Gheyi, H.R.; de Lima, G.S.; Pinheiro, F.W.A.; de Paiva, E.P.; Moreira, R.C.L.; Silva, L.D.A.; Fernandes, P.D. The right combination of NPK fertilization may mitigate salt stress in custard apple (*Annona squamosa* L.). *Acta Physiol. Plant.* **2021**, *43*, 59. [CrossRef]
27. Carvalho, F.E.L.; Lobo, A.K.M.; Bonifacio, A.; Martins, M.O.; Lima Neto, M.C.; Silveira, J.A.G. Aclimatação ao estresse salino em plantas de arroz induzida pelo pré-tratamento com H_2O_2. *Rev. Bras. Eng. Agric. Ambient.* **2011**, *15*, 416–423. [CrossRef]
28. Baxter, A.; Mittler, R.; Suzuki, N. EROS as key players in plant stress signalling. *JXB* **2014**, *65*, 1229–1240. [CrossRef]
29. Simões, W.L.; Coelho, D.S.; Mesquita, A.C.; Calgaro, M.; da Silva, J.S. Physiological and biochemical responses of sugarcane varieties to salt stress. *Rev. Caatinga* **2020**, *32*, 1069–1076. [CrossRef]
30. Dias, A.S.; De Lima, G.S.; Pinheiro, F.W.A.; Gheyi, H.R.; Soares, L.A.D.A. Gas exchanges, quantum yield and photosynthetic pigments of West Indian cherry under salt stress and potassium fertilization. *Rev. Caatinga* **2019**, *32*, 429–439. [CrossRef]
31. Silveira, A.G.; Silva, S.L.F.; Silva, E.V.; Viégas, R.V. Mecanismos biomoleculares envolvidos com a resistência ao estresse salino em plantas. In *Manejo da Salinidade na Agricultura: Estudos Básicos e Aplicados*, 2nd ed.; Gheyi, H.R., Dias, N.d.S., de Lacerda, C.F., Gomes Filho, É., Eds.; INCTSal: Fortaleza, Brazil, 2016; Chapter 13; pp. 181–197.
32. Hnilicková, H.; Hnilička, F.; Martinkova, J.; Kraus, K. Effects of salt stress on water status, photosynthesis and chlorophyll fluorescence of rocket. *Plant Soil Environ.* **2017**, *63*, 362–367. [CrossRef]
33. Turan, M.; Ekinci, M.; Kul, R.; Boynueyri, F.G.; Yildirim, E. Mitigation of salinity stress in cucumber seedlings by exogenous hydrogen sulfide. *J. Plant Res.* **2022**, *135*, 517–529. [CrossRef] [PubMed]
34. Manivannan, A.; Soundararajan, P.; Muneer, S.; Ko, C.H.; Jeong, B.R. Silicon mitigates salinity stress by regulating the physiology, antioxidant enzyme activities, and protein expression in *Capsicum annuum* 'Bugwang'. *BioMed Res. Int.* **2016**, *2016*, 3076357. [CrossRef]
35. Da Silva, A.A.R.; De Lima, G.S.; De Azevedo, C.A.V.; Gheyi, H.R.; Souza, L.D.P.; Veloso, L.L.D.S.A. Gas exchanges and growth of passion fruit seedlings under salt stress and hydrogen peroxide. *Pesq. Agropecu. Trop.* **2019**, *49*, e55671. [CrossRef]
36. Cattivelli, L.; Rizza, F.; Badeck, F.W.; Mazzucotelli, E.; Mastrangelo, A.M.; Francia, E.; Maré, C.; Tondelli, A.; Stanca, A.M. Drought tolerance improvement in crop plants: An integrated view from breeding to genomics. *Field Crops Res.* **2008**, *105*, 1–14. [CrossRef]
37. Monteiro, D.R.; De Melo, H.F.; Lins, C.M.T.; Dourado, P.R.M.; Santos, H.R.B.; De Souza, E.R. Chlorophyll a fluorescence in saccharine sorghum irrigated with saline water. *Rev. Bras. Eng. Agríc. Ambient.* **2018**, *22*, 673–678. [CrossRef]
38. Da Silva, A.A.R.; de Lima, G.S.; de Azevedo, C.A.V.; Gheyi, H.R.; Soares, L.A.D.A.; Veloso, L.L.D.S.A. Salicylic acid improves physiological indicators of soursop irrigated with saline water. *Rev. Bras. Eng. Agríc. Ambient.* **2022**, *26*, 412–419. [CrossRef]
39. Reis, F.O.; Campostrini, E. Microaspersão de água sobre a copa: Um estudo relacionado às trocas gasosas e à eficiência fotoquímica em plantas de mamoeiro. *Rev. Bras. Agroc.* **2011**, *17*, 284–295. [CrossRef]
40. Veloso, L.L.D.S.A.; da Silva, A.A.R.; Capitulino, J.D.; de Lima, G.S.; de Azevedo, C.A.V.; Gheyi, H.R.; Nobre, R.G.; Fernandes, P.D. Photochemical efficiency and growth of soursop rootstocks subjected to salt stress and hydrogen peroxide. *AIMS Agric. Food* **2020**, *5*, 1–13. [CrossRef]
41. Yadav, S.P.; Bharadwaj, R.B.; Nayak, H.; Mahto, R.; Singh, R.K.; Prasad, S.K. Impact of salt stress on growth, productivity and physicochemical properties of plants: A Review. *Int. J. Chem. Stud.* **2019**, *7*, 1793–1798.
42. Veloso, L.L.D.S.A.; De Lima, G.S.; De Azevedo, C.A.V.; Nobre, R.G.; Da Silva, A.A.R.; Capitulino, J.D.; Gheyi, H.R.; Bonifácio, B.F. Physiological changes and growth of soursop plants under irrigation with saline water and H_2O_2 in post-grafting phase. *Semin. Ciênc. Agrár.* **2020**, *41*, 3023–3038. [CrossRef]
43. Shobha, S.; Ashwani, K.; Sehrawat, N.; Kumar, K.; Kumar, N.; Lata, C.; Mann, A. Effect of saline irrigation on plant water traits, photosynthesis and ionic balance in durum wheat genotypes. *Saudi J. Biol. Sci.* **2021**, *28*, 2510–2517. [CrossRef]
44. Najar, R.; Aydi, S.; Sassi-Aydi, S.; Zarai, A.; Abdelly, C. Effect of salt stress on photosynthesis and chlorophyll fluorescence in Medicago truncatula. *Plant Biosyst.* **2019**, *153*, 88–97. [CrossRef]
45. Kilic, S.; Kahraman, A. The mitigation effects of exogenous hydrogen peroxide when alleviating seed germination and seedling growth inhibition on salinity-induced stress in barley. *Pol. J. Environ. Stud.* **2016**, *25*, 1053–1059. [CrossRef] [PubMed]

46. Da Silva, A.A.R.; De Lima, G.S.; De Azevedo, C.A.V.; Veloso, L.L.D.S.A.; Gheyi, H.R.; Soares, L.A.D.A. Salt stress and exogenous application of hydrogen peroxide on photosynthetic parameters of soursop. *Rev. Bras. Eng. Agríc. Ambient.* **2019**, *23*, 257–263. [CrossRef]
47. Nazir, F.; Fariduddin, Q.T.; Khan, A. Hydrogen peroxide as a signaling molecule in plants and its crosstalk with other plant growth regulators under heavy metal stress. *Chemosphere* **2020**, *252*, 126486. [CrossRef]
48. Farouk, S.; Amira, M.S.A.Q. Enhancing seed quality and productivity as well as physio-anatomical responses of pea plants by folic acid and/or hydrogen peroxide application. *Sci. Hortic.* **2018**, *240*, 29–37. [CrossRef]
49. Zhang, L.; Ma, H.; Chen, T.; Pen, J.; Yu, S.; Zhao, X. Morphological and physiological responses of cotton plants (*Gossypium hirsutum* L.) to salinity. *PLoS ONE* **2014**, *9*, e112807. [CrossRef]
50. Teixeira, P.C.; Donagemma, G.K.; Fontana, A.; Teixeira, W.G. *Manual de Métodos de Análise de Solo*, 3rd ed.; Embrapa Solos: Brasília, Brazil, 2017; p. 573.
51. Medeiros, J.F. Qualidade de Água de Irrigação e Evolução da Salinidade nas Propriedades Assistidas Pelo GAT nos Estados de RN, PB e CE. Master's Dissertation, Universidade Federal da Paraíba, Campina Grande, Brazil, 1992; 196p.
52. Richards, L.A. *Diagnosis and Improvement of Saline and Alkali Soils*; U.S. Department of Agriculture: Washington, DC, USA, 1954; 160p.
53. Cavalcante, F.J. *Recomendações de Adubação Para o Estado de Pernambuco*, 2nd ed.; IPA: Recife, Brazil, 2008; Volume 3, p. 212.
54. EMBRAPA. *A Cultura da Gravioleira*, 1st ed.; Embrapa Amazônia Ocidental: Manaus, Brazil, 1999; 19p.
55. Fernandes, E.A.; Soares, L.A.D.A.; de Lima, G.S.; Neta, A.M.D.S.S.; Roque, I.A.; da Silva, F.A.; Fernandes, P.D.; de Lacerda, C.N. Cell damage, gas exchange, and growth of *Annona squamosa* L. under saline water irrigation and potassium fertilization. *Semin. Ciências Agrárias* **2021**, *42*, 999–1018. [CrossRef]
56. Portella, C.R.; Marinho, C.S.; Amaral, B.D.; Carvalho, W.S.G.; Campos, G.S.; Silva, M.P.S.; Sousa, M.C. Desempenho de cultivares de citros enxertados sobre o trifoliateiro 'Flying Dragon' e limoeiro 'Cravo' em fase de formação do pomar. *Bragantia* **2016**, *1*, 70–75. [CrossRef]
57. Ferreira, D.F. SISVAR: A computer analysis system to fixed effects split plot type designs. *Rev. Bras. Biom.* **2019**, *37*, 529–535. [CrossRef]

Article

Horticultural Practices in Early Spring to Mitigate the Adverse Effect of Low Temperature on Fruit Set in 'Lapins' Sweet Cherry

Hao Xu *, Danielle Ediger and Mehdi Sharifi

Summerland Research and Development Centre, Agriculture and Agri-Food Canada, Summerland, BC V0H 1Z0, Canada
* Correspondence: hao.xu@agr.gc.ca

Abstract: Yield of sweet cherry (*Prunus avium* L.) is determined by fruit set, a developmental stage sensitive to variable spring environmental conditions. To sustain fruit production and enhance crop climate resilience, it is important to understand the impacts of abiotic stresses and the effectiveness of horticultural mitigations in the spring on the critical developmental processes during fruit set. In this study, flowering phenology, pistil browning and percent fruit set of 'Lapins' were monitored at five sites of different elevation and frost risk in the Okanagan Valley, British Columbia, Canada, in 2019 and 2022. At Site 1 in Summerland Research and Development Centre ("SuRDC1"), where a 'Lapins' on Krymsk 5 planting was located in a frost pocket where the crops were exposed to high risk of cold damage in the spring, a series of experiments were conducted to investigate the floral organ viability and percent fruit set under low temperatures, and under the effects of four spring horticultural mitigation measures. Installation of polyethylene sleeves and FAME spray (fatty acid methyl esters-based plant growth regulator, WAIKEN, SST Australia) were implemented in 2019; boric acid spray and postponed irrigation were tested in 2022. Low fruit set at SuRDC1 in both years was associated with severe pistil browning after night temperature dropped below $-4\,°C$ in late April. In 2019, the semi-enclosure of polyethylene sleeves led to an increase in the surface temperature (T_{surfae}) of floral buds by $2–4\,°C$, which prolonged the stage of first bloom, delayed petal fall and prevented frost damage on pistils, but led to the decrease in percent fruit set by 77%, due to ovule abortion or cessation of fruitlet development. The early and late sprays of FAME had no significant influence on either abundance of germinated pollen tubes or percent fruit set; however, the potential of late spray in improving pollen abundance and reducing pistil browning requires further investigation. In 2022, the spray of 0.01% boric acid solution led to a decrease in fruit set by 6.95%. Six-week postponement of irrigation starting from full bloom decreased soil moisture, but increased soil temperature and improved fruit set by 7.61%. The results improved our understanding about the damages of adverse spring air temperatures on pistils and ovules, and suggested the potential of irrigation adjustment in regulating soil moisture and temperature and improving fruit set in the cool and moist spring.

Keywords: anthesis; cold injury; fruit set; ovule abortion; pistil browning; pollen germination; soil temperature; warming

Citation: Xu, H.; Ediger, D.; Sharifi, M. Horticultural Practices in Early Spring to Mitigate the Adverse Effect of Low Temperature on Fruit Set in 'Lapins' Sweet Cherry. *Plants* **2023**, *12*, 468. https://doi.org/10.3390/plants12030468

Academic Editors: Alberto Soares De Melo and Hans Raj Gheyi

Received: 15 December 2022
Revised: 8 January 2023
Accepted: 12 January 2023
Published: 19 January 2023

1. Introduction

As a high-value fruit product, the sustainable production of sweet cherry (*Prunus avium* L.) is significant to the global market. Successful fruit set in the early growing season is a determinant for the yield of sweet cherry at harvest. Fruit set relies on the completion of a sequence of developmental processes in the spring, including bud break, anthesis (onset of flowering, pollination, fertilization) and initial fruitlet development [1–6]. Anthesis is a developmental stage that is particularly vulnerable to temperature stresses [7]. Low air temperature may damage floral organs during their differentiation and weaken pollen viability [3,4,8,9], thereby shortening the effective pollination period and reducing the probability of successful pollination and fertilization [6]. Adverse temperatures may also cause a decline in the population of pollinators [10], and disrupt their behaviors [11] and their interactions with the plants [12,13]. Low soil temperatures and waterlogging can inhibit root activity and affect the uptake and transport of water and mineral nutrients that are essential to floral organ viability and fruit set [14–16]. These edaphic issues are often a consequence of heavy rainfall events in the spring, which also cause physical injury of floral organs, and lead to the loss of pollen viability and dissemination capacity due to high humidity [17]. The disrupted developmental processes consequently trigger floral drop and poor fruit set [18].

Despite the difficulty of eliminating temperature stresses, some horticultural practices can be implemented to mitigate the fruit set issue. Pollinizers [19] and pollinators [20] can substantially increase pollen availability and improve the fruit set for the cultivars with pollen-related issues. In addition, a few plant growth regulators have been reported to be effective in tree fruits. For example, aminoethoxyvinylglycine (AVG, ReTain), an ethylene inhibitor, can reduce ethylene content in floral organs and extend ovule longevity [21]. Fatty acid methyl esters (FAME) can regulate the timing of bud break; therefore, they are a potential measure to schedule anthesis to avoid adverse temperatures [22]. Pollen viability can be improved by nutrient supplements that contain boron and calcium [23,24]. Dispersions of sprayable insulating coatings of cellulose nanocrystals and nanofibrils can protect floral organs from adverse temperatures during green tip–petal fall [25]. Furthermore, orchard moisture and temperature, and vegetative and reproductive growth of *Prunus* crops, can be influenced by irrigation [26–31] and protective structure [32,33]; the effects of these horticultural measures are profound and usually dependent on cultivars and orchard environmental conditions.

Under the changing climate, early spring weather conditions are becoming more unpredictable. Cherry flowering phenological rhythm and floral organ viability are affected by more frequent events of erratic and extreme temperatures. In the northern climate and at higher elevation where low temperature during anthesis is a predominant and frequent threat to sweet cherry fruit set, research is required to elucidate how floral organs are affected and which horticultural mitigations can effectively improve fruit set. In this study, floral buds and pistils were examined in a 'Lapins' sweet cherry/Krymsk 5 rootstock trial located in a frost pocket in Summerland in the Okanagan Valley, British Columbia, Canada (Site 1 at Summerland Research and Development Centre, "SuRDC1"), in comparison to four locations in the Valley with lower frost risk ("SuRDC2", "SuRDC3", "Oliver" and "Kelowna"; detailed description in Table 1 and Section 4.1), in the spring of 2019 and 2022. A series of experiments were conducted to investigate floral organ viability and percent fruit set under the effects of four spring horticultural mitigation measures, i.e., installation of polyethylene sleeves, FAME spray (WAIKEN, SST Australia), boric acid spray and postponed irrigation. This study aims to highlight the importance of site selection for spring frost avoidance, and to point out the necessity to evaluate the limitations of horticultural measures under specific growing conditions.

Table 1. Site description, phenology and percent fruit set of 'Lapins' sweet cherry at five monitored sites in the Okanagan Valley, British Columbia, Canada, in the spring of 2019 and 2022.

Sites	Elevation, Frost Risk	Rootstock, Canopy Structure	Replications for % Fruit Set	2019				2022				Year Effect
				First Bloom– Petal Fall	T_{min} (°C)	PP (mm)	% Fruit Set	First Bloom– Petal Fall	T_{min} (°C)	PP (mm)	% Fruit Set	
SuRDC1	487 m, High risk	Krymsk 5, TSA	6 plots, 3 trees/plot, 1–2 branches/tree	27 April –7 May	−4.10	3	40.9 ± 2.5 c	21 April –12 May	−4.12	12	23.0 ± 2.5 c *	F(1,53) = 20.06 $p < 0.001$
SuRDC2	416 m, Low risk	Krymsk 5, SSA	4 plots, 4 trees/plot, 1 trunk section/tree	24 April –3 May	−0.16	3	46.0 ± 1.9 bc	17 April –9 May	−1.72	12	35.8 ± 1.8 b *	F(1,31) = 8.90 $p = 0.004$
SuRDC3	424 m, Low risk	Mazzard, Open heart	9 trees in 4 rows, 3–4 branches	23 April –3 May	−0.98	3	59.2 ± 2.2 a	15 April –8 May	−2.10	12	66.4 ± 2.9 a	F(1,59) = 3.74 $p = 0.06$
Oliver	332 m, Low risk	Mazzard, Central leader	2 plots, 5 trees/plot, 3 branch/tree	17 April –30 April	−1.07	3	55.2 ± 0.3 a	11 April –4 May	−2.27	3	36.6 ± 1.8 b *	F(1,59) = 60.51 $p < 0.001$

Table 1. *Cont.*

Sites	Elevation, Frost Risk	Rootstock, Canopy Structure	Replications for % Fruit Set	2019				2022				Year Effect
				First Bloom–Petal Fall	T_{min} (°C)	PP (mm)	% Fruit Set	First Bloom–Petal Fall	T_{min} (°C)	PP (mm)	% Fruit Set	
Kelowna	483 m, Low risk	Mazzard, Central leader	2 plots, 5 trees/plot, 3 branch/tree	19 April –2 May	−1.27	1	53.1 ± 0.2 ab	12 April –6 May	−2.46	2	57.3 ± 3.5 a	$F_{(1,59)} = 0.84$ $p = 0.36$

Note: TSA and SSA stands for Tall Spindle Axes and Super Slender Axes canopy structures, respectively. Plots were randomized in each trial. T_{min} and PP stand for minimum air temperature and moisture deficit during full bloom–petal fall, respectively. Different letters in the same column of year I % fruit set represent significant difference among sites in each year; asterisks in the same row of site I %fruit set stand for significant difference between years at each site ($p < 0.05$, one-way ANOVA, Tukey's test for pairwise comparison, n = number of branches per site).

2. Results

2.1. Cold Injury on Floral Organs at SuRDC1

Phenological order of the five sites was consistent in 2019 and 2022: the earliest site was Oliver, followed by Kelowna, SuRDC3 and SuRDC2; located in a frost pocket at the foot of a hill, cooler weather delayed the anthesis at SuRDC1 (Table 1). Compared to 2019, first bloom was a week early, but anthesis was prolonged and petal fall was about 5 days late in 2022, due to a warmer March and a cooler April in the latter year (Table 2).

Table 2. The accumulated mean daily temperatures above 0 °C (T_{sum}) in March and April in 2019 and 2022 in Summerland, Oliver and Kelowna (British Columbia, Canada).

Sites	Weather Station Location	T_{sum} in March (°C)			T_{sum} in April (°C)		
		Historical Average	2019	2022	Historical Average	2019	2022
Summerland EC	Latitude: 49.5700°; Longitude: −119.6769°; Elevation: 454 m	143.62	108.2	192.4	269.9	275.5	215.8
Oliver Central	Latitude: 49.1575°; Longitude: −119.5514°; Elevation: 349 m	163.9	150.02	214.87	280.88	317.84	217.37
Kelowna East	Latitude: 49.8738°; Longitude: −119.4436°; Elevation: 432 m	148.83	144.93	182.84	268.54	278.88	213.7

Note: Data were acquired from the weather stations in adjacency to the cherry sites (Farmwest.com; accessed on 12 December 2022). The average of T_{sum} was the historical average of 61 years at Summerland EC station, and of 52 years at Oliver Central station and Kelowna East station.

SuRDC1 was the only site where night temperature dropped below −4 °C and frost risk was high after first bloom in both years. In early March of 2019, daily minimum temperature dropped to −15 °C, leading to the necrosis of floral bud primordia in about 10% of the examined 50 floral buds. Night temperature lower than −4 °C in late April of both years resulted in severe frost injury. Ratio of pistil browning in the examined flowers was 14.3% ± 4.5% on 1 May 2019 (n = 18, 30 flowers per tree), and 37.5% ± 1.8% on 6 May 2022 (n = 54, 30 flowers per tree). The percent fruit set in June was 40.9% in 2019 and 23% in 2022, significantly lower than the other four sites where no frost injury was observed on pistils in either year (Table 1; Site effect in 2019: $F_{(4,178)} = 10.22$, $p < 0.001$; Site effect in 2022: $F_{(4,169)} = 35.18$, $p < 0.001$). Percent fruit set in 2022 was significantly lower than in 2019 at SuRDC1, SuRDC2 and Oliver (Table 1; $p < 0.05$).

2.2. Effects of Elevated Temperature inside Polyethylene Sleeves

In the spring of 2019, polyethylene sleeves were wired around the bearing branches to increase ambient temperature around the floral buds (Figure 1A). The surface temperature of the floral buds surrounded by sleeves (Figure 1A,B) was 3.23 ± 0.16 °C higher than that of the untreated adjacent floral buds [Differential $T_{surface}$ mean ± SE, n = 18, $F_{(1,35)} = 146.5$, $p < 0.001$; data not graphed]. Under this setting, the stage of first bloom was prolonged and the stage of petal fall was delayed. It prevented frost damage on pistil, shown as no pistil browning in the flowers surrounded by the sleeves. However, the percent fruit set was 14.1% $\pm$ 5.4%, decreasing to about 1/3 of that on untreated branches [n = 18, F(1,35) = 20.28, p < 0.001; data not graphed], due to ovule abortion or cessation of fruitlet development (Figure 1C).

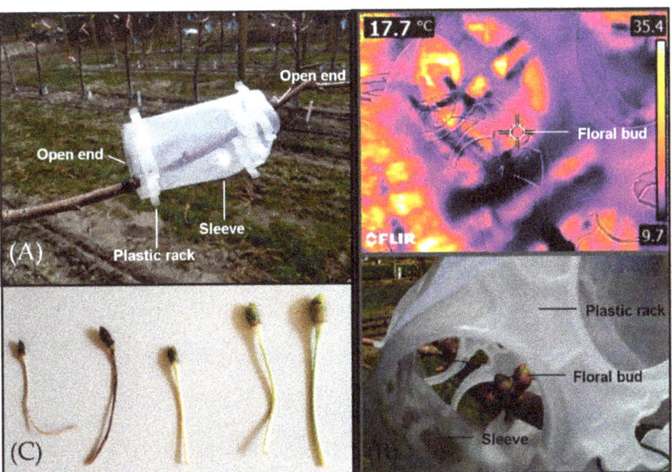

Figure 1. The installation of polyethylene sleeves to increase floral bud surface temperature ($T_{surface}$) in a 'Lapins'/Krymsk 5 trial at SuRDC1. (**A**) The set-up of polyethylene sleeves around floral buds. (**B**) Floral bud $T_{surface}$ measurement using an infrared imager (photo was taken through the open end of the sleeve; top: infrared channel, bottom: RGB channel). (**C**) Ovule abortion and cessation of fruitlet development under the setting.

2.3. Effects of FAME Spray

The early spray ("Early") and the late spray ("Late") of FAME were applied about 40 days and 20 days before the anticipated normal bud break, in order to advance or set back the bud break, respectively. Early spray did not change the timing of flowering compared to Control (no spray). Late spray postponed first bloom and sepal fall for two days in three out of six experimental plots. After sample preparation and Aniline Blue staining, pollens and pollen tubes were visualized in vivo successfully (Figure 2 photos). Under Control, abundance of pollen grains on the surface of stigma was significantly higher on 6 May than on 25 April (p = 0.01). Compared to Control, the median of pollen abundance was higher under early spray on 25 April, and was higher under late spray on 6 May; however, the mean was not significantly different (Figure 2A, n = 6; p = 0.35 on 25 April, p = 0.61 on 6 May). Late spray led to higher pollen abundance than early spray (p = 0.03). The treatments did not cause difference in the number of germinated pollen tubes within the style sections on the stigma side [Figure 2B, n = 6; F(4,22) = 0.31, p = 0.87, n = 6], or in the presence of pollen tubes on the ovary side [F(4,26) = 1.64, p = 0.19; data not graphed]. Pollen tubes were present in the ovary side in 1/3 of the styles sampled on 25 April and in 5/6 of those sampled on 6 May, showing a significant timing effect [F(1,23) = 7.62, p = 0.01; data not graphed].

The mean of pistil browning ratio was 9.5%, 14.3% and 20.6% under late spray, control and early spray treatments (Figure 3A). The mean of percent fruit set was 36.8%, 38.2% and 40.9% under early spray, late spray and control treatments (Figure 3B). Although the differences were not significant at $p \leq 0.05$ in either pistil browning ratio [F(2,52) = 2.10, p = 0.13] or percent fruit set [F(2,52) = 0.80, p = 0.46] (n = 18), the sprays shifted data distribution as shown in Figure 3.

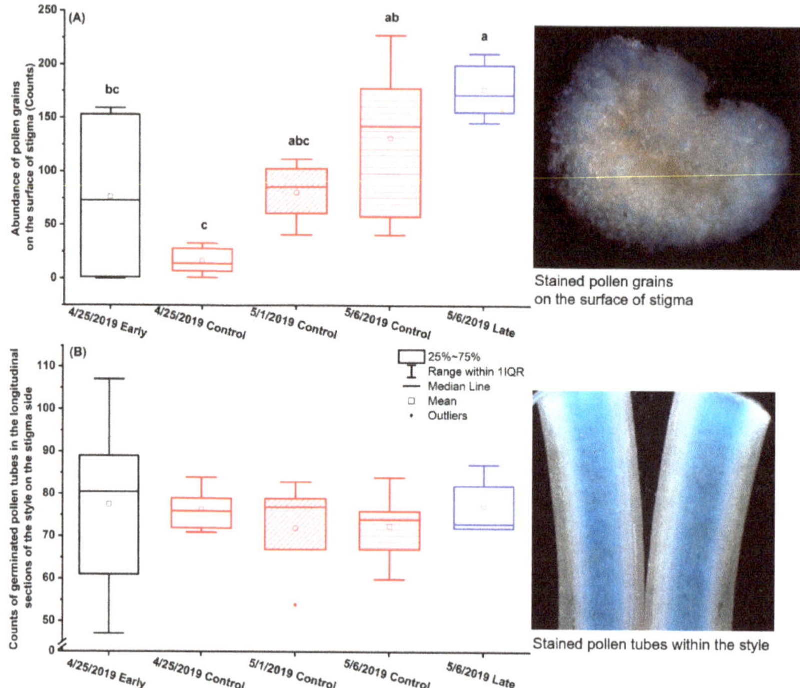

Figure 2. Impacts of fatty acid methyl esters (FAME) on (**A**) pollen abundance on the surface of stigma and (**B**) number of germinated pollen tubes in the longitudinal sections of the style on the stigma side of 'Lapins' sweet cherry in the spring of 2019 at SuRDC1. The early spray (Early) was applied on 7 March and flowers were sampled on 25 April. The late spray (Late) was applied on 1 April and flowers were sampled on 6 May. Flowers of Control treatment were sampled on 25 April, 1 May and 6 May. Pollen grains and pollen tubes were stained with Aniline Blue (n = 6 flowers). Boxplots show median (horizontal line) and interquartile ranges. In (**A**), the letters stand for significant difference at $p < 0.05$ [ANOVA, Tukey's test for pairwise comparison, $F_{(4,22)} = 6.97$, $p < 0.001$].

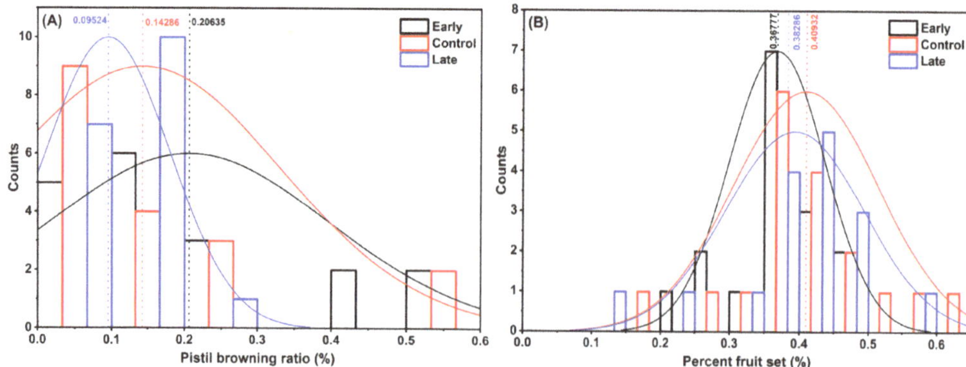

Figure 3. Impacts of FAME on (**A**) pistil browning ratio and (**B**) percent fruit set of 'Lapins' sweet cherry in the spring of 2019 at SuRDC1. The early spray (Early) and the late spray (Late) were applied on 7 March and 1 April, respectively. Solid lines in the histograms are distribution curves (n = 18 trees per treatment). Vertical dot lines stand for the mean for each data group.

2.4. Boric Acid Spray

The 0.01% boric acid spray led to the decrease in fruit set at SuRDC1 in the spring of 2022. Percent fruit set on the branches sprayed with 0.01% boric acid solution was 15.7% ± 0.3%, significantly lower than that of the untreated branches at 22.7% ± 0.3% [F(1,70) = 4.46, p = 0.04; n = 36; data not graphed].

2.5. Effects of Reduction in Spring Irrigation

Six-week postponement of irrigation starting from full bloom led to the decrease in soil VWC (Differential mean ± SD = 0.033 m^3/m^3 ± 0.012 m^3/m^3; range 0.006–0.067 m^3/m^3) (Figure 4A), the increase in soil temperature (Differential mean ± SD = 1.202 °C ± 0.523 °C; range 0.033–2.534 °C) (Figure 4B), a lower pistil browning ratio (mean ± SD 29.8% ± 1.8% versus 37.5% ± 1.8%) [Figure 4C; F(1,35) = 5.28, p = 0.025], and a higher percent fruit set (mean ± SD 22.4% ± 3.2% versus 14.8% ± 2.1%) [Figure 4D; F(1,35) = 4.17, p = 0.049].

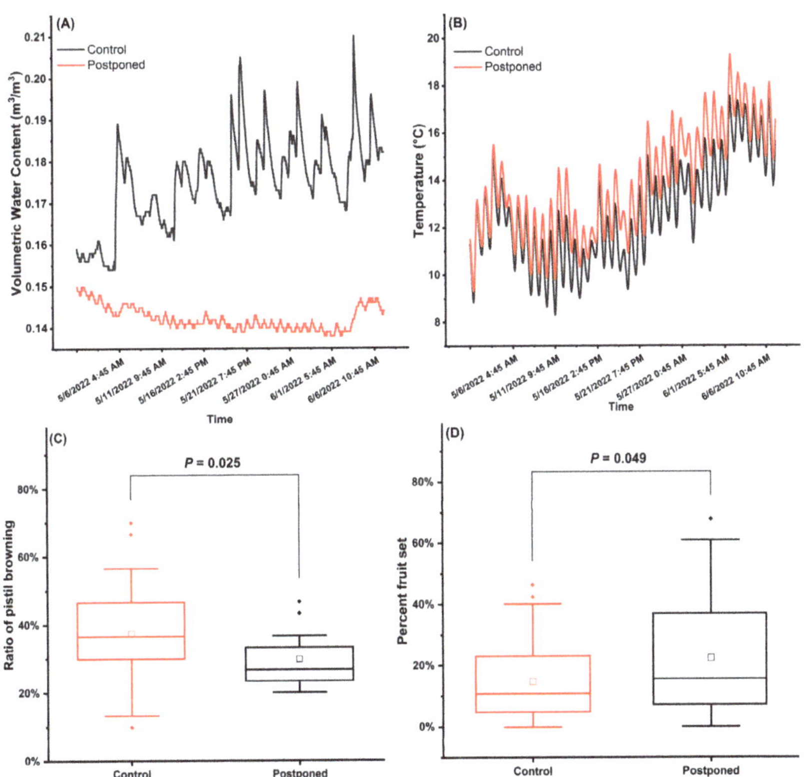

Figure 4. Postponed irrigation led to (**A**) the decrease in Volumetric Water Content (VWC) in topsoil, (**B**) the increase in soil temperature, (**C**) the decrease in pistil browning and (**D**) the increase in fruit set. Boxplots in (**C**) and (**D**) showed median (horizontal line), mean (square) and interquartile ranges; significant difference is shown as p value for each pair of comparison (AVONA, Tukey's test, p < 0.05, n = 18 trees per treatment).

3. Discussion

3.1. Vulnerability of Sweet Cherry Floral Organs under Variable Spring Conditions

In this study, fruit set in 'Lapins' sweet cherry at SuRDC1 was negatively affected by frost, precipitation and elevated ambient temperature during anthesis. Among the five sites, percent fruit set remained the lowest at SuRDC1 in both years, which was associated with pistil browning after daily minimum temperature (T_{min}) during anthesis dropped below $-4\,°C$. This suggested the importance of orchard location selection to avoid pistil injury due to spring frost. Compared to 2019, heavy precipitation and low air temperature in April 2022 at SuRDC1 led to a decrease in percent fruit set by nearly $\frac{1}{2}$ (Table 1). Rainfall events could disrupt pollinator activities [13], hinder anther dehiscence and pollen release [17], and affect stigma receptivity [3], causing the failure in fertilization. Concurring with rain, low temperature in April (Table 2) could hinder pollen tube elongation and initial ovary cell division and expansion, leading to ovule abortion and fruitlet drop.

On the other hand, an increase in floral bud $T_{surface}$ by 3.2 °C on average led to fruitlet development cessation (Figure 1C) and significantly lower fruit set on the branches surrounded by polyethylene sleeves, showing the negative consequence of simulated warming effect. Similar result was reported in Vignola and Sunburst sweet cherry, where an increase in maximum temperature by 5–7 °C resulted in an increase in average temperature by 1–3 °C and a drastic reduction in fruit set [6]. This could be attributed to rapid ovule senescence and reduced pistil viability in warm conditions [34]. This suggests that it is critical to characterize the ambient temperature dynamics and radiation conditions inside the surrounding protective mini-structures and facilities, and to avoid undesirable effects such as warming and reduced radiation when these structures are implemented in the spring. It also points out the potential impacts of global warming on cherry productivity and the necessity of implementing effective horticultural measures based on precise weather forecast to enhance fruit set resilience. In addition, it is important to define the cultivar-specific optimum temperature range for fruit set and use it to guide the site suitability evaluation prior to new cherry orchard establishment. Furthermore, compared to Mazzard rootstock, a lower fruit set was observed on Krymsk 5 rootstock (Table 1); such rootstock effect should be further investigated in continuous years in a multi-site trial that consists of different rootstocks at each location.

3.2. The Effectiveness of Horticultural Mitigations and Future Perspectives

The plant growth regulators that can change the readiness, viability and longevity of the floral organs can be used to improve pollination and fertilization and enhance the resilience of fruit set under adverse conditions. WAIKEN spray containing FAME was reported to effectively advance and delay the timing of bud break in the Tasmanian climate [22], suggesting its potential in scheduling anthesis to minimize the adverse temperature effect. In this study, however, neither the early spray nor the late spray altered the timing of bud break as expected. Furthermore, the sprays did not exert statistically significant effect on pollen abundance on the surface of stigma (Figure 2A), number of pollen tubes in the style sections (Figure 2B), pistil browning ratio (Figure 3A) or percent fruit set (Figure 3B). The ineffectiveness could be attributed to the low temperatures on 7 March 2019, the day of early spray application, which had affected the efficacy of the spray product. The daily T_{min} in early-mid March in the Okanagan region is usually close to 0 °C, which would be a persistent limitation to the use of the spray 35–50 days before bud break to advance anthesis. On the other hand, late spray showed potential in improving the pollen abundance (Figure 2A) and reducing pistil browning (Figure 3A). The statistical significance of late spray may have been underestimated due to the limited number of replications (n = 6 flowers), as statistical power depends on both effect size and sample size, and it is more likely to detect a smaller effect with a larger sample size. Within the recommended late spray threshold between 20 days before bud break and green tip (manufacturer's guide for WAIKEN Orchard Spray Emulsion Concentrate, SST Australia), sequential sprays at a 3-day incremental interval can be conducted to identify the best spray timing that can set back the bud break most effectively in the local climate of interest, as a measure to avoid frost and improve fruit set.

Boron is a critical element for the success of flowering and fruit set. Appropriate boron level can promote pollen germination and pollen tube elongation. Inappropriate concentration, formula, and application timing and method can cause deficiency, ineffectiveness or toxicity [35,36]. In this study, 0.01% boric acid spray negatively affected fruit set, suggesting that the acidic form in such concentration might not be a beneficial practice at full bloom, although it effectively promoted pollen germination and elongation in the in vitro study [23,24]. Tissue acidification [37] and high moisture retention [17] were reported to reduce pollen viability and anther dehiscence in fruit trees. Further study is required to elucidate whether the boric acid spray affects the viability of floral organs due to these issues, to investigate the appropriate concentration and form of boron, the timing, and the co-effects of calcium and nitrogen supplements in aerial spray, and to compare its effectiveness with fertigation.

Irrigation scheduling plays a critical role in determining the dynamics of water availability and soil temperature in the rhizosphere. Its effects on cherry tree hardiness, vigor and yield have been reported in several studies [38–40], whereas the effect on fruit set was yet to be elucidated. In the spring, irrigation can influence the root activity in transporting water and nutrients that are essential to flowering and fruit set, such as calcium, boron and zinc [41–45]. It can also affect the timing of dormancy break, the development of floral organs, the early vegetative growth and the abundance of carbohydrates for reproductive growth in woody plants [14,15]. Water stress can induce phytohormone imbalance in abscisic acid, auxin, gibberellin and ethylene, which dynamics can determine the formation of abscission zone at the base of pedicels, causing flower and fruit drop [46]. In this study, postponed irrigation during anthesis in the moist and cool spring of 2022 led to an increase in soil temperature without causing water deficit (Figure 4A,B), and an increase in percent fruit set which was associated with less pistil browning (Figure 4C,D). This is consistent with the study of Greer et al. which showed that warmer soil temperatures could promote bud break and early season development in apples, possibly attributed to higher root activity and better nutrient supply in the warmer rhizosphere [47]. However, Hammond and Seeley's earlier work showed that soil warming did not affect the spring bud development in the studied *Malus* or *Prunus* species [48]. This suggests that the root responses to warmer soil temperatures may be dependent on the interactions of other environmental factors that require more investigation. In addition, postponed irrigation in the wet spring may improve soil aeration in rhizosphere [16,49]. Research attention should also be drawn to the relation of water content and frost susceptibility of floral organs [50,51] under postponed irrigation. Irrigation scheduling shows great potential in improving fruit set, particularly in the early- and late-flowering cultivars grown at the geographic extremes which are more susceptible to the risks of adverse temperatures. However, it should be carefully tailored to the specific cultivars, the edaphic conditions of the orchard and the spring weather conditions that vary each year. Future studies should elucidate the soil water potential range during anthesis to achieve optimal fruit set.

4. Materials and Methods

4.1. Site Description and Fruit Set Monitoring

For treatment implementation, the main experimental site of 'Lapins' sweet cherry on 'Krymsk 5' rootstock was located in the frost pocket at the elevation of 487 m, in the experimental farm of Summerland Research and Development Centre, Agriculture and Agri-Food Canada, Summerland, the Okanagan Valley, British Columbia, Canada (Site "SuRDC1", 49.5657° N, 119.6365° W). The trial of 216 trees in 6 rows was established in 2015 in loamy sandy soil. Tree canopy was pruned to Tall Spindle Axes structure [52] and kept at the height of 2.2–2.5 m. Water was supplied through micro-sprinklers. Granular fertilizer (TerraLite 22-5-12) was casted under the canopy in June at the rate of 305 g per tree. Each plot (experimental unit) consisted of 3 sampling trees in the centre of the plot, with 1 buffer tree on each side of the plot; plots were distributed by completely randomized block design in the trial; treatment details of each horticultural mitigation trial were described in Section 4.2.

For phenology and fruit set monitoring, 18 trees were distributed in 6 control plots at SuRDC1 (3 trees/plot), in comparison to the other four monitored sites of lower frost risk, with 2 sites at lower elevation in the experimental farm ("SuRDC2", 49.5642° N, 119.6396° W, 16 trees; "SuRDC3", 49.5653° N, 119.6418° W, 9 trees), 1 in Oliver ("Oliver", 49.17° N, 119.56° W) and 1 in Kelowna ("Kelowna", 49.88° N, 119.37° W) (10 trees each site; precise geo-coordinates were omitted for growers' sites) (Table 1). Air temperature was monitored using HOBO sensors to detect daily T_{min} (MX2301A, Onset/HOBO). T_{sum} and precipitation data were acquired from farmwest.com (Summerland EC station for SuRDC sites, Olive Central for Oliver site, Kelowna East for Kelowna site; Tables 1 and 2.

Phenology of floral developmental stages from green tip (BBCH54) to sepal fall (Green Ovary, BBCH72) [53] was recorded on the tagged branches in each sampling tree (1 branch per tree in 2019 and 2 branches per tree in 2022 at SuRDC1, 1 trunk section per tree at SuRDC2, 3–4 branches per tree at SuRDC3, and 3 branches per tree at Kelowna and Olive; Table 1). At full bloom, flower samples were collected from adjacent untagged branches to examine pistil browning (5 flowers per tree). Flower counts at white tip—full bloom and fruitlet counts after shuck fall were recorded on the tagged branches; percent fruit set was estimated as fruitlet counts divided by flower counts for each branch and analyzed for statistical significance of site effect by ANOVA (n = number of branches per site, Tukey's pairwise comparison, $p \leq 0.05$). In addition, at SuRDC1, 50 random floral buds were sampled on 7 March 2019 and dissected to examine primordia, following the cold snap on 3–4 March (daily T_{min} below −15 °C).

4.2. Spring Horticultural Mitigations at SuRDC1

4.2.1. Installation of Polyethylene Sleeves to Increase Floral Bud $T_{surface}$

In the spring of 2019, 36 polyethylene sleeves (made of Uline 6 mil heavy duty poly tubing, about 0.152 mm thick) wrapped onto plastic racks were wired around the bearing branches to surround the floral buds, from bud break to sepal fall (2 sleeves on each tree, 18 trees; Figure 1A). The two ends of the polyethylene sleeves were kept open to allow air flow and pollinator activities. Two tagged branches on the same trees were monitored as the control. Thermal images of the floral buds were captured at 12:00–12:30 PM of 17 April, using FLIR E8 Infrared camera (FLIR® Systems Inc., Wilsonville, OR, USA; 7 mm focal lens, 320 × 240 IR resolution); On the surrounded branches, the photos were taken through the open end of the sleeve (Figure 1B). Infrared and RGB channels were separated in software FLIR Tools V. 6.4 (FLIR® Systems Inc.); the RGB image (Figure 1B bottom) was referred to locate the floral buds in the corresponding infrared image (Figure 1B top). $T_{surface}$ of floral buds was automatically computed in FLIR Tools; $T_{surface}$ of 3 buds was averaged to represent the mean bud $T_{surface}$ for each sleeved and tagged branch. Flowering stages, pistil browning and aborted ovule (Figure 1C) were recorded for each branch. For $T_{surface}$ and percent fruit set, the mean of the two branches per treatment per tree was calculated and analyzed for statistical significance of polyethylene sleeve effect by ANOVA (n = 18 trees, Tukey's test, $p \leq 0.05$).

4.2.2. FAME Spray to Regulate Bud Break

The efficacy of an orchard spray emulsion concentrate which contains FAME (WAIKEN, SST Australia; active constituent: 388 g/L FAME including 10–35% dibutyl phthalate, 10–20% ethoxylated nonylphenol, 1–12% oxirane methyl polymers), was tested as a run-off spray of 1:24 dilution (1 L of product diluted in 24 L water to make 25 L working spray; about 15.5 g/L of FAME in the working spray) in the spring of 2019 at two time points, to adjust the timing of bud break. The early spray ("Early") was applied on 7 March, about 40 days before the anticipated normal bud break, to advance bud break; T_{min} and T_{max} (maximum temperature) were −4.3 °C and 2.3 °C on this day. The late spray ("Late") was applied on 1 April, about 20 days before normal bud break, to set back the bud break; T_{min} and T_{max} were 1.3 °C and 15.8 °C on this day. No spray was applied to the control trees ("Control"). These treatments aimed to expose the flowers to different temperatures, and to investigate its impacts on pollen viability and percent fruit set. Each treatment was applied to 18 trees, with 3 trees per plot, 6 plots randomized in 6 rows (1 plot per row). On each tree, 2 branches were tagged for phenological observation and their mean percent fruit set was calculated; 7 flowers per tree were examined for pistil browning. Statistical significance of FAME spray effects was analyzed by ANOVA (n = 18 trees, Tukey's pairwise comparison, $p \leq 0.05$).

To examine the abundance of pollen grains landing on the stigma and the number of germinated pollen tubes in the style, six flowers were sampled for early spray on 25 April, for Control on 25 April, 1 May and 6 May each, and for late spray on 6 May respectively (n = 6 flowers). Pistils were preserved in Formaldehyde Alcohol Acetic Acid fixative (FAA; 50% ethanol, 5% (v/v) acetic acid, 3.7% (v/v) formaldehyde) at 4 °C until analysis. For sample preparation, stigma was cut apart from style using a sharp razor blade and set aside to analyze the pollen grains on its surface. The style was cut apart from the ovary, placed on a piece of double-sided tape to hold it down, and then pressed gently with a clean glass slide to flatten it slightly in order to make a straight cut; both ends of the style were cut off transversely about 0.5 mm from the ends and then dissected longitudinally for better dye penetration to stain the pollen tubes within the style. Both the stigma and the style were placed in Aniline Blue solution (0.1% Aniline Blue w/v in 0.1 mol L^{-1} KH_2PO_4, Sigma-Aldrich) for 30 min (modified based on Lu [54]). Using a transfer pipet, the stigma was positioned in the centre of a depression slide with the top pointing upwards. The four sections from each style (2 longitudinal sections from the stigma end and 2 from the ovule end) were placed on a coverslip facedown; the coverslip was then placed on the slide with the cut sections facing up. Sample images were acquired using Zeiss Axio Imager M2 wide-field microscope (Zeiss). Stigma images were acquired under 5× objective lens, with the microscope light source blocked out; an external warm white light was used to illuminate the sample from the side to provide better light diffusion and contrast. The images of style sections were acquired under 10× objective lens in green fluorescence channel under UV excitation. For both the stigma and the style images, multiple images were taken for each sample and stacked into one image with clear focus using Helicon Focus software (HeliconSoft.com). The counting tool in Image J [55] was used to count the number of pollen grains on the surface of stigma (Figure 2A), and the number of pollen tubes from both style sections in the stigma end which were summed up to estimate the number of germinated pollens (Figure 2B). In the ovary end of the style, the presence of pollen tubes was recorded. Statistical significance of FAME spray effects was analyzed by ANOVA (n = 6 flowers, Tukey's test, $p \leq 0.05$).

4.2.3. Boric Acid Spray

The 0.01% boric acid solution (0.099 g H_3BO_3 in 1 L of distilled water, Sigma-Aldrich) was sprayed to 1 branch per tree of 36 trees on 3 May 2022 at full bloom (with 3 trees per plot, 12 plots randomized in 6 rows, 2 plots per row). Flower counts at white tip–full bloom and fruitlet counts after shuck fall were recorded on treated branches and adjacent control branches of the same trees; percent fruit set was estimated as described in Section 4.1 and analyzed for statistical significance of boric acid spray effect by ANOVA (n = 36, Tukey's test, $p \leq 0.05$).

4.2.4. Postponed Irrigation and Monitoring of Soil Moisture and Temperature

Micro-sprinkler irrigation was scheduled for two hours every 72 h from full bloom to the end of September. When daily maximum temperature exceeded 30 °C in July and August, irrigation was scheduled for two hours every 48 h. In the moist and cool spring of 2022, irrigation was stopped between 28 April and 7 June (Postponed Irrigation) in and around 6 plots randomized in 6 rows (1 plot per row, 3 trees per plot, n = 18 trees). One control plot of 3 trees was located a plot apart from the Postponed Irrigation plot in the same row (n = 18 trees). The 5TM soil sensors (Meter Environment, Pullman, WA, USA) were installed in 2 Postponed Irrigation plots and 2 control plots, to monitor volumetric water content (VWC) and soil temperature in the topsoil of 20 cm depth at 30 min interval

(EM50 data loggers, Meter Environment). Flower counts, fruitlet counts and percent fruit set were recorded on 2 tagged branches per tree as described above.

4.3. Data Analysis

Statistical analysis and graphing were conducted using OriginPro 8.0 (OriginLab, Northampton, MA, USA). For the fruit set monitoring, significant difference among the five sites was assessed by ANOVA ($p \leq 0.05$, Tukey's test for pairwise comparison, site as the fixed effect). For each horticultural mitigation trial, significant difference between control and treatment was assessed by ANOVA ($p \leq 0.05$, Tukey's test, treatment as the fixed effect, outliers were excluded). Mean $\pm$ standard error were shown along with F[df (degree of freedom), N (total samples minus df)] and p values for statistical significance of the fixed effects in comparison. In the horticultural mitigation trials, individual flowers were analyzed as biological entity for the pollen abundance on stigma and the number of germinated pollen tubes (n = the number of sampled flowers). Sampling trees were analyzed as entity for ratio of floral organ damage, floral bud $T_{surface}$ and percent fruit set (n = the number of trees; the mean of branches under each treatment on each tree was calculated to represent the tree).

5. Conclusions

The fruit set in the 'Lapins'/Krymsk 5 planting located in a frost pocket was negatively affected by frost, precipitation and elevated ambient temperature during anthesis. Low percent fruit set was associated with pistil browning when T_{min} was below $-4\ ^{\circ}C$, and with fruitlet development cessation when floral bud $T_{surface}$ was elevated by about $3\ ^{\circ}C$. Postponed irrigation during anthesis in the moist and cool spring improved fruit set, suggesting irrigation adjustment as an effective measure to mitigate fruit set issue when there is excessive moisture. Further investigation on spray contents and timing is required to improve the efficacy of FAME and boron supplements. The FAME spray showed potential in delaying anthesis and preventing frost injury; its physiological mechanism and horticultural effects under the climate of interest wait to be elucidated with more replications using the in vivo pollen and pollen tube examination method documented in this study.

Author Contributions: Conceptualization, H.X.; methodology, D.E. and H.X.; validation, H.X.; formal analysis, H.X.; investigation, D.E. and H.X.; resources, H.X. and M.S.; data curation, H.X.; writing—original draft preparation, H.X.; writing—review and editing, H.X. and M.S.; visualization, H.X.; supervision, H.X.; project administration, H.X.; funding acquisition, H.X. All authors have read and agreed to the published version of the manuscript.

Funding: This research was funded by Agriculture and Agri-Food Canada and Canadian Agricultural Partnership Fund in collaboration with British Columbia Fruit Growers' Association (CAP; ASP-005 BCFGA Activity #7 J-002065).

Data Availability Statement: Data are available upon request.

Acknowledgments: The authors are grateful to Megan Lowe, Emily Payne and Bill Rabie for their technical support, to Tom Forge and Paige Munro for establishing the SuRDC1 trial, to Martyn Jones for providing the tested spray product, to Michael Weis for advising on microscopic work, to Amritpal Singh and Chris Pagliocchini for sharing the SuRDC3 site and their knowledge of cherry phenology, to Summerland Research and Development Centre Field Service for maintaining the SuRDC sites, to participating growers for supporting the research in their orchards, and to the reviewers for their constructive comments.

Conflicts of Interest: The authors declare no conflict of interest.

References

1. Mahmood, K.; Carew, J.G.; Hadley, P.; Battey, N.H. The effect of chilling and post-chilling temperatures on growth and flowering of sweet cherry (*Prunus avium* L.). *J. Hortic. Sci. Biotechnol.* **2000**, *75*, 598–601. [CrossRef]
2. Beppu, K.; Suehara, T.; Kataoka, I. High temperature and drought stress suppress the photosynthesis and carbohydrate accumulation in 'Satohnishiki'sweet cherry. *Acta Hortic.* **2002**, *618*, 371–377.
3. Hedhly, A.; Hormaza, J.I.; Herrero, M. The effect of temperature on stigmatic receptivity in sweet cherry (*Prunus avium* L.). *Plant Cell Environ.* **2003**, *26*, 1673–1680. [CrossRef]
4. Hedhly, A.; Hormaza, J.I.; Herrero, M. Effect of temperature on pollen tube kinetics and dynamics in sweet cherry, *Prunus avium* (Rosaceae). *Am. J. Bot.* **2004**, *91*, 558–564. [CrossRef] [PubMed]
5. Montiel, F.G.; Serrano, M.; Martinez-Romero, D.; Alburquerque, N. Factors influencing fruit set and quality in different sweet cherry cultivars. *Span. J. Agric. Res.* **2010**, *8*, 1118–1128. [CrossRef]
6. Hedhly, A.; Hormaza, J.I.; Herrero, M. Warm temperatures at bloom reduce fruit set in sweet cherry. *J. Appl. Bot. Food Qual.* **2007**, *81*, 158–164.
7. Kazan, K.; Lyons, R. The link between flowering time and stress tolerance. *J. Exp. Bot.* **2016**, *67*, 47–60. [CrossRef]
8. Postweiler, K.; Stösser, R.; Anvari, S.F. The effect of different temperatures on the viability of ovules in cherries. *Sci. Hortic.* **1985**, *25*, 235–239. [CrossRef]
9. Predieri, S.; Dris, R.; Sekse, L.; Rapparini, F. Influence of environmental factors and orchard management on yield and quality of sweet cherry. *Food Agric. Environ.* **2003**, *1*, 263–266.
10. Bosch, J.; Kemp, W.P. Effect of wintering duration and temperature on survival and emergence time in males of the orchard pollinator *Osmia lignaria* (Hymenoptera: Megachilidae). *Environ. Entomol.* **2003**, *32*, 711–716. [CrossRef]
11. Vicens, N.; Bosch, J. Weather-dependent pollinator activity in an apple orchard, with special reference to *Osmia cornuta* and *Apis mellifera* (Hymenoptera: Megachilidae and Apidae). *Environ. Entomol.* **2000**, *29*, 413–420. [CrossRef]
12. Holzschuh, A.; Dudenhöffer, J.H.; Tscharntke, T. Landscapes with wild bee habitats enhance pollination, fruit set and yield of sweet cherry. *Biol. Conserv.* **2012**, *153*, 101–107. [CrossRef]
13. Forrest, J.R. Plant–pollinator interactions and phenological change: What can we learn about climate impacts from experiments and observations? *Oikos* **2015**, *124*, 4–13. [CrossRef]
14. Kaufmann, M.R. Water Deficits and Reproductive Growth. In *Water Deficit and Plant Growth. Volume III Plant Responses and Control of Water Balance*; Kozlowski, T.T., Ed.; Academic Press: New York, NY, USA; London, UK, 1972; pp. 91–124.
15. Pallardy, S.G. *Physiology of Woody Plants*; Academic Press: New York, NY, USA, 2010; Volume 231–233, pp. 244–252.
16. Bradford, K.J.; Yang, S.F. Physiological responses of plants to waterlogging. *HortScience* **1981**, *16*, 25–30. [CrossRef]
17. Gradziel, T.M.; Weinbaum, S.A. High relative humidity reduces anther dehiscence in apricot, peach, and almond. *HortScience* **1999**, *34*, 322–325. [CrossRef]
18. Herrero, M.; Rodrigo, J.; Wünsch, A. Flowering, fruit set and development. In *Cherries—Botany, Production and Uses*; Quero-Garcia, J., Iezzoni, A., Puławska, J., Lang, G., Eds.; CABI: Oxfordshire, UK, 2017; pp. 23–29.
19. Beyhan, N.; Karakaş, B. Investigation of the fertilization biology of some sweet cherry cultivars grown in the Central Northern Anatolian Region of Turkey. *Sci. Hortic.* **2009**, *121*, 320–326. [CrossRef]
20. Mateos-Fierro, Z.; Garratt, M.P.; Fountain, M.T.; Ashbrook, K.; Westbury, D.B. Wild bees are less abundant but show better pollination behaviour for sweet cherry than managed pollinators. *J. Appl. Entomol.* **2022**, *146*, 361–371. [CrossRef]
21. Bound, S.A.; Close, D.C.; Jones, J.E.; Whiting, M.D. Improving fruit set of 'Kordia' and 'Regina' sweet cherry with AVG. *Acta Hortic.* **2014**, *1042*, 285–292. [CrossRef]
22. Bound, S.A.; Miller, P. Manipulating time of bud break, flowering and crop development of sweet cherry with the dormancy breaker Waiken®. *Acta Hortic.* **2016**, *1130*, 285–292. [CrossRef]
23. Muengkaew, R.; Chaiprasart, P.; Wongsawad, P. Calcium-boron addition promotes pollen germination and fruit set of mango. *Int. J. Fruit Sci.* **2017**, *17*, 147–158. [CrossRef]
24. Ureta Ovalle, A.; Atenas, C.; Larraín, P. Application of an *Ecklonia maxima* seaweed product at two different timings can improve the fruit set and yield in 'Bing' sweet cherry trees. *Acta Hortic.* **2019**, *1235*, 319–326. [CrossRef]
25. Alhamid, J.O.; Mo, C.; Zhang, X.; Wang, P.; Whiting, M.D.; Zhang, Q. Cellulose nanocrystals reduce cold damage to reproductive buds in fruit crops. *Biosyst. Eng.* **2018**, *172*, 124–133. [CrossRef]
26. Feldmane, D. Effects of irrigation and woodchip mulch on growth and habit of sour cherries. In Proceedings of the Annual 15th International Scientific Conference Research for Rural Development, Aguascalientes, Mexico, 18–21 June 2009; pp. 64–70.
27. Feldmane, D. Precocity of sour cherry cultivars influenced by using woodchip mulch and drip irrigation. In Proceedings of the Research for Rural Development Annual 16th International Scientific Conference Proceedings, Jelgava, Latvia, 19–21 May 2010; pp. 48–54.
28. Yin, X.; Long, L.E.; Huang, X.L.; Jaja, N.; Bai, J.; Seavert, C.F.; le Roux, J. Transitional effects of double-lateral drip irrigation and straw mulch on irrigation water consumption, mineral nutrition, yield, and storability of sweet cherry. *HortTechnology* **2012**, *22*, 484–492. [CrossRef]
29. Long, L.E.; Yin, X.; Huang, X.L.; Jaja, N. Responses of sweet cherry water use and productivity and soil quality to alternate groundcover and irrigation systems. *Acta Hortic.* **2014**, *1020*, 331–338. [CrossRef]

30. Neilsen, G.H.; Neilsen, D.; Kappel, F.; Forge, T. Interaction of irrigation and soil management on sweet cherry productivity and fruit quality at different crop loads that simulat those occurring by environmental extremes. *HortScience* **2014**, *49*, 215–220. [CrossRef]

31. Blanco, V.; Blaya-Ros, P.J.; Torres-Sánchez, R.; Domingo, R. Influence of regulated deficit irrigation and environmental conditions on reproductive response of sweet cherry trees. *Plants* **2020**, *9*, 94. [CrossRef]

32. Lang, G.A.; Sage, L.; Wilkinson, T. Ten years of studies on systems to modify sweet cherry production environments: Retractable roofs, high tunnels, and rain-shelters. *Acta Hortic.* **2016**, *1130*, 83–90. [CrossRef]

33. Martínez-Gómez, P.; Rahimi Devin, S.; Salazar, J.A.; López-Alcolea, J.; Rubio, M.; Martínez-García, P.J. Principles and Prospects of *Prunus* Cultivation in Greenhouse. *Agronomy* **2021**, *11*, 474. [CrossRef]

34. Zhang, L.; Ferguson, L.; Whiting, M.D. Temperature effects on pistil viability and fruit set in sweet cherry. *Sci. Hortic.* **2018**, *241*, 8–17. [CrossRef]

35. Wojcik, P.; Wojcik, M. Effect of boron fertilization on sweet cherry tree yield and fruit quality. *J. Plant Nutr.* **2006**, *29*, 1755–1766. [CrossRef]

36. Botelho, R.V.; Müller, M.M.L.; Umburanas, R.C.; Laconski, J.M.O.; Terra, M.M. Boron in fruit crops: Plant physiology, deficiency, toxicity, and sources for fertilization. In *Boron in Plants and Agriculture*; Academic Press: New York, NY, USA, 2022; pp. 29–50.

37. Chagas, E.A.; Barbosa, W.; Saito, A.; Pio, R.; Feldberg, N.P. Temperature, pH and development period on in vitro pollen germination in *Pyrus calleryana. Acta Hortic.* **2008**, *800*, 521–526. [CrossRef]

38. Feldmane, D.; Ruisa, S.; Rubauskis, E.; Kaufmane, E. Winter hardiness of sour cherries influenced by cultivar and soil moisture treatment. *Acta Hortic.* **2016**, *1130*, 111–115. [CrossRef]

39. Neilsen, G.H.; Neilsen, D.; Forge, T. Environmental limiting factors for cherry production. In *Cherries—Botany, Production and Uses*; Quero-Garcia, J., Iezzoni, A., Puławska, J., Lang, G., Eds.; CABI: Oxfordshire, UK, 2017; pp. 205–213.

40. Houghton, E.A.; Bevandick, K.; Neilsen, D.; Hannam, K.D.; Nelson, L.M. Effects of postharvest deficit irrigation on sweet cherry (*Prunus avium*) in five Okanagan Valley, Canada orchards: I. Tree water status, photosynthesis, and growth. *Can. J. Plant Sci.* **2022**. Just-In. [CrossRef]

41. DeMoranville, C.J.; Deubert, K.H. Effect of commercial calcium-boron and manganese-zinc formulations on fruit set of cranberries. *J. Hortic. Sci.* **1987**, *62*, 163–169. [CrossRef]

42. Usenik, V.; Stampar, F. Effect of foliar application of zinc plus boron on sweet cherry fruit set and yield. *Acta Hortic.* **2001**, *594*, 245–249. [CrossRef]

43. Hepler, P.K. Calcium: A central regulator of plant growth and development. *Plant Cell* **2005**, *17*, 2142–2155. [CrossRef] [PubMed]

44. Satya, S.; Pitchai, J.G.; Indirani, R. Boron nutrition of crops in relation to yield and quality—A review. *Agric. Rev.* **2009**, *30*, 139–144.

45. Hassan, G.; Mir, M.; Khachi, B.; Slathia, S. Effect of calcium and boron on reproductive performance of sweet cherry cv Bigarreau Noir Grossa. *Int. J. Farm Sci.* **2016**, *6*, 120–127.

46. van Doorn, W.G.; Stead, A.D. Abscission of flowers and floral parts. *J. Exp. Bot.* **1997**, *48*, 821–837. [CrossRef]

47. Greer, D.H.; Wünsche, J.N.; Norling, C.L.; Wiggins, H.N. Root-zone temperatures affect phenology of bud break, flower cluster development, shoot extension growth and gas exchange of 'Braeburn'(*Malus domestica*) apple trees. *Tree Physiol.* **2006**, *26*, 105–111. [CrossRef]

48. Hammond, M.W.; Seeley, S.D. Spring bud development of *Malus* and *Prunus* species in relation to soil temperature1. *J. Am. Soc. Hortic. Sci.* **1978**, *103*, 655–657. [CrossRef]

49. Friedman, S.P.; Naftaliev, B. A survey of the aeration status of drip-irrigated orchards. *Agric. Water Manag.* **2012**, *115*, 132–147. [CrossRef]

50. Rodrigo, J. Spring frosts in deciduous fruit trees—Morphological damage and flower hardiness. *Sci. Hortic.* **2000**, *85*, 155–173. [CrossRef]

51. Kappel, F. Sweet cherry cultivars vary in their susceptibility to spring frosts. *HortScience* **2010**, *45*, 176–177. [CrossRef]

52. Long, L.E.; Lang, G.A.; Whiting, M.D.; Musacchi, S. *Cherry Training Systems*; PNW 667. A Pacific Northwest Extension Publication; Oregon State University: Corvallis, OR, USA; Washington State University: Pullman, WA, USA; University of Idaho: Perimeter, MO, USA; Michigan State University: East Lansing, MI, USA, 2015; pp. 43–56.

53. Fadón, E.; Herrero, M.; Rodrigo, J. Flower development in sweet cherry framed in the BBCH scale. *Sci. Hortic.* **2015**, *192*, 141–147. [CrossRef]

54. Lu, Y. Arabidopsis pollen tube aniline blue staining. *Bio-Protocol* **2011**, *1*, e88. [CrossRef]

55. Schneider, C.A.; Rasband, W.S.; Eliceiri, K.W. NIH Image to ImageJ: 25 years of image analysis. *Nat. Methods* **2012**, *9*, 671–675. [CrossRef]

Article

Effect of Salinity and Silicon Doses on Onion Post-Harvest Quality and Shelf Life

Jefferson Bittencourt Venâncio [1], Nildo da Silva Dias [1,*], José Francismar de Medeiros [1], Patrícia Lígia Dantas de Morais [1], Clístenes Williams Araújo do Nascimento [2], Osvaldo Nogueira de Sousa Neto [3], Luciara Maria de Andrade [1], Kleane Targino Oliveira Pereira [1], Tayd Dayvison Custódio Peixoto [1], Josinaldo Lopes Araújo Rocha [4], Miguel Ferreira Neto [1] and Francisco Vanies da Silva Sá [1,*]

[1]   Center for Agrarian Sciences, Federal Rural University of the Semi-Arid Region, Mossoró 59625-900, Brazil
[2]   Department of Agronomy, Federal Rural University of Pernambuco, Recife 52171-900, Brazil
[3]   Department of Engineering, Federal Rural University of the Semi-Arid Region, Angicos 59515-000, Brazil
[4]   Center for Agro-Food Science and Technology, Federal University of Campina Grande, Pombal 58840-000, Brazil
*   Correspondence: nildo@ufersa.edu.br (N.d.S.D.); vanies_agronomia@hotmail.com (F.V.d.S.S.); Tel.: +55-(84)-99684-4875 (N.d.S.D.); +55-(84)-99651-3164 (F.V.d.S.S.)

Abstract: Salt stress during pre-harvest limits the shelf life and post-harvest quality of produce; however, silicon nutrition can mitigate salt stress in plants. Thus, we evaluated the effects of salinity and fertilization with Si, in pre-harvest, on the morpho-physiological characteristics of onion bulbs during shelf life. The experiment was set up in randomized complete blocks, with treatments arranged in split-split plots. The plots had four levels of electrical conductivity of irrigation water (0.65, 1.7, 2.8, and 4.1 dS m^{-1}). The subplots had five fertilization levels with Si (0, 41.6, 83.2, 124.8, and 166.4 kg ha^{-1}). The sub-sub plots had four shelf times (0, 20, 40, and 60 days after harvest). Irrigation water salinity and shelf time reduced firmness and increased the mass loss of onion bulbs during shelf life. Salt stress reduced the contents of sugars and total soluble solids of onion bulbs during storage; however, Si supply improved the contents of these variables. Salinity, Si supply, and shelf time increased the concentrations of pyruvic and ascorbic acids in onion bulbs during shelf life. Si doses between 121.8 and 127.0 kg ha^{-1} attenuated the impacts caused by moderate salinity, increasing the synthesis of metabolites and prolonging the onion bulbs' shelf life.

Keywords: *Allium cepa* L.; mineral nutrition; horticulture; soil fertility; diatomaceous earth; *Melosira granulate*

Citation: Venâncio, J.B.; Dias, N.d.S.; Medeiros, J.F.d.; Morais, P.L.D.d.; Nascimento, C.W.A.d.; Sousa Neto, O.N.d.; Andrade, L.M.d.; Pereira, K.T.O.; Peixoto, T.D.C.; Rocha, J.L.A.; et al. Effect of Salinity and Silicon Doses on Onion Post-Harvest Quality and Shelf Life. *Plants* **2022**, *11*, 2788. https://doi.org/10.3390/plants11202788

Academic Editor: Pedro Diaz-Vivancos

Received: 17 September 2022
Accepted: 13 October 2022
Published: 20 October 2022

Publisher's Note: MDPI stays neutral with regard to jurisdictional claims in published maps and institutional affiliations.

1. Introduction

Onion (*Allium cepa* L.) is a vegetable appreciated worldwide for its food, nutritional and medicinal characteristics [1,2]. Onions are in high demand all year, so the bulbs are usually stored due to logistical limitations and the seasonality of crop harvests [3].

Pre-harvest aspects of crop management techniques, such as mineral nutrition, irrigation, cultivar, or other agronomic conditions, affect the post-harvest conservation parameters, processing, and quality of onion bulbs. Post-harvest factors contributing to storage performance include the method and duration of curing, grading, packaging method, and storage environment [4]. The significant post-harvest losses of onions are mainly caused due to bulb sprouting and rotting, which contribute to loss in storage life and quality [5]. The market value is predominantly related to bulb firmness and dry matter content [6]. To supply customers and processors with high-quality, firm onions with a high dry matter content devoid of sprouts, secondary roots, and illnesses, mature bulbs are cured, dried, and stored in cool rooms between seasons [7]. Sugars and organic acids contribute to the organoleptic test and distinct flavor and aroma [8]. Biotic and abiotic stresses affect post-harvest conservation, such as pathogen incidence, exposure to temperature extremes, solar radiation, light, winds, and salinity [9,10].

Salt stress through acclimatization programs regulate plant development, physiology, and metabolism [11]. Redox metabolism and cellular osmoregulation are altered at the gene expression level to produce antioxidant and compatible molecules. Studies have examined that plants under salt stress increase their sugars, proline, betaine, glycine, polyamines, ascorbate, glutathione, tocopherols, carotenoids, thiols, and flavonoid levels in tissues; on the other hand, they decrease fresh bulb weight, production of large bulbs, bulb yield, the quality of onion bulbs, bulb firmness and bulb pH [12–17]. The literature reveals that salt stress alters production levels of proline, phenolic compounds, and pyruvic acid precursors in onions [17–19].

There is a need to evaluate salt stress effects on metabolite production related to onion bulb quality and conservation, such as pH, ascorbic acid, sugars, soluble solids, and titratable acidity.

Silicon is an element considered beneficial to plants [20]. Si improves plant acclimatization to multiple stressors, such as salt stress [21]. In salt stress, Si fertilization improves the essential plant nutrient uptake, antioxidant defense systems, and solute and plant hormone production [20]. In onions, ref. [22] showed that fertilization with Si, in interactions with zeolite and selenium (Se), improved some bulbs' qualitative characteristics, such as the large-bulb production, bulb-dry matter, soluble solids, and protein content. Refs. [17,22] observed that Si improved the physiological (enzymatic activity, chlorophyll levels, and photosynthetic activity) and nutritional (increase in nitrogen, nitrate, and potassium concentrations; decrease in sodium concentration) of the quality of onion bulbs, and plant salt tolerance under conditions of saline soil and brackish irrigation water. However, Si fertilization effects on shelf life and post-harvest characteristics of onion bulbs grown under saline stress are less well known.

We hypothesized that Si fertilization improves the post-harvest quality and shelf life of onion bulbs grown under saline stress. In this study, we examined the effects of Si fertilization on the post-harvest quality and shelf life of onion bulbs grown under increasing salinity levels of irrigation water.

2. Results

2.1. Bulb Firmness and Mass Loss

The general analysis of variance (ANOVA) showed an interaction between the salinity of irrigation water (EC) and shelf life (SL), and a non-significant effect of the Si dose (SD) on the variables bulb firmness (BF) and mass loss (ML).

The BF of 'Rio das Antas' onions was significantly reduced with increases in EC and SL, being described by the nonlinear parabolic model (Figure 1A). The minimum BF of 30.07 N was reached at 28.4 shelf days under an irrigation water salinity of 4.0 dS m^{-1}. The ML of onion bulbs, on the other hand, increased significantly with increases in EC and SL, being described by the nonlinear flat model (Figure 1C). The maximum ML of 9.23% was reached at 60 shelf days, with an EC of 4.0 dS m^{-1}.

Fertilization with Si did not significantly influence onion BF and ML. However, the significant effects of the SL factor on these variables allowed fits of three-dimensional parabolic and flat models, respectively, to the SD × SL interaction analysis data (Figure 1B,D). Minimum BF values of 33.47 and 35.07 N were reached at 28.4 shelf days with Si doses ranging from 0 to 166.4 kg ha^{-1} (Figure 1B). Maximum ML values of 8.27 and 8.60% were reached at 60 shelf days with the same Si doses (Figure 1D).

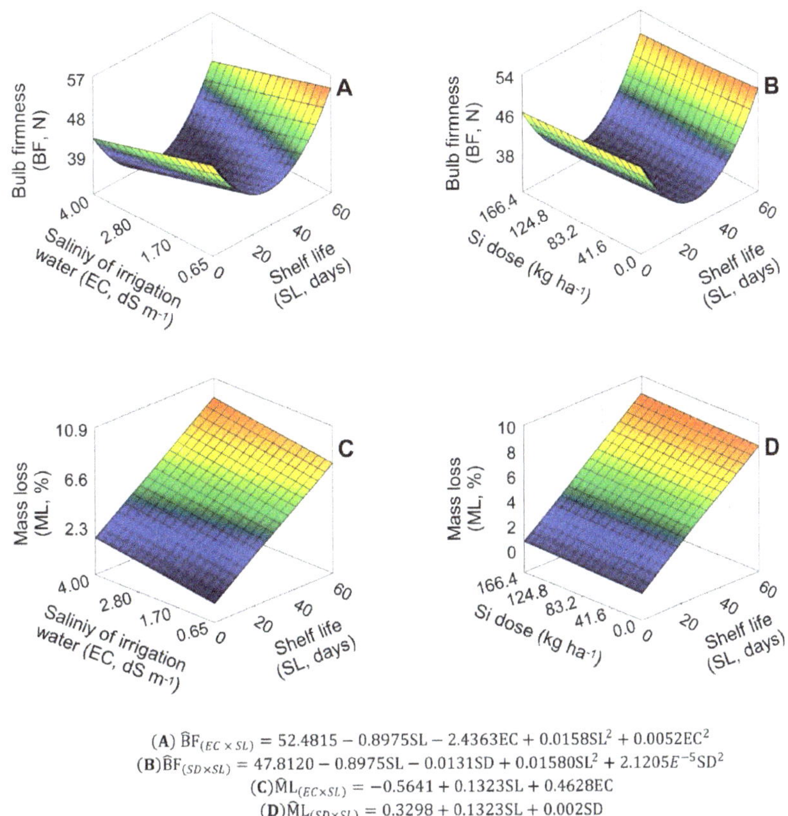

$$\textbf{(A)} \; \hat{B}F_{(EC \times SL)} = 52.4815 - 0.8975SL - 2.4363EC + 0.0158SL^2 + 0.0052EC^2$$
$$\textbf{(B)} \hat{B}F_{(SD \times SL)} = 47.8120 - 0.8975SL - 0.0131SD + 0.01580SL^2 + 2.1205E^{-5}SD^2$$
$$\textbf{(C)} \hat{M}L_{(EC \times SL)} = -0.5641 + 0.1323SL + 0.4628EC$$
$$\textbf{(D)} \hat{M}L_{(SD \times SL)} = 0.3298 + 0.1323SL + 0.002SD$$

Figure 1. Unfolding of the treatment factors of salinity of irrigation water (EC) and silicon doses in the soil (SD), with shelf life (SL), for the variables: bulb firmness (FB, in N) (**A**,**B**); and mass loss (ML, %) (**C**,**D**), in 'Rio da Antas' onion bulbs.

2.2. Tunic Color

Irrigation water salinity and Si fertilization failed to the color of onion bulbs. However, shelf time affected bulb color, altering the CIE color parameters L*, a*, b*, C*, and °H*.

For the EC and SD factors, the CIE parameters L*, a*, b*, C*, and °H were 70.50, 9.71, 32.05, 33.78, and 72.96, respectively (Figure 2A,B). For SL, for each storage day, there was a linear reduction of 0.2328, 0.3945, 0.3142, and 0.2565% in the CIE parameters L*, b*, C*, and H*, respectively, in addition to a linear increase of 0.8643% in the CIE parameter a* (Figure 2C).

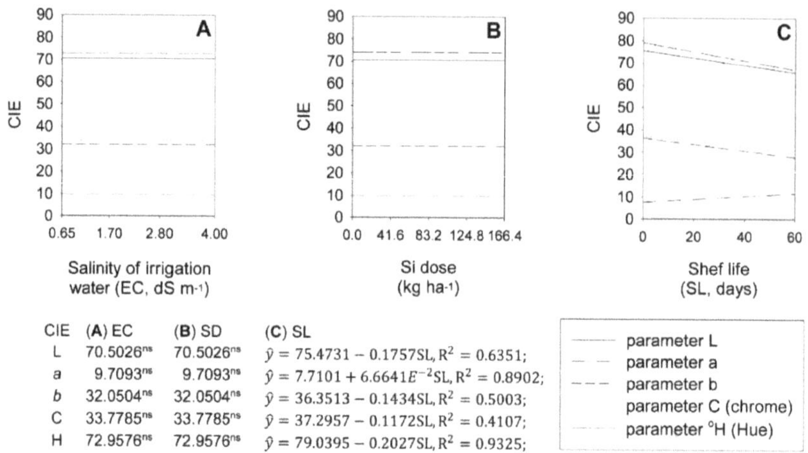

Figure 2. Effect of increasing salinity (EC) (**A**), silicon doses (SD) (**B**), and shelf life (SL) (**C**) on the coloration of 'Rio das Antas' onion bulbs, parameterized according to CIE standard L*, a*, b*, C*, and °H.

2.3. Soluble Sugars and Total Soluble Solids

There was a significant main effect of the treatment factors EC, SD, and SL on soluble sugars (SSg). Regression analyses showed that EC, SD, and SL differently affected the metabolism of soluble sugars during shelf life, with fits of linear, cubic, and cubic models, respectively (Figure 3).

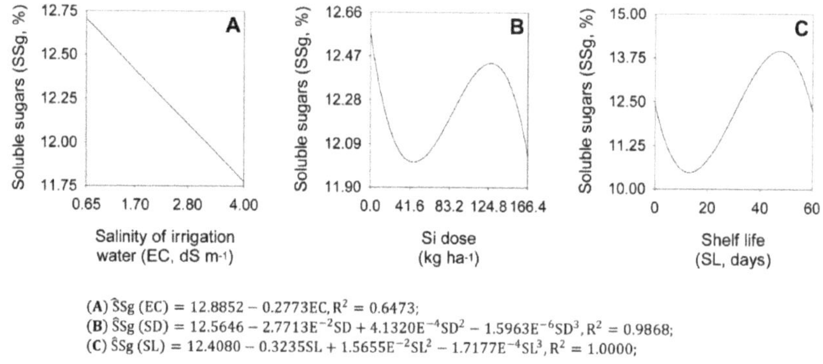

(**A**) $\hat{S}Sg$ (EC) $= 12.8852 - 0.2773EC, R^2 = 0.6473;$
(**B**) $\hat{S}Sg$ (SD) $= 12.5646 - 2.7713E^{-2}SD + 4.1320E^{-4}SD^2 - 1.5963E^{-6}SD^3, R^2 = 0.9868;$
(**C**) $\hat{S}Sg$ (SL) $= 12.4080 - 0.3235SL + 1.5655E^{-2}SL^2 - 1.7177E^{-4}SL^3, R^2 = 1.0000;$

Figure 3. Effect of salinity levels (EC) (**A**), silicon doses (SD) (**B**), and shelf life (SL) (**C**) on total soluble sugars (SSg) in 'Rio das Antas' onion bulbs.

Salinity decreases onion bulb SSg content during shelf life. Starting from 12.89% SSg, the increase in salinity caused a decrease at a rate of 2.15% for each EC unit added to the irrigation water (Figure 3A). In Si fertilization, the minimum SSg content was 12.01%, obtained with a Si dose of 45.6 kg ha^{-1}, while the maximum was 12.44%, obtained with a Si dose of 127.0 kg ha^{-1} (Figure 3B). In shelf life, the minimum SSg content was 10.47% at 13.2 shelf-life days, while the maximum was 13.95% at 47.6 days (Figure 3C).

Figure 4 shows the main effect of the EC, SD, and SL on the total soluble solids (SS). The salinity of irrigation water decreased onion bulb SS content during shelf life. Based on the initial SS content of 6.47 °Brix, there was a decrease of 1.09% for each EC unit added to the irrigation water (Figure 4A). However, Si fertilization caused cubic variation in the onion bulb SS content during shelf life, being: 6.40 °Brix, in the control treatment; 6.24 °Brix, with the Si dose of 40.4 kg ha^{-1}; 6.40 °Brix, with the Si dose of 121.8 kg ha^{-1}; and 6.20 °Brix, with the Si dose of 166.4 kg ha^{-1} (Figure 4B). For shelf life, the increase in the exposure time of the bulbs caused cubic variation in SS, being: 7.00 °Brix, at the beginning of shelf life; 5.99 °Brix, at 16.7 shelf days; 6.62 °Brix, at 43.7 shelf days; and 5.65 °Brix, at 60 shelf days (Figure 4C).

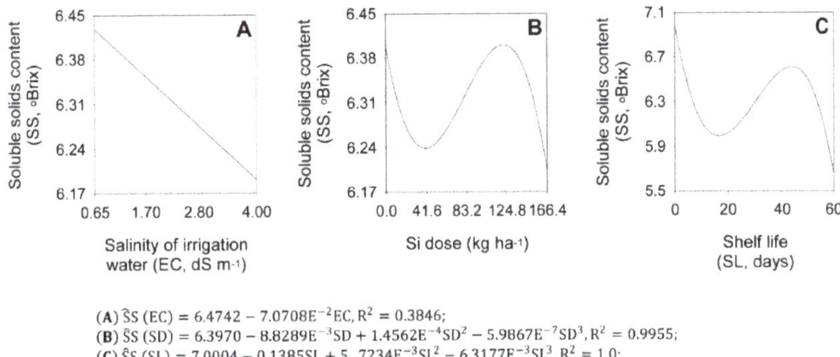

(**A**) $\hat{S}S$ (EC) = 6.4742 − 7.0708E^{-2}EC, R^2 = 0.3846;
(**B**) $\hat{S}S$ (SD) = 6.3970 − 8.8289E^{-3}SD + 1.4562E^{-4}SD^2 − 5.9867E^{-7}SD^3, R^2 = 0.9955;
(**C**) $\hat{S}S$ (SL) = 7.0004 − 0.1385SL + 5 .7234E^{-3}SL^2 − 6.3177E^{-3}SL^3, R^2 = 1.0;

Figure 4. Effect of salinity levels (EC) (**A**), silicon doses (Si) (**B**), and shelf life (SL) (**C**) on soluble solids content (SS) variable in 'Rio das Antas' onion bulbs.

2.4. pH, Titratable Acidity, and SS/TA Ratio in Bulbs

Figure 5 shows pH responses to EC × SL and SD × SL interactions. To EC × SL interaction, the basal pH was 5.09. EC increase caused a pH decrease of 0.2336% for each EC unit. SL increase caused a pH increase of 0.0412% for each shelf-life day (Figure 5A). For SD × SL interaction, the basal pH was 5.07. SD did not change pH significantly, but the SL increase caused an increase of 0.04114% for each shelf-life day (Figure 5B).

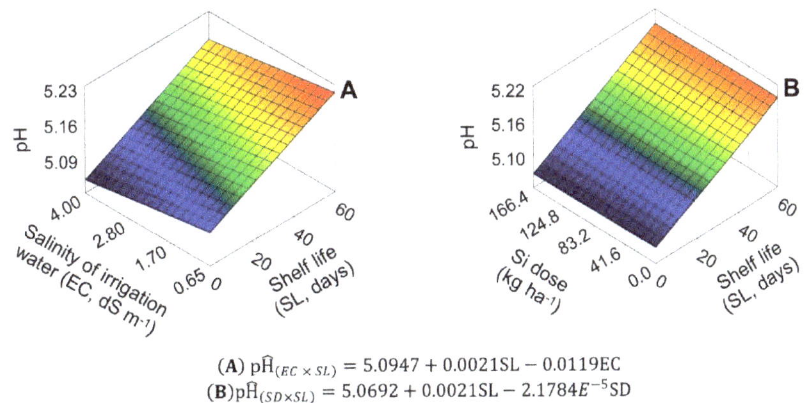

(**A**) pĤ$_{(EC \times SL)}$ = 5.0947 + 0.0021SL − 0.0119EC
(**B**) pĤ$_{(SD \times SL)}$ = 5.0692 + 0.0021SL − 2.1784E^{-5}SD

Figure 5. Unfolding of the treatment factors of salinity of irrigation water (EC) (**A**) and silicon dose (SD) (**B**), with shelf life (SL), for the hydrogenic potential (pH) variable in 'Rio da Antas' onion bulbs.

The variable TA and SS/TA ratio showed significant effects for the parameters related to EC and SL, but there was no parameterization for SD (Figures 6–8). For EC × SL interaction, basal TA was 3.01%. The salinity increase caused a TA increase of 3.57% for each EC unit increased, and the SL increase caused a TA decrease of 0.16% for each day of shelf life (Figure 6A). For SD × SL interaction, the initial TA was 3.28%. SD did not change TA significantly, but the SL increase caused a TA decrease of 0.15% for each shelf-life day (Figure 6B).

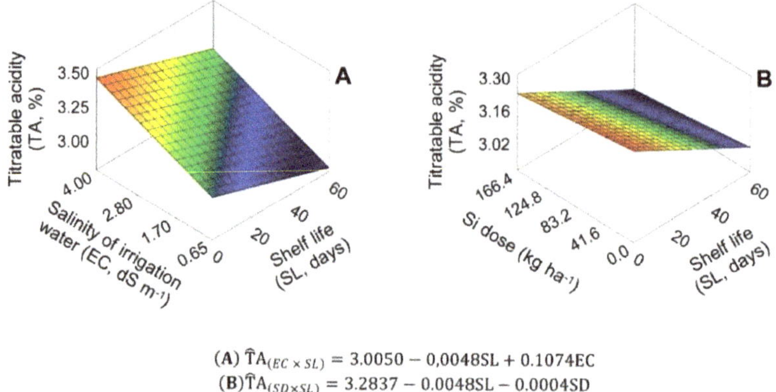

$$(\textbf{A}) \ \widehat{T}A_{(EC \times SL)} = 3.0050 - 0{,}0048SL + 0.1074EC$$
$$(\textbf{B}) \widehat{T}A_{(SD \times SL)} = 3.2837 - 0.0048SL - 0.0004SD$$

Figure 6. Unfolding of the treatment factors of salinity of irrigation water (EC) (**A**) and silicon dose (Si) (**B**), with shelf life (SL), for the variable titratable acidity (TA, %), in 'Rio da Antas' onion bulbs.

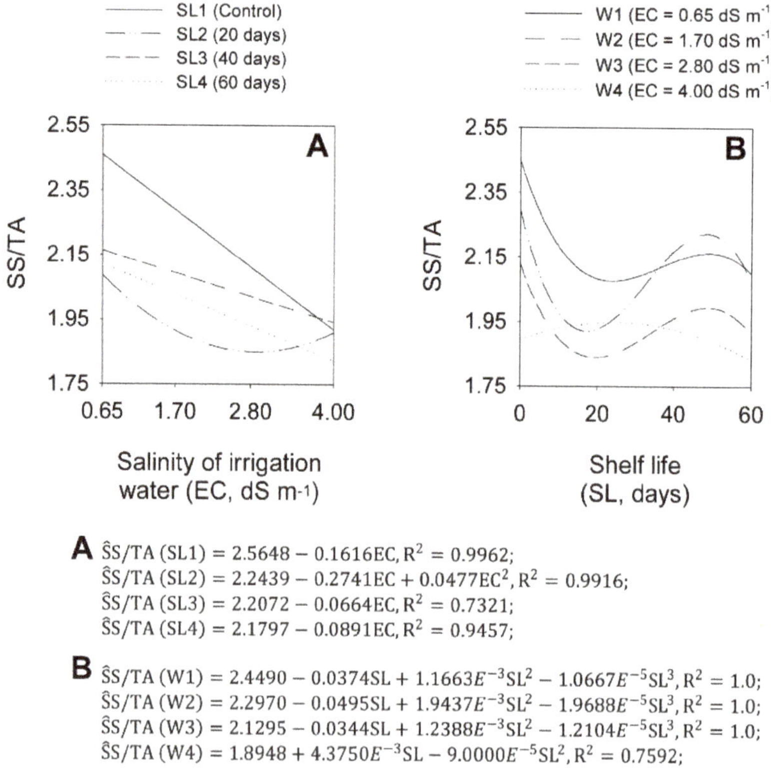

$$\textbf{A} \quad \widehat{S}S/TA \ (SL1) = 2.5648 - 0.1616EC, R^2 = 0.9962;$$
$$\widehat{S}S/TA \ (SL2) = 2.2439 - 0.2741EC + 0.0477EC^2, R^2 = 0.9916;$$
$$\widehat{S}S/TA \ (SL3) = 2.2072 - 0.0664EC, R^2 = 0.7321;$$
$$\widehat{S}S/TA \ (SL4) = 2.1797 - 0.0891EC, R^2 = 0.9457;$$

$$\textbf{B} \quad \widehat{S}S/TA \ (W1) = 2.4490 - 0.0374SL + 1.1663E^{-3}SL^2 - 1.0667E^{-5}SL^3, R^2 = 1.0;$$
$$\widehat{S}S/TA \ (W2) = 2.2970 - 0.0495SL + 1.9437E^{-3}SL^2 - 1.9688E^{-5}SL^3, R^2 = 1.0;$$
$$\widehat{S}S/TA \ (W3) = 2.1295 - 0.0344SL + 1.2388E^{-3}SL^2 - 1.2104E^{-5}SL^3, R^2 = 1.0;$$
$$\widehat{S}S/TA \ (W4) = 1.8948 + 4.3750E^{-3}SL - 9.0000E^{-5}SL^2, R^2 = 0.7592;$$

Figure 7. Effect of the interaction between salinity of irrigation water levels (EC) (**A**) and shelf life (SL) (**B**) on the variable SS/AT in 'Rio da Antas' onion bulbs.

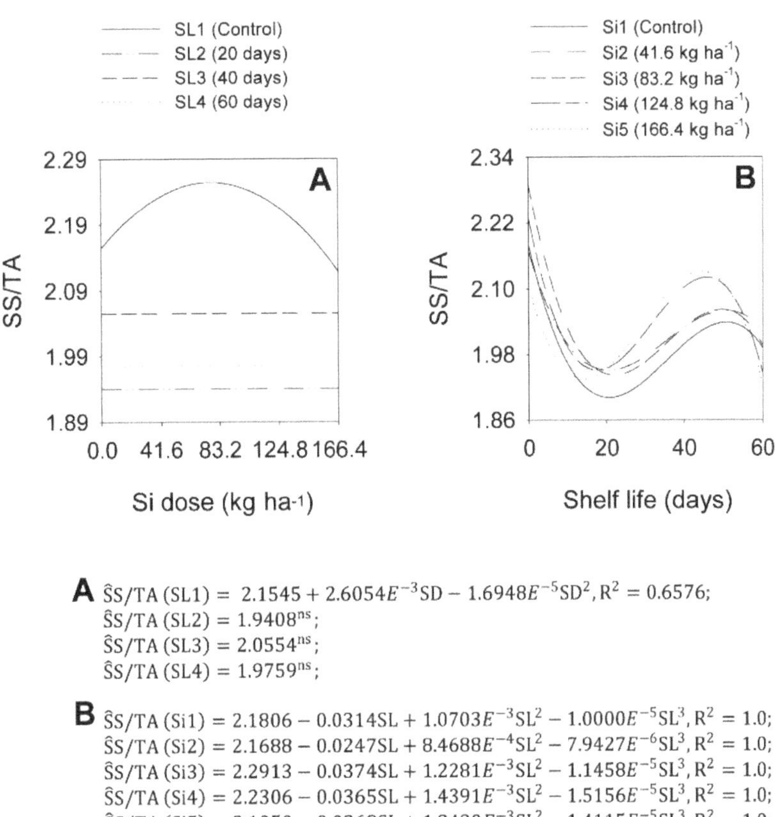

A $\hat{S}S/TA\,(SL1) = 2.1545 + 2.6054E^{-3}SD - 1.6948E^{-5}SD^{2}, R^{2} = 0.6576;$
$\hat{S}S/TA\,(SL2) = 1.9408^{ns};$
$\hat{S}S/TA\,(SL3) = 2.0554^{ns};$
$\hat{S}S/TA\,(SL4) = 1.9759^{ns};$

B $\hat{S}S/TA\,(Si1) = 2.1806 - 0.0314SL + 1.0703E^{-3}SL^{2} - 1.0000E^{-5}SL^{3}, R^{2} = 1.0;$
$\hat{S}S/TA\,(Si2) = 2.1688 - 0.0247SL + 8.4688E^{-4}SL^{2} - 7.9427E^{-6}SL^{3}, R^{2} = 1.0;$
$\hat{S}S/TA\,(Si3) = 2.2913 - 0.0374SL + 1.2281E^{-3}SL^{2} - 1.1458E^{-5}SL^{3}, R^{2} = 1.0;$
$\hat{S}S/TA\,(Si4) = 2.2306 - 0.0365SL + 1.4391E^{-3}SL^{2} - 1.5156E^{-5}SL^{3}, R^{2} = 1.0;$
$\hat{S}S/TA\,(Si5) = 2.1050 - 0.0268SL + 1.2430E^{-3}SL^{2} - 1.4115E^{-5}SL^{3}, R^{2} = 1.0;$

Figure 8. Effect of the interaction between Si dose levels (SD) (**A**) and shelf life (SL) (**B**) on the variable SS/AT in 'Rio da Antas' onion bulbs.

EC analyses at each SL showed that the salinity increase decreased the onion bulb SS/TA ratio, depending on shelf life. At SL1, starting from the initial SS/TA ratio of 2.56, each EC unit increase caused a linear decrease of 6.30%. At SL2, the minimum SS/TA ratio value of 1.85 occurred in the EC of 2.87 dS m^{-1}. At SL3 and SL4, each EC unit increase caused a linear decrease of 3.00 and 4.09%, starting from the initial SS/TA ratio of 2.21 and 2.18, respectively (Figure 7A).

SL analyses at each EC showed that the SL increase affected the onion bulb SS/TA ratio (Figure 7B). At salinity levels W1, W2, and W3, the SS/TA ratio was at its maximum at the beginning of the shelf period, being 2.45, 2.30, and 2.13, decreasing to 2.08, 1.92, and 1.84, at 23.8, 17.3 and 19.4 shelf life days, increasing to 2.16, 2.22 and 1.99 at 49.1, 48.6 and 48.8 days, and decreasing to minimum values of 2.1, 2.07 and 1.91 at 60 days, respectively. At W4, the initial SS/TA ratio was 1.89, reaching a maximum of 1.95 at 24.3 shelf-life days and decreasing to 1.83 at 60 days.

SD analyses at each SL showed that the SD increase effects on the onion bulb SS/TA ratio were significant at SL1 and not significant at SL2, SL3, and SL4. At SL1, the maximum SS/TA ratio value of 2.25 occurred at a Si dose of 76.9 kg ha^{-1}. At SL2, SL3, and SL4 levels, average SS/TA ratio values were 1.94, 2.06, and 1.98, respectively (Figure 8A). SL analyses at each SD showed that the SL increase decreased the onion bulb SS/TA ratio, depending on the Si dose (Figure 8B). At Si1, Si2, Si3, Si4, and Si5 levels, the SS/TA ratios were at a maximum at the beginning of the shelf life, being 2.18, 2.17, 2.29, 2.23, and 2.11, decreasing to 1.90, 1.95, 1.94, 1.95, and 1.93 at 20.6, 20.4, 21.5, 17.6, and 14.2 shelf life days, and increasing to

2.04, 2.06, 2.06, 2.12, and 2.13 at 50.7, 50.6, 49.5, 45.7, and 44.5 days, decreasing to minimum SS/TA ratio values of 1.99, 2.02, 2.00, 195, and 1.92 at 60 days, respectively.

2.5. Concentrations of Pyruvic and Ascorbic Acids

The PyA responses to EC $\times$ SL and DS $\times$ SL interaction were described by a three-dimensional parabolic model, with maximum PyA concentrations of 6.93 and 6.75 $\mu M\ g^{-1}$ of FM at EC and SD of 2.81 dS m^{-1} and 78.3 kg ha^{-1}, respectively, both at 22.1 shelf life days (Figure 9A,B).

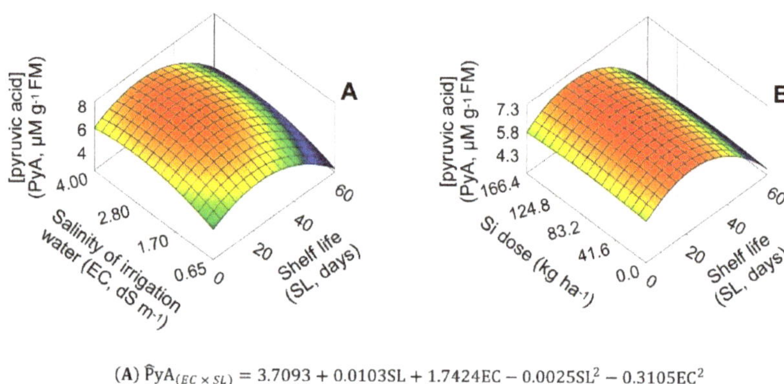

(A) $\hat{P}yA_{(EC \times SL)} = 3.7093 + 0.0103SL + 1.7424EC - 0.0025SL^2 - 0.3105EC^2$
(B) $\hat{P}yA_{(SD \times SL)} = 5.5206 + 0.1103SL + 0.0041SD - 0.025SL^2 - 2.6177E^{-5}SD^2$

Figure 9. Unfolding of the treatment factors of salinity of irrigation water (EC) (**A**) and silicon dose (SD) (**B**), with shelf life (SL), for the variable pyruvic acid concentration (PyA, $\mu M\ g^{-1}$ FM) in 'Rio da Antas' onion bulbs.

EC analyses at each SL showed that the EC increases increased onion bulb AsA concentration, depending on the onion shelf life (Figure 10A). At SL1, there was no significant effect of the increase in EC on onion bulb AsA concentration, which was 33.23 mg 100 g^{-1} of FM. At SL2, starting from the initial AsA of 42.21 mg 100 g^{-1} of FM, each EC unit increase caused a linear increase of 24.43% in the onion bulb AsA concentration. At SL3, starting from the initial AsA of 36.76 mg 100 g^{-1} of FM, each EC unit increase caused a linear increase of 4.99% in the onion bulb AsA concentration. At SL4, starting from the initial AsA of 54.44 mg 100 g^{-1} of FM, each EC unit increase caused a linear increase of 4.50% in the onion bulb AsA concentration.

SL analysis at each EC showed that the SL increases increased the onion bulb AsA concentration, depending on the salinity level (Figure 10B). At W1, starting from the initial AsA of 34.40 mg 100 g^{-1} of FM, the SL increase caused a linear increase of 0.91% in the concentration. At W2, the increase in EC caused a quadratic increase in AsA, which reached a maximum of 54.13 mg 100 g^{-1} of FM at 50 shelf-life days. At W3, the increase in EC caused a quadratic increase in AsA, with a maximum value of 57.59 mg 100 g^{-1} of FM at 47.7 shelf life days. At W4, the EC increase caused a quadratic increase in AsA, which reached a maximum AsA value of 67.63 mg 100 g^{-1} of FM at 36.4 shelf-life days.

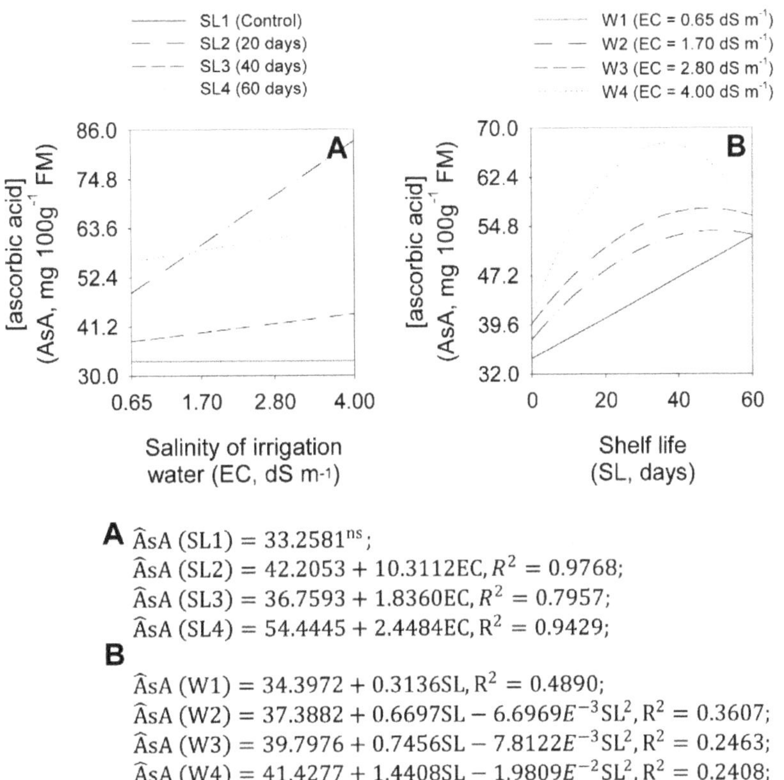

A $\hat{A}sA\ (SL1) = 33.2581^{ns};$

$\hat{A}sA\ (SL2) = 42.2053 + 10.3112EC, R^2 = 0.9768;$

$\hat{A}sA\ (SL3) = 36.7593 + 1.8360EC, R^2 = 0.7957;$

$\hat{A}sA\ (SL4) = 54.4445 + 2.4484EC, R^2 = 0.9429;$

B

$\hat{A}sA\ (W1) = 34.3972 + 0.3136SL, R^2 = 0.4890;$

$\hat{A}sA\ (W2) = 37.3882 + 0.6697SL - 6.6969E^{-3}SL^2, R^2 = 0.3607;$

$\hat{A}sA\ (W3) = 39.7976 + 0.7456SL - 7.8122E^{-3}SL^2, R^2 = 0.2463;$

$\hat{A}sA\ (W4) = 41.4277 + 1.4408SL - 1.9809E^{-2}SL^2, R^2 = 0.2408;$

Figure 10. Effect of the interaction between salinity of irrigation water levels (EC) (**A**) and shelf life (SL) (**B**) on ascorbic acid (AsA) concentration in 'Rio da Antas' onion bulbs.

SD analyses at each SL showed that the SD increases increased onion bulb AsA concentration, depending on the shelf life (Figure 11A). At SL1, there was no significant effect of SD increase on onion bulb AsA concentration, on average, 33.23 mg 100 g^{-1} of FM. At SL2, starting from the initial AsA of 61.38 mg 100 g^{-1} of FM, each SD increase caused a linear increase of 0.0863% in the onion bulb AsA concentration. At SL3, starting from the initial AsA of 38.45 mg 100 g^{-1} of FM, each SD increase caused a linear increase of 0.0785% in the onion bulb AsA concentration. At SL4, the SD increase increased onion bulb AsA concentration, up to a maximum AsA of 61.17 mg 100 g^{-1} of FM, with 71.3 kg ha^{-1}.

SL analysis at each SD showed that the SL increase quadratically increased the onion bulb AsA concentration, depending on the Si dose (Figure 11B). At Si1, the SL increase caused a quadratic increase in AsA, which reached a maximum AsA value of 53.56 mg 100 g^{-1} of FM at 57.5 shelf-life days. At Si2, the SL increase caused a quadratic increase in AsA, with a maximum AsA of 57.84 mg 100 g^{-1} of FM at 54.2 shelf-life days. At Si3, the SL increase caused a quadratic increase in AsA, with a maximum AsA of 57.60 mg 100 g^{-1} of FM, at 53.2 shelf-life days. At Si4, the SL increase caused a quadratic increase in AsA, with a maximum AsA of 55.95 mg 100 g^{-1} of FM at 41.8 shelf-life days. At Si5, the SL increase caused a quadratic increase in AsA, with a maximum AsA of 59.97 mg 100 g^{-1} of FM at 38.3 shelf-life days.

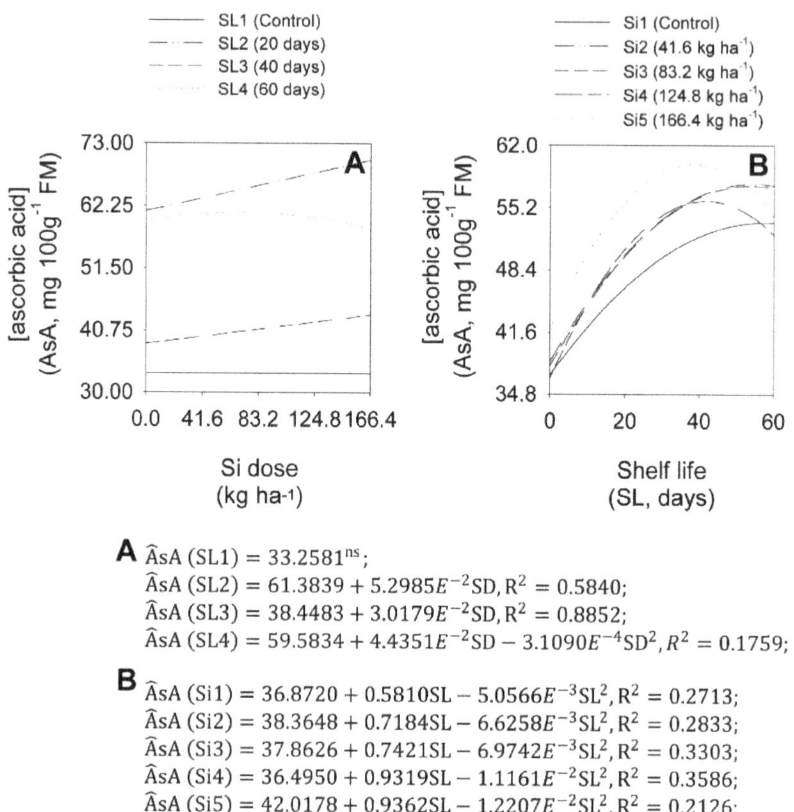

A $\hat{A}sA\ (SL1) = 33.2581^{ns};$

$\hat{A}sA\ (SL2) = 61.3839 + 5.2985E^{-2}SD, R^2 = 0.5840;$

$\hat{A}sA\ (SL3) = 38.4483 + 3.0179E^{-2}SD, R^2 = 0.8852;$

$\hat{A}sA\ (SL4) = 59.5834 + 4.4351E^{-2}SD - 3.1090E^{-4}SD^2, R^2 = 0.1759;$

B $\hat{A}sA\ (Si1) = 36.8720 + 0.5810SL - 5.0566E^{-3}SL^2, R^2 = 0.2713;$

$\hat{A}sA\ (Si2) = 38.3648 + 0.7184SL - 6.6258E^{-3}SL^2, R^2 = 0.2833;$

$\hat{A}sA\ (Si3) = 37.8626 + 0.7421SL - 6.9742E^{-3}SL^2, R^2 = 0.3303;$

$\hat{A}sA\ (Si4) = 36.4950 + 0.9319SL - 1.1161E^{-2}SL^2, R^2 = 0.3586;$

$\hat{A}sA\ (Si5) = 42.0178 + 0.9362SL - 1.2207E^{-2}SL^2, R^2 = 0.2126;$

Figure 11. Effect of the interaction between Si dose levels (SD) (**A**) and shelf life (SL) (**B**) on ascorbic acid concentration (AsA) in 'Rio da Antas' onion bulbs.

3. Discussion

Salt stress affects different parameters in onions. The literature shows that irrigated onion yield with 0.65 dS m^{-1} water is 99.6 Mg ha^{-1} and decreases to 64.4 Mg ha^{-1} when irrigated with 4.00 dS m^{-1}, representing a yield loss of 35.3% [17]. The authors showed that Si does not increase irrigated onion yield with 4.00 dS m^{-1} water but improves the irrigated onion yield with 1.7–2.8 dS m^{-1} water, mainly producing of Class 3 bulbs ($50 \leq \Phi < 70$ mm). Onions irrigated with 1.7 dS m^{-1} and fertilized with 166.4 kg ha^{-1} of Si produce 93.8 t ha^{-1}. However, onions irrigated with water of 2.8 dS m^{-1} only responded up to 78.5 kg ha^{-1} of Si with a production of 81.2 t ha^{-1} [17]. The onions fertilized with Si and irrigated with water of 1.7–2.8 dS m^{-1} decreased the yield by 5.8–18.5% [17]. The salinity of 1.7–4.0 dS m^{-1} reduced onion yield, but few studies verify the effect of salinity on onion shelf life. There is a need to evaluate salt stress effects on the metabolite production related to onion bulb quality and conservation, such as pH, ascorbic acid, sugars, soluble solids, and titratable acidity. We examined the effects of Si fertilization on the post-harvest quality and shelf life of onion bulbs grown under increasing salinity levels of irrigation water.

Our results showed that irrigation with saline water decreased the firmness of onion bulbs. Onion bulbs have high firmness at harvest because of the endogenous uronic acid concentrations and the relationship between total and water-soluble pectins [23]. However, salt stress caused oxidative damage to the aerial part of the onion and decreased the relative water content and membrane stability index in leaf and bulb tissues [1,24]. These changes could decrease the firmness levels of onion bulbs. Membrane stability loss promotes electrolyte leakage and oxidative and hydrolyzing reactions by contact between enzymes and substrates [9]. Low relative water content in bulb tissues causes pressure and intercellular cohesion loss, detaching the middle lamella from the cell walls by shear forces [25].

In storage, an increase in onion BF occurred after the 28th shelf day due to the elastic properties of the epidermal tissues of the bulb cataphylls [26], which may increase the resistance to penetration in onion bulbs due to the decrease in turgor cells resulting from the loss of moisture from the bulbs during their shelf life.

Mass loss (ML) in onions occurs due to transpiration water loss from the bulb tissues [27,28]. We observed that an increase in irrigation water salinity significantly increased the ML of 'Rio das Antas' onions, compromising its shelf life. Salinity-induced osmotic stress in the pre-harvest caused a water deficiency in onion bulbs after harvest, leading to an increase in the rate of cell membrane dehydration during shelf life [29]. In general, salinity increases the production of ROS in intra- and intercellular spaces, resulting in cell death, electrolyte leakage, senescence, and increased dehydration of plant tissues in fresh food products [29,30]. Although salinity increased the ML of onion bulbs during conservation, the maximum ML of 9.23%, reached with EC of 4.00 dS m^{-1} at the end of 60 days of storage, was slightly lower than the maximum tolerable mass loss of 10% before the onion is considered non-marketable [29]. Water loss reduces the shelf life of vegetables, revealing changes in qualitative characteristics related to texture, such as softening and visible wilting, which are essential in indicating the deterioration of fresh products [6,29,31].

Onion bulbs can be stored for long periods (up to 8 months) when kept under refrigeration conditions (temperature, 2 °C) and high relative humidity (RH, 98%) [7,28]. In this experiment, we verified that each conservation day increased by 0.1323% in ML under an average air temperature of 29.5 $\pm$ 0.7 °C and RH of 67 $\pm$ 5%. Since transpiration is directly proportional to the water vapor gradient between the surface of the plant and the surrounding air [29], storage environment conditions influenced the ML.

Si-accumulating plants deposit Si in the root endoderm, leaf epidermis, and leaf cuticle in the form of hydrated amorphous silica ($SiO_2.nH_2O$), close to the cell walls to form alternative polyphenolic complexes to lignin [32–34]. Thus, the remarkable properties of Si in plants confer its potential capacity to increase the resistance of their cell walls [20] and reduce the transpiration processes of the tissues by improving interfaces of resistance to transpiration close to the cuticular layers [31]. In this study, the Si increase in soil fertilization did not significantly alter the resistance (BF) and ML of onion bulbs. The beneficial effects of Si on shoot tissues may not occur in onion bulbs because shoots of this species fail to accumulate Si even at high levels in the soil [32]. Ref. [22] observed that the increase in Si fertilization did not significantly alter the concentrations of Si in the aerial part of onion plants and, according to [33], the processes of silicification of the cell walls in the aerial part of the plants depend on the species and are specific to each type of cellular tissue. So far, we only know that onions respond to fertilization with Si, increasing its content and improving the anatomy of the roots [32].

The L*, b*, and C* reductions indicate a darkening yellow color tone decrease and saturation loss of five onion bulbs along shelf life. On the other hand, the increase in a* and hue angle (°H) reduction indicates that the shelf time increased the red color tone of the bulbs. The b* and C* reductions in the bulbs suggest degradation of quercetin phenolic compounds during onion shelf life. Ref. [35] reported that increased concentrations of flavonoids such as quercetin 7,4-diglucoside, quercetin 3,4-diglucoside, quercetin 3-glucoside and quercetin 4-glucoside are responsible for conferring yellow color to the onion tunic. Quercetin compartmentalization in the transition zones between the living and dead cells of the epidermis suberization process and flavonol aglycone degradation occurs by a self-catalytic oxidation process triggered by the generation of radical molecules of quercetin and superoxide radical ($O_2{}^{\bullet-}$) [36].

The °H angle reduction and a* increase occur due to the onion tunic darkening observed during the bulb self-life. We found that the cured yellow onions show color changes in the tunic of the bulbs, from light whitish brown to dark reddish-brown, concomitantly with the gradual reduction in the hue angle (°H) [28]. Nonetheless, [37] found that the hue angle of the tunic is negatively correlated with the contents of anthocyanin and total flavonoids in red onions ('Red Baron'). However, this correlation fails to be observed in the yellow onion cultivars ('Sherpa' and 'Wellington') [37].

Salinity increases mitochondrial ROS production and the generation of free radical molecules in harvested products [30]. In contrast, the soluble sugar degradation can feed the NADPH-producing metabolic pathways, as in the oxidative way of phosphate pentoses, contributing to the elimination of ROS and regulation of cellular redox [38,39] in several pathways of the antioxidant defense system, such as the recovery of enzymatic cofactors of the Asada-Halliwell cycle and the feeding of the proline biosynthesis pathway [40–42]. Thus, higher soluble sugar consumption and production in plants grown under high salinity conditions occur due to their multifaceted function in the physiological responses of plants to abiotic stress [43]. After harvest, however, sink organs are disconnected from their respective sources of sugars and begin the processes of respiratory degradation of carbohydrates and senescence [44,45]. However, in shelf life, products grown under pre-harvest salinity conditions, in addition to exhibiting the natural catabolism of sugars in senescence and respiration processes, exhibit exacerbated energy consumption due to the regulation of osmoprotective metabolism, antioxidant (ROS eliminator) metabolism, and balance of cellular redox [12,30,39].

Carbohydrate metabolism responses to the nutritional Si supply vary widely with the species, genetic material within the species, specific Si nutritional habits, and environmental conditions [46–50]. Refs. [46,47] found that Si increased the concentration of soluble sugars in plants under salinity conditions, reducing cellular stresses caused by osmotic and ionic stresses. This effect occurs due to the lower catabolic rate of soluble sugars in plants treated with Si than in untreated plants. Although our results showed oscillation in SSg content as a function of Si doses.

The reduction of SSg in the first shelf-life days is possibly due to respiratory consumption and ethylene production, which may have been initially accelerated in response to stimuli related to the regeneration of lesions that occurred in the harvesting process [44,51]. In addition, the conditions of the atmosphere of the storage system (average air temperature of 29.5 ± 0.7 °C and relative humidity of 67 ± 5%) were favorable to trigger enzymatic activities and, consequently, cellular energy consumption [44], which possibly contributed to the higher initial consumption of SSg during the onion's shelf life.

The increase in the SSg content of bulbs between 13.2 and 47.6 shelf-life days may be related to increased activity of cellulase and enzymes that solubilize peptic substances, as well as other enzymes that degrade carbohydrates, converging to cellulose degradation and conversion into glucose [44]. Ref. [23] found a 40% reduction in cellulose concentrations and changes in the activities of wall-modifying enzymes (polygalacturonase and pectin methyl esterase) in onion strains under storage conditions, which was related to decreased cell wall resistance. In addition, there was a significant increase in soluble solids content concomitantly with the reduction of cellulose content in onion strain M87-WOPL, indicating probable degradation of cellulose into sugars in stored bulbs.

SS is the solid cellular compound found in a plant juice aliquot, detectable by diffractometry calibrated using a sucrose solution. Although there is a significant correlation between the obtained values and the sucrose solution, solids include several organic and inorganic components, such as carbohydrates, organic acids, proteins, fats, and minerals. Ref. [52] found that the increase in salinity in the cultivation solution did not cause significant changes in the SS content of onion bulbs after harvest. On the other hand, Ref. [53] found that the SS content of the bulbs increased quadratically as a function of decreasing periods of salt stress duration during the phenological cycle of the plant. Concerning fertilization with Si, Ref. [22] reported an increase in the SS content of bulbs after harvesting due to increased doses of Si in the field. On the other hand, in the post-harvest life, fluctuations in the SS content of onion bulbs vary considerably between genetic materials and according to factors related to technology and pre- and post-harvest handling [4,5,22,23,52–55]. Ref. [4] suggested that the variations in SS content in onion bulbs during post-harvest storage were due to fluctuations in respiratory rates and endogenous content of soluble sugars.

The increase in pH by increasing salinity occurred due to increased intra- and intercellular concentrations of cations, especially Na^+ [19]. In general, high cellular concentrations of cations trigger programs of synthesis for organic anions in the leaves for the buffering of excess cations absorbed by the roots [56]. In bulb formation, cations can be translocated from the leaves to the bulbs in ionic pairs with synthesized anions (especially malate), causing the pH reduction. On the other hand, the increase in bulb pH by the effect of shelf time is possibly due to the respiratory catabolism of organic acids using terminal oxidation to CO_2 and H_2O [51].

The increase in TA by increasing salinity occurred due to increased vacuolar concentrations of organic acids since only this analysis method identifies protonated forms of organic acids [56]. Ref. [57] reported that photorespiratory conditions and a high level of reducing pressure in the photosynthesis electron transfer system indicate the partial flow of the tricarboxylic acid (TCA) cycle, inducing the synthesis of organic acids and the consequent export of these acids from mitochondria to the vacuoles. Organic acids can supply electrons to the mitochondrial electron transport chain or can be accumulated for long periods as osmolytes or provide redox energy when needed [58]. Organic acids exhibit greater metabolic flexibility than large coenzymes, such as NADH and NADPH, and can transfer electrons and protons through membranes to other compartments.

In shelf life, the TA reduction in onion bulbs occurred due to the respiratory catabolism of organic acids to obtain the energy necessary for the processes related to the ripening and senescence of the bulbs using terminal oxidation of organic acids to CO_2 and H_2O, which also corroborates the increase in pH [51].

By increasing irrigation water salinity, the onion bulbs SS/TA ratio reduction occurred due to more soluble sugar catabolism than total titratable acids. In general, onion bulbs prefer to use soluble sugars as energy reserve forms to meet the higher ATP requirements, maintenance of redox balance, and enzymatic detoxification of lipid radical molecules and ROS caused by salt stress [12,30,38,39,45]. However, from the initial zero shelf-time (ST1) level to the subsequent levels ST2, ST3, and ST4, saline stress decreased rates of SS/AT ratio reduction, showing an increase in organic acid contribution to respiratory and redox metabolism of onion bulbs during their shelf life. The literature indicates the participation of organic acids in respiratory processes and mitochondrial redox regulation [51,57,58].

In the first shelf-life days, the salinity levels A2 and A3 led to a lower SS/TA ratio compared to the values observed at the control salinity level A1, indicating higher sugar consumption compared to titratable acids in these treatments. At the salinity level A4, we observed values lower than the control. The shelf time caused an increase in the SS/TA ratio of the onion, suggesting a preponderance in the titratable organic acid consumption compared to the sugars that make up the SS under high salinity. Therefore, these results show the organic acid consumption regulating the cellular redox of onion bulbs under salt stress.

The increase in the SS/TA ratio of the bulbs, as a function of Si supply, under salinity conditions, at ST1 possibly occurred because of improvements in an onion plants' carbohydrate metabolism and osmotic adjustment. Previous studies have reported an increase in the soluble sugar concentration and osmotic regulation in plants treated with Si under salt and water stresses [46,49,50]. However, our results reveal that Si supply effect on the SS/TA ratio of onion bulbs occurred exclusively at ST1 (harvest). However, the absence of a significant effect of Si on the SS/TA ratio at ST2, ST3, and ST4 leads us to speculate on a possible transient accumulation mechanism of soluble sugars mediated by Si in onion bulbs for osmotic regulation of plants during the reproductive phenological period. Interestingly, [49] found that the increase in the soluble sugar concentration in cucumber plants occurred in the root system, accompanied by positive regulation of the expression of the aquaporin gene mediated by Si. However, further studies are needed to fully understand the effect of Si on the mechanism regulation related to sugar metabolism and osmotic control in onions pre- and post-harvest.

The increases in PyA in onion bulbs after harvest show significant changes in sulfur (S) metabolism in onion plants due to pre-harvest conditions of increases in salinity and Si fertilization. The modulation of activities and gene expression of rate-controlling enzymes in the metabolism of assimilation and absorption of S in plants by salinity stress have been previously reported [59]. We identified that Si supply alters the metabolism of S in different plant species, causing increases in S absorption and regulating the synthesis of amino acids and polyamines involved in stress response and tolerance [60,61]. Antioxidant defense related to the AsA-GSH cycle depends on the metabolic pathways of S [62]. Molecular domains containing thiol are, directly or indirectly, oxidized by ROS, generating relatively more stable oxidation products with modified physical conformations and biochemical activities. In addition, oxidized S-cysteine (S-Cys) chemical species, including sulfenic acid, glutathionylated Cys, sulfanilamide groups, and metal-sulfur bonds, are significant in redox signaling and regulation of ROS metabolism, causing direct effects on protein molecules, transcription factors and gene expression levels [12]. Interestingly, our results showed that salinity and Si supply significantly increased the concentration of sulfurous compounds in onion bulbs after harvest, up to a certain point, as shown by the increase in PyA concentration.

The increase of PyA in the first shelf life days and its subsequent reduction after 22.1 days may be due to the effects of concentration and degradation of thio-compounds that are precursors of pyruvic acid in onion bulbs, since we observed the following correlations: negative, between the PyA content and the mass loss of the bulbs; and positive, between the PCA content and bulb's content of SS and TA.

AsA is a non-enzymatic antioxidant and redox buffer of many biological processes, including photosynthesis, ROS detoxification, and elimination of radical molecules [12,63–69]. There is strong evidence that adverse environmental conditions regulate AsA biosynthesis, including salinity stress [15,65,66]. AsA can directly eliminate the molecules of $O_2^{\bullet-}$, $^{\bullet}OH$, $^1O_2$ or react with biologically generated radicals such as tocopheroxyl (that is, regeneration of tocopherols) and alkoxyl/peroxyl. In addition, AsA can reduce H_2O_2 molecules to H_2O, acting as a cofactor in Mehler's peroxidative reactions catalyzed by ascorbate peroxidase (APX). The AsA acts on the quenching of excess energy from the molecules of $^3Chl^*$ and $^1O_2$ as a cofactor in the catalytic reaction of violaxanthin de-epoxidase (VDE), which converts violaxanthin into zeaxanthin [70–74]. The reason for the absence of a significant response of AsA to the increase in salinity at ST1 was possibly its dilution in bulbs still bloated in the harvest phase, not allowing the detection of significant variances by the adopted test. However, the means obtained at ST1 showed a slight increase in AsA (not significant) as a function of the increase in salinity. At the same time, the concentrations of AsA observed at ST2, ST3, and ST4 showed linear increases. Nevertheless, AsA content showed a positive correlation with the mass loss of the bulbs.

Storage time seems to concentrate AsA in onion bulbs treated with Si in pre-harvest without irrigation water salinity. The AsA concentration effect was possibly due to the increase in the mass loss of bulbs during storage and occurred exclusively in the control treatment (A1). This response shows that the AsA pool produced was sufficient to support the reactions involved in the natural processes of ripening and senescence. However, despite the increases in the AsA reservoirs in A2, A3, and A4 treatments under salt stress, in the first days of storage, AsA degradation occurred at the end of the onion's shelf life. The increase in salinity reduced the shelf life of AsA, evidencing that salinity increased AsA consumption, possibly due to the increase of reactions related to the metabolism of cellular redox maintenance and catalysis of ROS and radical molecules generated biologically.

Studies conducted with *Acacia gerrardii* Benth and *Triticum aestivum*, under salinity stress and Si supply, showed significant improvements in AsA levels due to the supply of Si [75,76]. According to [76], the improvement in Si-mediated antioxidant defense occurs by modulation in enzymatic activity (including SOD, POD, CAT, APX, and GR) and biosynthesis of non-enzymatic antioxidants, including ascorbic acid, proline, and glycine betaine. We found that the increase in Si doses caused an increase in the AsA concentration of onion bulbs only at shelf times of 20, 40, and 60 days.

Although the AsA levels of onion bulbs, as a function of shelf time, reached higher peaks with the elevation of SD levels, we observed that the AsA peaks occurred at increasingly shorter shelf times as the SD levels increased. In many species, AsA regulates plant tolerance against factors of multiple abiotic stresses [65,77]. Ref. [65] reported that the redox state of ascorbate and the level of the apoplastic AsA pool affect the hormonal balance of plants, MAPK signaling cascades, and antioxidant enzymatic activities, which is therefore vital in the perception of environmental stress, redox homeostasis, and regulation of oxidative stress and plant physical-biochemical responses. However, the upstream mechanisms by which salinity and Si supply can regulate the activities and expression of the components of the antioxidant defense system in onions under storage conditions and the processes of senescence and ripening are not yet known. However, other studies, such as the description of proteomic and metabolomic profiles, gene identification, and PCR over time, are still needed to better understand the signaling mechanisms and responses to pre-harvest conditions of salt stress and Si supply in onion bulbs under storage conditions.

4. Materials and Methods

4.1. Location and Experimental Design

The experiment was conducted at the Rafael Fernandes Experimental Farm, belonging to the Federal Rural University of the Semi-Arid Region (UFERSA), located in the district of Alagoinha, Mossoró-RN, Brazil ($5°03'37''$ S; $37°23'50''$ W and altitude of 72 m). The region recorded an average annual rainfall of 625 mm, with a dry season between June and January. The average annual air temperature has been recorded at a minimum of $21.3\,°C$ and a maximum of $34.5\,°C$. The climate is classified as semi-arid (BSh), dry, and very hot, according to Köppen's classification system [78]. The soil in the experimental area is an ARGISSOL, with a loamy-clay-sandy texture (726 g kg^{-1} of sand, 48 g kg^{-1} of silt, and 226 g kg^{-1} of clay, in the diagnostic B horizon). Table 1 shows a physicochemical analysis of the soil in the 0–20 cm layer.

Table 1. Physicochemical soil of the experimental area in the 0–20 cm layer.

pH	Ca^{2+}	Mg^{2+}	K$^+$	Na$^+$	Al^{3+}	H$^+$+Al^{3+}	P	M.O.	Si	Clay	Silt	Sand
				cmol$_c$dm^{-3}			mg dm^{-3}	g kg^{-1}	mg dm^{-3}		g kg^{-1}	
7.30	1.60	0.41	0.14	0.07	0.00	0.44	31.08	4.65	0.10	29	21	950

pH in water (1:2.5); Ca^{2+}, Mg^{2+}, and Al^{3+}: 1 mol L^{-1} KCl extractor; K$^+$ and P: Mehlich^{-1} extractor; H$^+$+Al^{3+}: SMP extractor; M.O.: organic matter by Walkey–Black; Si: 0.05 mol L^{-1} acetic acid extractor.

The experimental design was in randomized blocks, arranged in a split-split plot scheme, with four replicates. The plots had four salinity levels of irrigation water (EC: 0.61; 1.74; 2.87, and 4.0 dS m^{-1}), and the subplots had five levels of fertilization with silicon (SD: 0; 41.6; 83.2; 124.8, and 166.4 kg ha^{-1} of Si) and sub-sub plots lots had four conservation times (0, 20, 40, and 60 days after harvest), totaling 320 experimental units. The sample consisted of five onion bulbs per experimental unit, totaling 1600 bulbs [Classes 3 ($50 \leq \Phi < 70$ mm) and 4 ($70 \leq \Phi < 90$ mm)]. The samples were labeled and taken to the laboratory to evaluate the post-harvest quality of the bulbs within their respective conservation times. Figure 12 shows the details of Si-fertilization and onion storage.

Figure 12. Si-fertilization (**A**) and onion storage (**B**,**C**).

4.2. Plant Material, Pre-Harvest Treatments, and Post-Harvest Storage

The experiment was carried out with onions (*Allium cepa* L.) sown on 16 July 2019. The genetic material used was the hybrid 'Rio das Antas', a common cultivar in the northeastern semi-arid region of Brazil. Cultivation, fertilization, and irrigation were performed according to local recommendations, crop needs, and availability of water and nutrients in the soil [79–82].

The treatments were of four salinity levels of irrigation water (0.61; 1.74, 2.87, and 4.0 dS m^{-1}) and five levels of fertilization with silicon (0; 41.6; 83.2, 124.8, and 166.4 kg ha^{-1} of Si), arranged in plots and subplots, respectively. Water salinity was elevated by the addition of sodium chloride, calcium chloride, and magnesium sulfate salts in a molar charge ratio of 7:2:1. Fertilization with Si was carried out at planting, with a fertilizer based on natural diatomaceous earth derived from the species Melosira granulata—Agrisilica^{TM} with 2 mm diameter (26% Si; 0.07% N; 0.02% P; 0.08% K; 0.09% S; 1.4% Ca; 1.1% Mg; 1.3% Fe; 219 mg kg^{-1} Mn; < 5 mg kg^{-1} B; 22 mg kg^{-1} Cu; 18 mg kg^{-1} Zn; and 2.1 mg kg^{-1} Mo), produced by Agripower Australia Limited.

Harvest was carried out on 16 December 2019, after a natural curing process of the bulbs, for 10 days, under field conditions. The beginning of the curing process was defined when 70% of the plant population was physiologically mature (leaves fallen over) when the suspension of the water supply by irrigation was also determined. The bulbs harvested were selected according to the class of commercial diameter, between 50 and 70 mm, and then placed on shelves, in polyethylene nets (10 mm mesh), for 0, 20, 40, and 60 days, under ambient storage conditions, at an average air temperature of 29.5 ± 0.7 °C and relative humidity of 67 ± 5%, were obtained by digital thermo-hygrometer (Jprolab®, São José dos Pinhais, Brazil).

4.3. Evaluated Characteristics

The morphological analyses of the plants were bulb firmness (BF), mass loss (ML), and color (CLR) of onion bulbs, determined at 0, 20, 40, and 60 days after harvest (DAH). BF and CLR analyses were performed separately in each of the bulbs in the sample, recording the mean of the five bulbs for each treatment replicate. ML was determined by percentage ratio, recording the initial and final weights of each sample set of five bulbs.

The post-harvest quality of the bulbs was evaluated by determining the levels of soluble sugar (SSg), total soluble solids (SS), total titratable acidity (TA), and SS/TA ratio, as well as pH and the concentrations of pyruvic (PyA) and ascorbic (AsA) acids. The evaluations were performed in homogenized juice from the five onion bulbs of the sample, obtained using a centrifugal juice extractor 700 W with stainless steel blades (Philips Walita®, Brazil).

4.3.1. Evaluation of Bulb Firmness and Mass Loss

Bulb firmness, measured in Newton force (N), was evaluated using a penetrometer (Lutron® PTR-300, Taiwan), with a tip of 8 mm in diameter and penetration at a depth of 7 mm. The readings were performed in the middle equatorial portion of the bulbs at two equidistant points, on opposite sides, after removing the dry tunic from the bulbs.

Mass loss was obtained using the percentage ratio between the weight of onion bulbs at the initial shelf time (0 DAH) and the weight of the bulbs at their respective shelf times, at the moment of evaluation, according to the following equation:

$$ML\ (\%) = \left(\frac{M_i}{M_f} - 1 \right) \times 100 \times (-1) \tag{1}$$

where;

ML is the mass loss, in %;

M_i is the mass of fresh matter of the sample at the beginning of storage, in g; and

M_f is the mass of fresh matter of the sample at the evaluated shelf time, in g.

4.3.2. Evaluation of Bulb Color

Bulb color was evaluated from the color sensation recommended by the International Lighting Commission (*Commission Internationale de l'Eclairage*—CIE), quantified by the determination of color parameters L* (lightness or brightness), a* (color variation between green and red), b* (color variation between blue and yellow), C* (chromaticity or saturation) and °H (hue or hue angle). The CIE coordinates L*, a*, and b* were obtained by a colorimeter (Konica Minolta® CR-410, Japan) (variation of Y reflectance: from 0.01% to 160.00%), with a silicon photocell detector and light source in the form of a xenon flashtube, using a wide-area illumination color measuring head, reflecting directly in the upper polar region of the onion bulbs, covering a circular area of 50 mm in diameter. The standard observer corresponds to Standard 2° CIE 1931 ($\bar{x}2\lambda$, $\bar{y}\lambda$, $\bar{y}\lambda$).

The CIE color spaces C* and H* were calculated from the CIE coordinates L*, a*, and b* using the following mathematical applications [83]:

$$C^* = \sqrt{a^{*2} + b^{*2}} \tag{2}$$

$$h^* = 180 + \left[\frac{\left(\arctan\frac{b^*}{a^*} \right)}{6.2832} \right] \times 360, \text{ when } a^* < 0 \tag{3}$$

$$h^* = \left[\frac{\left(\arctan\frac{b^*}{a^*} \right)}{6.2832} \right] \times 360, \text{ when } a^* > 0 \tag{4}$$

4.3.3. Evaluation of Soluble Sugars (SSg), Soluble Solids (SS), Titratable Acidity (TA), SS/TA Ratio, and Hydrogen Potential (pH)

The soluble sugar (SSg) and titratable acidity (TA) content were evaluated in onion bulbs according to the methods described by [84,85], respectively. SSg was extracted from 1 g samples of juice from fresh bulbs and dissolved in 100 mL of pure demineralized water. The determinations were carried out by colorimetry from 50 μL aliquots taken from the extracts. For the readings, the extract aliquots (50 μL) were mixed with 950 μL of pure water and 2000 μL of anthrone reagent (2 g L^{-1} of H_2SO^4), heated (water bath at 100 °C for 8 min), and cooled (ice bath). The readings were recorded in the spectral range of 620 nm, and SSg content was calculated from a standard glucose curve. TA was determined from 1 g samples of onion bulb juice dissolved in 50 mL of pure demineralized water by titration with 0.1 N NaOH and phenolphthalein indicator at 1%.

The soluble solids (SS) content and the hydrogen potential (pH) of onion bulbs were determined directly in the juice using a portable digital Brix refractometer (Dbr-92, China), converted to the Brix scale (%), and an electrical potentiometer (mV), converted to the pH scale. The device used to measure SS was the DBR45 digital refractometer (refractive index of 1.3330–1.4098), with automatic temperature compensation, while the device used to measure pH was the benchtop pH meter (Hanna® Instruments HI 2221, United States), with HI-1131B pH electrode and HI 7662 automatic temperature compensation probe.

SS/TA ratio was obtained through the mathematical ratio between the observed values of SS and TA.

4.3.4. Evaluation of Pyruvic Acid and Ascorbic Acid Concentrations

Pyruvic acid (PyA) and ascorbic acid (AsA) concentrations were evaluated in onion bulbs according to the methods described by [86,87], respectively. PyA was extracted from 0.5 g samples of juice from fresh bulbs, dissolved in 1.5 mL of 5% trichloroacetic acid (TCA) and 18 mL of pure demineralized water. The determinations were carried out by colorimetry, using 1 mL aliquots taken from the extracts. For readings, the extract aliquots (1 mL) were mixed with 1 mL of 2,4-dinitrophenylhydrazine (2.4-DNPH), 0.125 g L^{-1} of 2 N HCl, and 1 mL of pure water and then heated (water bath at 37 °C for 10 min), cooled (ice bath) and mixed with 5 mL of 0.6 N NaOH. Readings were recorded in the spectral range of 420 nm, and the PyA content was calculated from a standard sodium pyruvate curve. AsA was extracted from 1 g samples of onion bulb juice, dissolved in 50 mL of 0.5% oxalic acid. The determinations were carried out by titration with Tillman solution (2,6-dichlorophenol indophenol) at 0.02% (refrigerated) from 5 mL of the extract dissolved in a volume of 50 mL based on pure demineralized water. The content was calculated by proportional ratios based on the titration of a standard ascorbic acid solution.

4.4. Statistical Analysis

The results obtained were subjected to analysis of variance and polynomial regression tests, and then polynomial regression equations were fitted using linear and nonlinear regression models. The criteria for selecting the models were biological meaning, the significance of the estimators of regression parameters, and R^2 values.

5. Conclusions

The present research results show that the increase in salinity of irrigation water and shelf time reduces the quality of onion bulbs, causing reduced firmness and significant mass loss. At the same time, fertilization with Si did not contribute significantly to improving these variables. On the other hand, salinity and Si do not considerably affect the tunic color of onion bulbs, showing that onions grown under such environmental conditions have appeared for sale. In addition, our studies have shown that increased salinity causes a reduction in the content of sugars and total soluble solids of onion bulbs, as well as pH and SS/TA ratios, increasing the concentrations of titratable acids and pyruvic acid along with shelf life. Onion bulbs grown under salinity conditions may exhibit a more astringent and acidic flavor to the consumer, especially in prolonged permanence under shelf conditions.

In contrast, the increase in Si fertilization with doses between 121.8 and 127.0 kg ha^{-1} contributes to increasing the content of sugars and total soluble solids and reducing the SS/TA ratio, especially at 20 shelf-life days, with only a slight increase in pyruvic acid concentration. Si fertilization can improve the flavor qualities of onions grown under salinity, promoting a better balance between the astringency and sweetness of bulbs after cooking or improving the aromatic background of salads and culinary dishes by increasing pungency expression. Although salinity results in losses of palatable quality in onion bulbs, it can promote the increased concentration of ascorbic acid during shelf life until approximately 36 days of storage. Additionally, the increase in fertilization with Si, up to the maximum dose of 166.4 kg ha^{-1}, also promotes an increased concentration of ascorbic acid in bulbs during shelf life until approximately 40 days of storage. Thus, the increase in irrigation water salinity and the management of Si fertilization promotes the biofortification of onions with vitamin C. Therefore, our results can help define guidelines for fertilization with Si in the cultivation of onions under salinity conditions, focusing on developing new markets based on the quality of products and the quality of life for consumers.

Author Contributions: Conceptualization, methodology, statistical analysis, investigation, data curation, writing—original draft preparation, graphic arts, J.B.V.; conceptualization, resources and writing—review and editing, N.d.S.D.; methodology and resources, J.F.d.M.; conceptualization and laboratory resources, P.L.D.d.M.; conceptualization and field resources, C.W.A.d.N.; conceptualization and resources, O.N.d.S.N.; redaction and writing—review and editing, L.M.d.A., K.T.O.P., T.D.C.P., J.L.A.R., M.F.N. and F.V.d.S.S. All authors have read and agreed to the published version of the manuscript.

Funding: This research was funded by Coordenação de Aperfeiçoamento de Pessoal de Nível Superior, grant number 001.

Data Availability Statement: All data are presented in the paper.

Conflicts of Interest: The authors declare no conflict of interest.

References

1. Rady, M.O.A.; Semida, W.M.; Abd El-Mageed, T.A.; Hemida, K.A.; Rady, M.M. Up-regulation of antioxidative defense systems by glycine betaine foliar application in onion plants confer tolerance to salinity stress. *Sci. Hortic.* **2018**, *240*, 614–622. [CrossRef]
2. Ricciardi, L.; Mazzeo, R.; Marcotrigiano, A.R.; Rainaldi, G.; Iovieno, P.; Zonno, V.; Pavan, S.; Lotti, C. Assessment of Genetic Diversity of the "Acquaviva Red Onion" (*Allium cepa* L.) Apulian Landrace. *Plants* **2020**, *9*, 260. [CrossRef] [PubMed]
3. Ohanenye, I.C.; Alamar, M.C.; Thompson, A.J.; Terry, L.A. Fructans redistribution prior to sprouting in stored onion bulbs is a potential marker for dormancy break. *Postharvest Biol. Technol.* **2019**, *149*, 221–234. [CrossRef] [PubMed]
4. Sharma, P.; Sharma, S.R.; Dhall, R.K.; Mittal, T.C. Effect of γ-radiation on post-harvest storage life and quality of onion bulb under ambient condition. *J. Food Sci. Technol.* **2020**, *57*, 2534–2544. [CrossRef]
5. Sharma, K.; Ko, E.Y.; Assefa, A.D.; Nile, S.H.; Park, S.W. A comparative study of anaerobic and aerobic decomposition of quercetin glucosides and sugars in onion at an ambient temperature. *Front. Life Sci.* **2015**, *8*, 117–123. [CrossRef]
6. Chávez-Mendonza, C.; Vega-García, M.O.; Guevara-Aguilar, A.; Sánchez, E.; Alvarado-González, M.; Flores-Córdova, M. Effect of prolonged storage in controlled atmospheres on the conservation of the onion (*Allium cepa* L.) quality. *Emir. J. Food Agric.* **2016**, *28*, 842–852. [CrossRef]
7. Islam, M.N.; Wang, A.; Pedersen, J.S.; Edelenbos, M. Microclimate Tools to Monitor Quality Changes in Stored Onions. *Acta Hortic.* **2017**, *1154*, 229–234. [CrossRef]
8. Dhumal, K.; Datir, S.; Pandey, R. Assessment of bulb pungency level in different Indian cultivars of onion (*Allium cepa* L.). *Food Chem.* **2007**, *100*, 1328–1330. [CrossRef]
9. Chitarra, M.I.F.; Chitarra, A.B. *Pós-Colheita de Frutas e Hortaliças: Fisiologia e Manuseio*, 2nd ed.; UFLA: Lavras, Brazil, 2005.
10. Valverde-Miranda, D.; Díaz-Pérez, M.; Gómez-Galán, M.; Callejón-Ferre, Á.J. Total soluble solids and dry matter of cucumber as indicators of shelf life. *Postharvest Biol. Technol.* **2021**, *180*, 111603. [CrossRef]
11. Hasegawa, P.M. Sodium (Na$^+$) homeostasis and salt tolerance of plants. *Environ. Exp. Bot.* **2013**, *92*, 19–31. [CrossRef]
12. Foyer, C.H.; Noctor, G. Redox homeostasis and antioxidant signaling: A metabolic interface between stress perception and physiological responses. *Plant Cell* **2005**, *17*, 1866–1875. [CrossRef] [PubMed]
13. Ahmad, P.; Ahanger, M.A.; Alyemeni, M.N.; Wijaya, L.; Alam, P.; Ashraf, M. Mitigation of sodium chloride toxicity in *Solanum lycopersicum* L. by supplementation of jasmonic acid and nitric oxide. *J. Plant Interact.* **2018**, *13*, 64–72. [CrossRef]
14. Takahashi, Y.; Tahara, M.; Yamada, Y.; Mitsudomi, Y.; Koga, K. Characterization of the polyamine biosynthetic pathways and salt stress response in *Brachypodium distachyon*. *J. Plant Growth Regul.* **2018**, *37*, 625–634. [CrossRef]
15. Wang, Y.; Zhao, H.; Qin, H.; Li, Z.; Liu, H.; Wang, J.; Zhang, H.; Quan, R.; Huang, R.; Zhang, Z. The synthesis of ascorbic acid in rice roots plays an important role in the salt tolerance of rice by scavenging ROS. *Int. J. Mol. Sci.* **2018**, *19*, 3347. [CrossRef] [PubMed]
16. Chen, S.; Wu, F.; Li, Y.; Qian, Y.; Pan, X.; Li, F.; Wang, Y.; Wu, Z.; Fu, C.; Lin, H.; et al. NtMYB4 and NtCHS1 are critical factors in the regulation of flavonoid biosynthesis and are involved in salinity responsiveness. *Front. Plant Sci.* **2019**, *10*, 178. [CrossRef]
17. Venâncio, J.B.; Dias, N.S.; Medeiros, J.F.; Morais, P.L.D.; Nascimento, C.W.A.; Sousa Neto, O.N.; Sá, F.V.S. Yield and Morphophysiology of onion grown under salinity and fertilization with silicon. *Sci. Hortic.* **2022**, *301*, 111095. [CrossRef]
18. Amal, A.; Aly, A. Alteration of some secondary metabolites and enzymes activity by using exogenous antioxidant compound in onion plants growth under seawater salt stress. *Am.-Eurasian J. Sci. Res.* **2008**, *3*, 139–146.
19. Hanci, F.; Cebeci, E.; Uysal, E.; Dasgan, H. Effects of salt stress on some physiological parameters and mineral element contents of onion (*Allium cepa* L.) plants. *Acta Hortic.* **2016**, *1143*, 179–186. [CrossRef]
20. Rizwan, M.; Ali, S.; Ibrahim, M.; Farid, M.; Adrees, M.; Bharwana, S.A.; Zia-ur-Rehman, M.; Qayyum, M.F.; Abbas, F. Mechanisms of silicon-mediated alleviation of drought and salt stress in plants: A review. *Environ. Sci. Pollut. Res.* **2015**, *22*, 15416–15431. [CrossRef]

21. Etesami, H.; Jeong, B.R. Silicon (Si): Review and future prospects on the action mechanisms in alleviating biotic and abiotic stresses in plants. *Ecotoxicol. Environ. Saf.* **2018**, *147*, 881–896. [CrossRef]

22. Bybordi, A.; Saadat, S.; Zargaripour, P. The effect of zeolite, selenium and silicon on qualitative and quantitative traits of onion grown under salinity conditions. *Arch. Agron. Soil Sci.* **2018**, *64*, 520–530. [CrossRef]

23. Coolong, T.W.; Randle, W.M.; Wicker, L. Structural and chemical differences in the cell wall regions in relation to scale firmness of three onion (*Allium cepa* L.) selections at harvest and during storage. *J. Sci. Food Agric.* **2008**, *88*, 1277–1286. [CrossRef]

24. Semida, W.M.; Abdelkhalik, A.; Rady, M.O.A.; Marey, R.A.; Abd El-Mageed, T.A. Exogenously applied proline enhances growth and productivity of drought stressed onion by improving photosynthetic efficiency, water use efficiency and up-regulating osmoprotectants. *Sci. Hortic.* **2020**, *272*, 109580. [CrossRef]

25. Safar, A.; Wyatt, H.L.; Mihai, L.A. Debonding of cellular structures under shear deformation. In Proceedings of the 25th Conference of the UK Association for Computational Mechanics, University of Birmingham, Birmingham, UK, 11–13 April 2017.

26. Ng, A.; Parker, M.L.; Parr, A.J.; Saunders, P.K.; Smith, A.C.; Waldron, K.W. Physicochemical Characteristics of Onion (*Allium cepa* L.) Tissues. *J. Agric. Food Chem.* **2000**, *48*, 5612–5617. [CrossRef] [PubMed]

27. Maw, B.W.; Mullinix, B.G. Moisture loss of sweet onions during curing. *Postharvest Biol. Technol.* **2005**, *35*, 223–227. [CrossRef]

28. Eshel, D.; Teper-Bamnolker, P.; Vinokur, Y.; Saad, I.; Zutahy, Y.; Rodov, V. Fast curing: A method to improve postharvest quality of onions in hot climate harvest. *Postharvest Biol. Technol.* **2014**, *88*, 34–39. [CrossRef]

29. Díaz-Pérez, J.C. Transpiration. In *Postharvest Physiology and Biochemistry of Fruits and Vegetables*; Yahia, E.M., Ed.; Woodhead Publishing: Sawston, UK, 2018; pp. 157–173. [CrossRef]

30. Madani, B.; Mirshekari, A.; Imahori, Y. Physiological Responses to Stress. In *Postharvest Physiology and Biochemistry of Fruits and Vegetables*; Yahia, E.M., Ed.; Woodhead Publishing: Sawston, UK, 2019; pp. 405–423. [CrossRef]

31. Galindo, F.G.; Herppich, W.; Gekas, V.; Sjöholm, I. Factors Affecting Quality and Postharvest Properties of Vegetables: Integration of Water Relations and Metabolism. *Crit. Rev. Food Sci. Nutr.* **2004**, *44*, 139–154. [CrossRef]

32. Fleck, A.T.; Schulze, S.; Hinrichs, M.; Specht, A.; Waßmann, F.; Schreiber, L.; Schenk, M.K. Silicon Promotes Exodermal Casparian Band Formation in Si-Accumulating and Si-Excluding Species by Forming Phenol Complexes. *PLoS ONE* **2015**, *10*, e0138555. [CrossRef]

33. Kumar, S.; Soukup, M.; Elbaum, R. Silicification in grasses: Variation between different cell types. *Front. Plant Sci.* **2017**, *8*, 438. [CrossRef]

34. Ferrón-Carrillo, F.; Urrestarazu, M. Effects of Si in nutrient solution on leaf cuticles. *Sci. Hortic.* **2021**, *278*, 109863. [CrossRef]

35. Ko, E.Y.; Nile, S.H.; Sharma, K.; Li, G.H.; Park, S.W. Effect of different exposed lights on quercetin and quercetin glucoside content in onion (*Allium cepa* L.). *Saudi J. Biol. Sci.* **2015**, *22*, 398–403. [CrossRef] [PubMed]

36. Takahama, U. Oxidation of vacuolar and apoplastic phenolic substrates by peroxidase: Physiological significance of the oxidation reactions. *Phytochem. Rev.* **2004**, *3*, 207–219. [CrossRef]

37. Downes, K.; Chope, G.A.; Terry, L.A. Effect of curing at different temperatures on biochemical composition of onion (*Allium cepa* L.) skin from three freshly cured and cold stored UK-grown onion cultivars. *Postharvest Biol. Technol.* **2009**, *54*, 80–86. [CrossRef]

38. Hu, M.; Shi, Z.; Zhang, Z.; Zhang, Y.; Li, H. Effects of exogenous glucose on seed germination and antioxidant capacity in wheat seedlings under salt stress. *Plant Growth Regul.* **2012**, *68*, 177–188. [CrossRef]

39. Obata, T.; Geigenberger, P.; Fernie, A.R. Redox control of plant energy metabolism: The complex intertwined regulation of redox and metabolism in plant cells. *Biochemist* **2015**, *37*, 14–18. [CrossRef]

40. Meloni, D.A.; Oliva, M.A.; Martinez, C.A.; Cambraia, J. Photosynthesis and activity of superoxide dismutase, peroxidase and glutathione reductase in cotton under salt stress. *Environ. Exp. Bot.* **2003**, *49*, 69–76. [CrossRef]

41. Sharma, P.; Jha, A.B.; Dubey, R.S.; Pessarakli, M. Reactive oxygen species, oxidative damage, and antioxidative defense mechanism in plants under stressful conditions. *J. Bot.* **2012**, 17037. [CrossRef]

42. Khanna-Chopra, R.; Semwal, V.K.; Lakra, N.; Pareek, A. 5 proline—A key regulator conferring plant tolerance to salinity and drought. In *Plant Tolerance to Environmental Stress: Role of Phytoprotectants*; Hasanuzzaman, M., Fujita, M., Oku, H., Islam, T., Eds.; CRC Press Taylor & Francis Group: Boca Raton, FL, USA, 2019; pp. 59–80.

43. Sami, F.; Yusuf, M.; Faizan, M.; Faraz, A.; Hayat, S. Role of sugars under abiotic stress. *Plant Physiol. Biochem.* **2016**, *109*, 54–61. [CrossRef]

44. Yahia, E.M.; Carrillo-López, A.; Bello-Perez, L.A. Chapter 9–Carbohydrates. In *Postharvest Physiology and Biochemistry of Fruits and Vegetables*; Yahia, E.M., Ed.; Woodhead Publishing: Sawston, UK, 2019; pp. 175–205. [CrossRef]

45. Foyer, C.H.; Ruban, A.V.; Noctor, G. Viewing oxidative stress through the lens of oxidative signaling rather than damage. *Biochem. J.* **2017**, *474*, 877–883. [CrossRef]

46. Pei, Z.F.; Ming, D.F.; Liu, D.; Wan, G.L.; Geng, X.X.; Gong, H.J.; Zhou, W.J. Silicon Improves the Tolerance to Water-Deficit Stress Induced by Polyethylene Glycol in Wheat (*Triticum aestivum* L.) Seedlings. *J. Plant Growth Regul.* **2010**, *29*, 106–115. [CrossRef]

47. Yin, L.; Wang, S.; Li, J.; Tanaka, K.; Oka, M. Application of silicon improves salt tolerance through ameliorating osmotic and ionic stresses in the seedling of Sorghum bicolor. *Acta Physiol. Plant.* **2013**, *35*, 3099–3107. [CrossRef]

48. Zhu, Y.; Gong, H. Beneficial effects of silicon on salt and drought tolerance in plants. *Agron. Sustain. Dev.* **2014**, *34*, 455–472. [CrossRef]

49. Zhu, Y.-X.; Xu, X.-B.; Hu, Y.-H.; Han, W.-H.; Yin, J.-L.; Li, H.-L.; Gong, H.-J. Silicon improves salt tolerance by increasing root water uptake in *Cucumis sativus* L. *Plant Cell Rep.* **2015**, *34*, 1629–1646. [CrossRef] [PubMed]

50. Zhu, Y.-X.; Gong, H.-J.; Yin, J.-L. Role of Silicon in Mediating Salt Tolerance in Plants: A Review. *Plants* **2019**, *8*, 147. [CrossRef] [PubMed]

51. Saltveit, M.E. Chapter 4—Respiratory Metabolism. In *Postharvest Physiology and Biochemistry of Fruits and Vegetables*; Yahia, E.M., Ed.; Woodhead Publishing: Sawston, UK, 2019; pp. 73–91. [CrossRef]

52. Coca, C.A.; Carranza, C.E.; Miranda, D.; Rodríguez, M.H. Efecto del NaCl sobre los parámetros de crecimiento, rendimiento y calidad de la cebolla de bulbo (*Allium cepa* L.) bajo condiciones controladas (NaCl effects on growth, yield and quality parameters in the onion (*Allium cepa* L.) under controlled conditions). *Rev. Colomb. Ciências Horticolas* **2012**, *6*, 196–212. [CrossRef]

53. Chang, P.-T.; Randle, W.M. Sodium Chloride Timing and Length of Exposure Affect Onion Growth and Flavor. *J. Plant Nutr.* **2005**, *28*, 1755–1766. [CrossRef]

54. Chope, G.A.; Terry, L.A.; White, P.J. Effect of controlled atmosphere storage on abscisic acid concentration and other biochemical attributes of onion bulbs. *Postharvest Biol. Technol.* **2006**, *39*, 233–242. [CrossRef]

55. Sohany, M.; Sarker, M.K.U.; Mahmud, M.S. Physiological changes in red onion bulbs at different storage temperature. *World J. Eng. Technol.* **2016**, *4*, 261–266. [CrossRef]

56. Vallarino, J.G.; Osorio, S. Chapter 10—Organic Acids. In *Postharvest Physiology and Biochemistry of Fruits and Vegetables*; Yahia, E.M., Ed.; Woodhead Publishing: Sawston, UK, 2019; pp. 207–224. [CrossRef]

57. Igamberdiev, A.U.; Eprintsev, A.T. Organic Acids: The Pools of Fixed Carbon Involved in Redox Regulation and Energy Balance in Higher Plants. *Front. Plant Sci.* **2016**, *7*, 1042. [CrossRef]

58. Igamberdiev, A.U.; Bykova, N.V. Role of organic acids in integrating cellular redox metabolism and mediation of redox signalling in photosynthetic tissues of higher plants. *Free Radic. Biol. Med.* **2018**, *122*, 74–85. [CrossRef]

59. Khan, N.A.; Khan, M.I.R.; Asgher, M.; Fatma, M.; Masood, A.; Syeed, S. Salinity tolerance in plants: Revisiting the role of sulfur metabolites. *J. Plant Biochem. Physiol.* **2014**, *2*, 2–8. [CrossRef]

60. Ali, N.; Schwarzenberg, A.; Yvin, J.-C.; Hosseini, S.A. Regulatory Role of Silicon in Mediating Differential Stress Tolerance Responses in Two Contrasting Tomato Genotypes Under Osmotic Stress. *Front. Plant Sci.* **2018**, *9*, 1475. [CrossRef] [PubMed]

61. Maillard, A.; Ali, N.; Schwarzenberg, A.; Jamois, F.; Yvin, J.-C.; Hosseini, S.A. Silicon transcriptionally regulates sulfur and ABA metabolism and delays leaf senescence in barley under combined sulfur deficiency and osmotic stress. *Environ. Exp. Bot.* **2018**, *155*, 394–410. [CrossRef]

62. Réthoré, E.; Ali, N.; Yvin, J.-C.; Hosseini, S.A. Silicon Regulates Source to Sink Metabolic Homeostasis and Promotes Growth of Rice Plants under Sulfur Deficiency. *Int. J. Mol. Sci.* **2020**, *21*, 3677. [CrossRef] [PubMed]

63. Do, H.; Kim, I.-S.; Jeon, B.W.; Lee, C.W.; Park, A.K.; Wi, A.R.; Shin, S.C.; Park, H.; Kim, Y.-S.; Yoon, H.-S.; et al. Structural understanding of the recycling of oxidized ascorbate by dehydroascorbate reductase (OsDHAR) from *Oryza sativa* L. *japonica*. *Sci. Rep.* **2016**, *6*, 19498. [CrossRef] [PubMed]

64. Sharma, S.; Sehrawat, A.; Deswal, R. Asada-Halliwell pathway maintains redox status in Dioscorea alata tuber which helps in germination. *Plant Sci.* **2016**, *250*, 20–29. [CrossRef]

65. Akram, N.A.; Shafiq, F.; Ashraf, M. Ascorbic acid—A potential oxidant scavenger and its role in plant development and abiotic stress tolerance. *Front. Plant Sci.* **2017**, *8*, 613. [CrossRef]

66. Cao, S.; Du, X.H.; Li, L.H.; Liu, Y.D.; Zhang, L.; Pan, X.; Li, Y.; Li, H.; Lu, H. Overexpression of Populus tomentosa cytosolic ascorbate peroxidase enhances abiotic stress tolerance in tobacco plants. *Russ. J. Plant Physiol.* **2017**, *64*, 224–234. [CrossRef]

67. Sharma, S.; Kiran, U.; Sopory, S.K. Developing stress-tolerant plants by manipulating components involved in oxidative stress. In *Plant Biotechnology: Principles and Applications*; Abdin, M.Z., Kiran, U., Kamaluddin, A.A., Eds.; Springer: Singapore, 2017; pp. 233–248. [CrossRef]

68. Yoshimura, K.; Ishikawa, T. Chemistry and metabolism of ascorbic acid in plants. In *Ascorbic Acid in Plant Growth, Development and Stress Tolerance*; Hossain, M.A., Munné-Bosch, S., Burritt, D.J., Diaz-Vivancos, P., Fujita, M., Lorence, A., Eds.; Springer International Publishing: Cham, Switzerland, 2017; pp. 1–23. [CrossRef]

69. Ishikawa, T.; Maruta, T.; Yoshimura, K.; Smirnoff, N. Biosynthesis and regulation of ascorbic acid in plants. In *Antioxidants and Antioxidant Enzymes in Higher Plants*; Gupta, D.K., Palma, J.M., Corpas, F.J., Eds.; Springer International Publishing: Cham, Switzerland, 2018; pp. 163–179. [CrossRef]

70. Eskling, M.; Arvidsson, P.-O.; Åkerlund, H.-E. The xanthophyll cycle, its regulation and components. *Physiol. Plant.* **1997**, *100*, 806–816. [CrossRef]

71. Giorgia, S.; Giorgetti, A.; Fufezan, C.; Giacometti, G.M.; Bassi, R.; Morosinotto, T. Mutation analysis of violaxanthin de-epoxidase identifies substrate-binding sites and residues involved in catalysis. *J. Biol. Chem.* **2010**, *285*, 23763–23770. [CrossRef]

72. Chang, H.-Y.; Lin, S.-T.; Ko, T.-P.; Wu, S.-M.; Lin, T.-H.; Chang, Y.-C.; Huang, K.-F.; Lee, T.-M. Enzymatic characterization and crystal structure analysis of Chlamydomonas reinhardtii dehydroascorbate reductase and their implications for oxidative stress. *Plant Physiol. Biochem.* **2017**, *120*, 144–155. [CrossRef] [PubMed]

73. Smirnoff, N. Ascorbic acid metabolism and functions: A comparison of plants and mammals. *Free Radic. Biol. Med.* **2018**, *122*, 116–129. [CrossRef] [PubMed]

74. García-Caparrós, P.; Hasanuzzaman, M.; Lao, M.T. Oxidative Stress and Antioxidant Defense in Plants under Salinity. In *Reactive Oxygen, Nitrogen and Sulfur Species in Plants*; Hasanuzzaman, M., Fotopoulos, V., Nahar, K., Fujita, M., Eds.; John Wiley & Sons Ltd.: Hoboken, NJ, USA, 2019; pp. 291–309. [CrossRef]

75. Saqib, M.; Zörb, C.; Schubert, S. Silicon-mediated improvement in the salt resistance of wheat (*Triticum aestivum*) results from increased sodium exclusion and resistance to oxidative stress. *Fun. Plant Biol.* **2008**, *35*, 633–639. [CrossRef] [PubMed]
76. Al-Huqail, A.A.; Alqarawi, A.A.; Hashem, A.; Malik, J.A.; Abd_Allah, E.F. Silicon supplementation modulates antioxidant system and osmolyte accumulation to balance salt stress in *Acacia gerrardii* Benth. *Saudi J. Biol. Sci.* **2019**, *26*, 1856–1864. [CrossRef]
77. Farooq, A.; Bukhari, S.A.; Akram, N.A.; Ashraf, M.; Wijaya, L.; Alyemeni, M.N.; Ahmad, P. Exogenously Applied Ascorbic Acid-Mediated Changes in Osmoprotection and Oxidative Defense System Enhanced Water Stress Tolerance in Different Cultivars of Safflower (*Carthamus tinctorious* L.). *Plants* **2020**, *9*, 104. [CrossRef]
78. Carmo Filho, F.; Oliveira, O.F. *Mossoró: Um Município do Semi-Árido Nordestino, Caracterização Climática e Aspecto Florístico*; ESAM: Mossoró, Brazil, 1995; p. 62.
79. Chauhan, C.P.S.; Shisodia, P.K.; Minhas, P.S.; Chauhan, R.S. Response of onion (*Allium cepa*) and garlic (*Allium sativum*) to irrigation with different salinity waters with or without mitigating salinity stress at seedling establishment stage. *Indian J. Agric. Sci.* **2007**, *77*, 483–485.
80. Aguiar Neto, P.; Grangeiro, L.C.; Mendes, A.M.S.; Costa, N.D.; Marrocos, S.T.P.; Sousa, V.F.L. Growth and accumulation of macronutrients in onion crop in Baraúna (RN) and Petrolina (PE). *Rev. Bras. Eng. Agrícola E Ambient.* **2014**, *18*, 370–380. [CrossRef]
81. Behairy, A.G.; Mahmoud, A.R.; Shafeek, M.; Ali, A.H.; Hafez, M.M. Growth, yield and bulb quality of onion plants (*Allium cepa* L.) as affected by foliar and soil application of potassium. *Middle East J. Agric. Res.* **2015**, *4*, 60–66.
82. Marrocos, S.T.; Grangeiro, L.C.; Sousa, V.F.L.; Ribeiro, R.M.P.; Cordeiro, C.J. Potassium fertilization for optimization of onion production. *Rev. Caatinga* **2018**, *31*, 379–384. [CrossRef]
83. Petropoulos, S.A.; Fernandes, Â.; Barros, L.; Ferreira, I.C.F.R.; Ntatsi, G. Morphological, nutritional and chemical description of "Vatikiotiko", an onion local landrace from Greece. *Food Chem.* **2015**, *182*, 156–163. [CrossRef]
84. Yemm, E.W.; Willis, A.J. The estimation of carbohydrates in plant extracts by anthrone. *Biochem. J.* **1954**, *57*, 508–514. [CrossRef] [PubMed]
85. Instituto Adolfo Lutz. *Métodos Físico-Químicos Para Análise de Alimentos*, 4th ed.; Instituto Adolfo Lutz: São Paulo, Brasil, 2008.
86. Schwimmer, S.; Weston, W.J. Onion Flavor and Odor, Enzymatic Development of Pyruvic Acid in Onion as a Measure of Pungency. *J. Agric. Food Chem.* **1961**, *9*, 301–304. [CrossRef]
87. Strohecker, R.; Zaragoza, F.M.; Henning, H.M. *Analisis de Vitaminas: Métodos Comprovados*; Editorial Paz Montalvo: Madrid, Spain, 1967.

Article

Production and Physiological Quality of Seeds of Mini Watermelon Grown in Substrates with a Saline Nutrient Solution Prepared with Reject Brine

Tatianne Raianne Costa Alves [1], Salvador Barros Torres [1], Emanoela Pereira de Paiva [1], Roseane Rodrigues de Oliveira [1], Renata Ramayane Torquato Oliveira [1], Afonso Luiz Almeida Freires [1], Kleane Targino Oliveira Pereira [1], Douglas Leite de Brito [2], Charline Zaratin Alves [2], Alek Sandro Dutra [3], Clarisse Pereira Benedito [1], Alberto Soares de Melo [4], Miguel Ferreira-Neto [1], Nildo da Silva Dias [1] and Francisco Vanies da Silva Sá [1,*]

[1] Department of Agronomic and Forest Science, Federal Rural University of the Semi-Arid—UFERSA, Mossoró 59625-900, Brazil
[2] Department of Agronomy, Federal University of Mato Grosso do Sul—UFMS, Chapadão do Sul 79560-000, Brazil
[3] Department of Fitotecnia, Federal University of Ceará—UFC, Fortaleza 60356-001, Brazil
[4] Department of Biology, State University of Paraíba—UEPB, Campina Grande 58429-500, Brazil
* Correspondence: vanies_agronomia@hotmail.com or vanies@ufersa.edu.br; Tel.: +55-(83)9-9861-9267

Citation: Alves, T.R.C.; Torres, S.B.; de Paiva, E.P.; de Oliveira, R.R.; Oliveira, R.R.T.; Freires, A.L.A.; Pereira, K.T.O.; de Brito, D.L.; Alves, C.Z.; Dutra, A.S.; et al. Production and Physiological Quality of Seeds of Mini Watermelon Grown in Substrates with a Saline Nutrient Solution Prepared with Reject Brine. *Plants* 2022, 11, 2534. https://doi.org/10.3390/plants11192534

Academic Editors: Alberto Gianinetti and Pedro Diaz-Vivancos

Received: 26 August 2022
Accepted: 23 September 2022
Published: 27 September 2022

Publisher's Note: MDPI stays neutral with regard to jurisdictional claims in published maps and institutional affiliations.

Abstract: The economically profitable production of crops is related, among other factors, to seed quality, the production system, and the water used in irrigation or preparation of nutrient solutions. Therefore, the objective was to evaluate the phenology, production, and vigor of seeds of mini watermelons grown in saline nutrient solution and different substrates. In the fruit and seed production phase, the experiment occurred in a greenhouse with five electrical conductivities of water for nutrient solution preparation, ECw (0.5, 2.4, 4.0, 5.5, and 6.9 dS m^{-1}), and two growing substrates (coconut fiber and sand). We evaluated the physiological quality of seeds previously produced under the five electrical conductivities of water and two substrates. High salinities for the hydroponic cultivation of the mini watermelon cultivar 'Sugar Baby' accelerated fruit maturation and crop cycle, decreasing fruit size. However, in both substrates, the seed production of mini watermelons, seed viability, and seed vigor occurred adequately with a reject brine of 6.9 dS m^{-1} in the hydroponic nutrient solution. The seed production of 'Sugar Baby' mini watermelons using reject brine in a hydroponic system with coconut fiber and sand substrates is viable in regions with water limitations.

Keywords: *Citrullus lanatus*; germination; salinity; hydroponics; seed vigor

1. Introduction

Low rainfall and high temperatures characterize the Brazilian semi-arid region for most of the year and high evapotranspiration rates [1]. Along with these factors, soil salinization is prevalent, causing losses in crop yield and quality [2].

Salinity is one of the main factors threatening agriculture and food security globally, mainly due to salts' osmotic and ionic effects. These effects can interfere with cell stability and reduce plant water uptake, resulting in ion toxicity and changes in the cell's physiological and metabolic processes [2,3]. Salt stress also alters hormone-regulated responses, limiting stomatal opening and photosynthesis [4]. These changes caused phenological changes resulting in reduced leaf expansion and photosynthetic area, lower plant growth rates, and smaller fruits and seeds [5].

Seeds' physiological quality and vigor are essential for crop establishment [6]. Technical and environmental conditions of the production phase influenced the physiological seed quality [7], particularly stressful conditions such as salt stress. Excess salt can cause losses in the transport and accumulation of reserves, leading to low seed viability and vigor and an accelerated deterioration process [8].

The limited availability of good quality water for agriculture encourages strategies to use brackish water in agriculture. Brackish water desalination by reverse osmosis systems has guaranteed access to potable water for many communities in the Brazilian semi-arid region. However, water treatment by reverse osmosis generates brine waste, which must be used in fish farming or hydroponics to avoid soil contamination [5]. In this regard, using brackish water to prepare nutrient solutions for hydroponic crops is noteworthy as they do not present matric potential, and the roots are in a constant state of saturation, thus reducing the deleterious effects of salinity on the plants [9].

Agricultural producers are increasingly turning to hydroponic cultivation in protected environments, which enables intensive production and a continuous supply of products. Compared to conventional crops, the production cycle in hydroponics is shorter and more productive, and the products present a better quality [10]. In protected environment agriculture, hydroponics has been cultivated in substrates to provide the plant with excellent stability and support, supply oxygen, and promote carbon dioxide exchange between the roots and the external air [11]. Using coconut fiber substrate has demonstrated promising results in hydroponic cultivation with saline water [12,13].

Watermelon (*Citrullus lanatus* Schrad.), which belongs to the Cucurbitaceae family, is one species that thrives in hydroponic systems. It is one of the most important crops grown in Brazil. The Northeast region is the largest producer of watermelons in Brazil, accounting for 38% of national production [14]. In hydroponic cultivation, mini watermelon plants present great productive potential in all-year seasons, depending on management practices [15,16]. However, there are no studies on the effects of saline water in hydroponic cultivation of this species for phenology, production, and the physiological quality of seeds. This study aimed to evaluate the phenology, production, and vigor of seeds of mini watermelon grown in a nutrient solution prepared with reject brine and two growing substrates.

2. Results

During the mini watermelon cycle, there was a significant interaction ($p < 0.05$) between salinity and substrates. Single effect of salinity occurred only for the days to flowering—DTF ($p < 0.001$), and for the days from flowering to fruit maturity—DFFTFM ($p < 0.001$). There were also single effects of substrates on DTF ($p < 0.001$) and DFFTFM ($p < 0.01$). Mini watermelon flowering was delayed by eight days (Figure 1A) and there was a reduction of three days from flowering to fruit maturity (Figure 1C) for plants cultivated in the sand compared to those cultivated in coconut fiber.

Flowering was delayed by approximately one day for each unit increment in ECw, with a difference of 6.2 days when irrigation was performed with water of the highest salinity (6.9 dS m^{-1}) compared to the control (0.5 dS m^{-1}) (Figure 1B). The days from flowering to fruit maturity were reduced by 0.62 days for each increment of 1 dS m^{-1} in irrigation water, corresponding to a difference of 3.97 days in the treatment of the highest ECw (6.9 dS m^{-1}) compared to the control (0.5 dS m^{-1}) (Figure 1D).

The plant growth cycle in coconut fiber lasted an average of 81 days, regardless of salinity. The plant growth cycle in sand lasted from 83.6 to 88.6 days between salinity levels of 0.5 and 6.9 dS m^{-1}, and the highest plant growth cycle occurred in the ECw of 4.9 dS m^{-1} (Figure 2). The cycles of plants cultivated in the sand were 2.6, 7.0, 7.0, 5.3, and 7.0 days longer than those grown in coconut fiber, at salinity levels of 0.5, 2.4, 4.0, 5.5, and 6.9 dS m^{-1}, respectively (Figure 2).

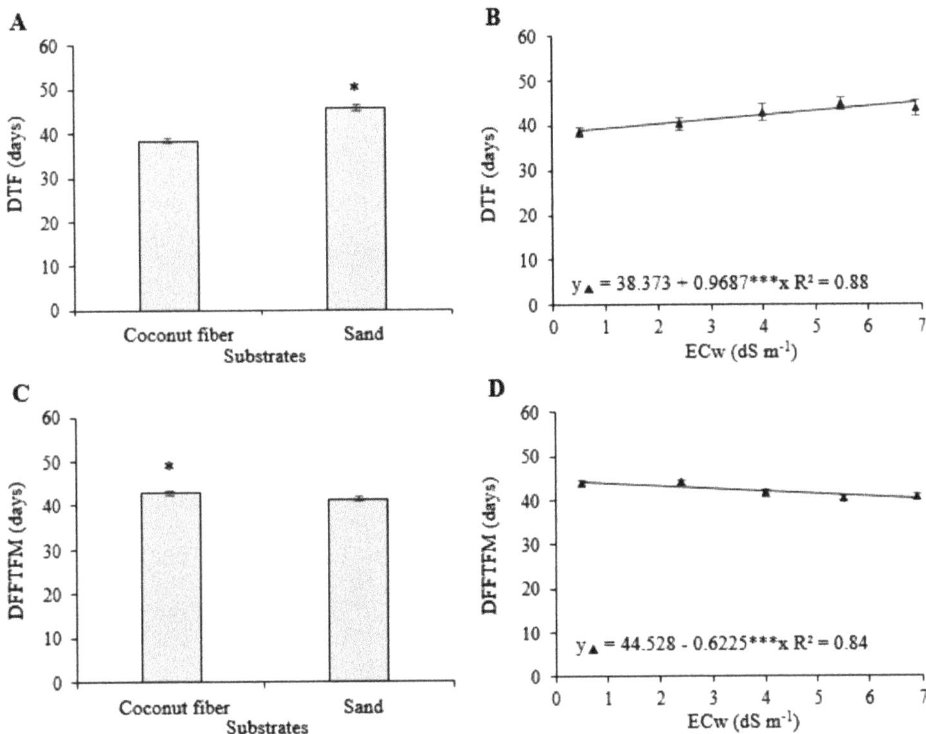

Figure 1. Regression analysis and means comparison test (Tukey, $p < 0.05$ and SD, $n = 20$) for the variables days to flowering—DTF (**A,B**) and days from flowering to fruit maturity—DFFTFM (**C,D**) of mini watermelon in hydroponic cultivation using a nutrient solution prepared with reject brine and different substrates. *** and * significant at 0.001 and 0.05 probability levels, respectively.

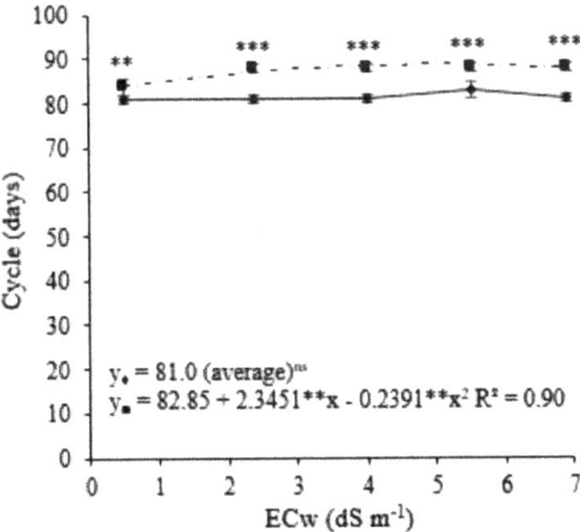

Figure 2. Regression analysis and means comparison test (Tukey, $p < 0.05$ and SD, $n = 4$) for the cycle of mini watermelon in hydroponic cultivation using a nutrient solution prepared with reject brine and different substrates. (♦ Coconut fiber; ■ Sand). ***, **, and ns significant at 0.001, 0.01 probability levels, and not significant, respectively.

The factors salinity, substrate, and time had significant interaction ($p < 0.001$) for longitudinal fruit diameter (LFD) and transverse fruit diameter (TFD) of a mini watermelon. In coconut fiber cultivation, the severe effects of salinity started five days after anthesis (DAA) for longitudinal and transverse diameters (Figure 3A,C). For longitudinal diameter, salinities S1 and S2 did not differ during the 45 DAA, but salinities S3, S4, and S5 differed from S1 and S2 from 20, 15, and 5 DAA, respectively (Figure 3A). At 45 days, the lowest longitudinal diameters occurred in salinity S5, with a reduction of 17.17%, followed by S4 and S3, with reductions of 4.37 and 7.74% compared to S1, respectively (Figure 3A). For transverse diameter, salinities S2, S3, S4, and S5 differed from S1 from 30, 15, 5, and 5 DAA, respectively (Figure 3C). At 45 days, reductions of 3.9, 5.6, 10.2, and 12.9% in the transverse diameter were seen under salinities S2, S3, S4, and S5 compared to S1, respectively (Figure 3C).

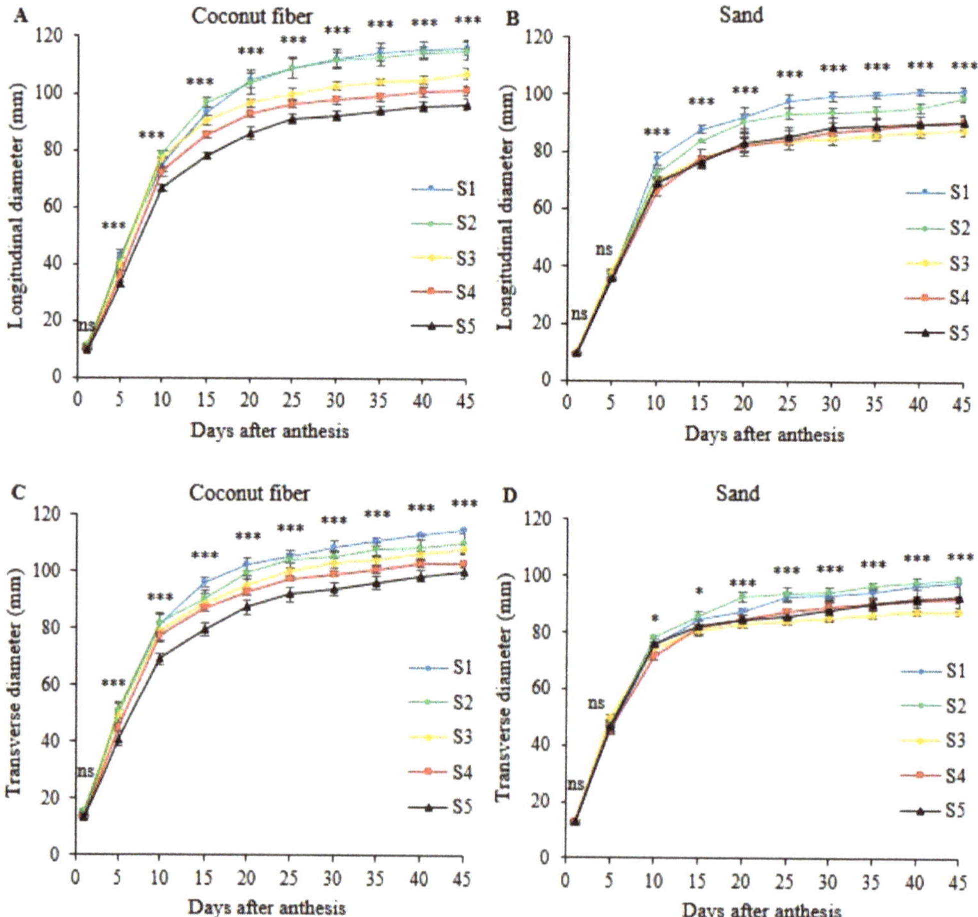

Figure 3. Means comparison test (Tukey, $p < 0.05$ and SD, $n = 4$) for longitudinal diameter (**A,B**) and transverse diameter (**C,D**) of mini watermelon fruits as a function of days after anthesis, in hydroponic cultivation using nutrient solution prepared with reject brine (S1—0.5 dS m^{-1}, S2—2.4 dS m^{-1}, S3—4.0 dS m^{-1}, S4—5.5 dS m^{-1}, S5—6.9 dS m^{-1}) and substrates (Coconut fiber and Sand). ***, * and ns significant at 0.001, 0.05 probability levels, and non-significant, respectively.

In sand cultivation, the effects of salinity began at 10 DAA for longitudinal and transverse diameters (Figure 3B,D). S1 was superior to the other salinities for longitudinal diameter on all days after anthesis, and S2 was superior to S3, S4, and S5 from 15 DAA (Figure 3B). At 45 days after anthesis, there were reductions of 2.35, 13.32, 10.44, and 10.83% in longitudinal diameter under salinities S2, S3, S4, and S5 compared to S1, respectively (Figure 3B). Salinities S3, S4, and S5 were similar for longitudinal diameter on all days after anthesis (Figure 3B). For transverse diameter, salinities S1 and S2 did not differ during the 45 DAA and were superior to S3, S4, and S5 from 15 DAA (Figure 3C). At 40 DAA, the transverse fruit diameters of plants under salinities S3, S4, and S5 decrease by 12.06, 7.56, and 7.10% compared to S1, respectively (Figure 3C). There was no significant difference in the transverse diameter of the fruits for those cultivated under salinities S4 and S5 (Figure 3C).

The interaction between salinity and growing substrates had a significant effect ($p < 0.05$) on fruit weight (FW) and hundred-seed weight (HSW). The substrate factor was significant for the weight of seeds per fruit (WSF) ($p < 0.01$) and for seed thickness (ST) ($p < 0.001$). The number of seeds per fruit (NSF), seed length (SL), and seed width (SW) were not significantly ($p > 0.05$) affected by the factors studied.

The increase in irrigation water salinity reduced the weight of mini watermelon fruits by 50.5 and 29.5% for plants grown in coconut fiber and sand substrates compared with those obtained at high (6.9 dS m^{-1}) and low (0.5 dS m^{-1}) salinity levels (Figure 4A).

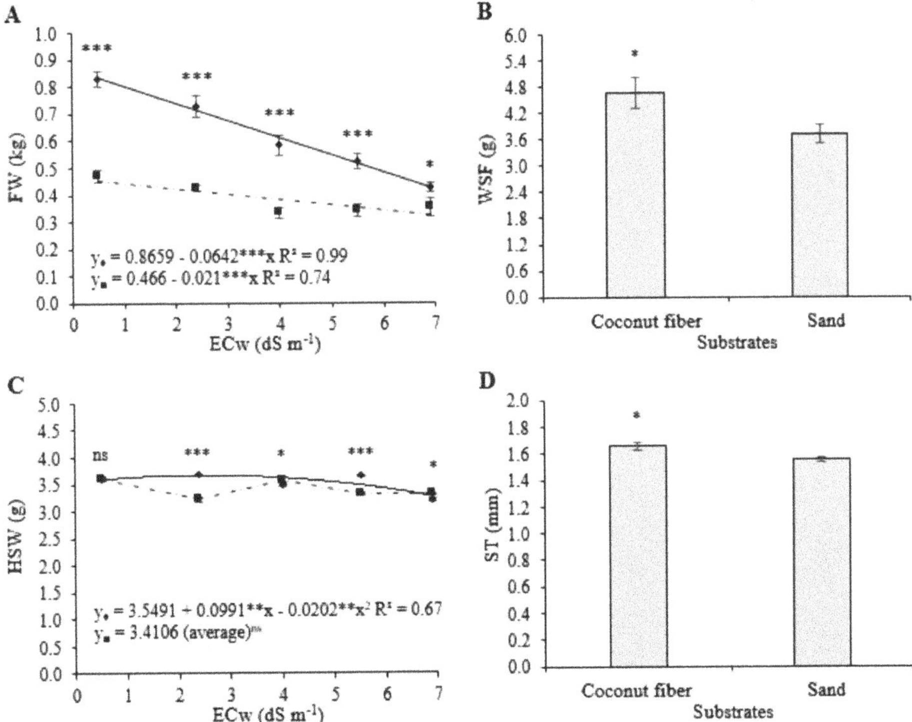

Figure 4. Regression analysis and means comparison test (Tukey, $p < 0.05$ and SD, $n = 20$) for the variables fruit weight–FW (**A**), the weight of seeds per fruit–WSF (**B**), hundred-seed weight–HSW (**C**), and seed thickness –ST (**D**) of mini watermelon in hydroponic cultivation using a nutrient solution prepared with reject brine and different substrates (♦ Coconut fiber; ■ Sand). ***, **, *, and ns significant at 0.001, 0.01, 0.05 probability levels, and not significant, respectively.

The weight of mini watermelon fruits obtained under coconut fiber cultivation was higher than that obtained under sand cultivation for all salinities, precisely 75.9, 71.0, 75.6, 53.5, and 20.2% higher at the salinities of 0.5, 2.4, 4.0, 5.5, and 6.9 dS m^{-1}, respectively (Figure 4A). The weight of mini watermelon seeds per fruit cultivated in coconut fiber was 60.5% (2.3 g) higher than that obtained for sand cultivation (Figure 4B).

The average hundred-seed weight of plants grown in the sand was 3.41 g, regardless of salinity. In coconut fiber cultivation, the hundred-seed weight ranged from 3.27 to 3.67 g between salinities of 0.5 and 6.9 dS m^{-1}, and the estimated salinity of 2.45 dS m^{-1} obtained the highest value (Figure 4C). For the hundred-seed weight, the substrates did not differ at the salinity of 0.5 dS m^{-1}. At salinities of 2.4 and 5.5 dS m^{-1}, the hundred-seed weight of watermelon cultivated in coconut fiber was 14.2 and 10.0% higher than the values obtained with sand cultivation, respectively. At the salinity levels of 4.0 and 6.9 dS m^{-1}, the hundred-seed weight of plants grown in the sand was 2.5 and 3.2% higher than those obtained with coconut fiber cultivation, respectively (Figure 4C).

The thickness of seeds of mini watermelon cultivated in coconut fiber was 7.0% (0.11 mm) higher than that obtained with sand cultivation (Figure 4D). Fruits produced in coconut fiber substrate obtained higher seed weight. Therefore, this result is not related to the number of seeds but their weight due to their greater thickness.

For the variables that indicate the physiological quality of the seeds, there was a significant interaction between the factors of salinity and substrates for germination ($p < 0.05$), electrical conductivity ($p < 0.001$), and accelerated aging ($p < 0.01$). The salinity factor significantly affected the emergence ($p < 0.01$).

The germination of mini watermelon seeds from coconut fiber cultivation was not influenced by water salinity, with an average of 100% (Figure 5A). Seeds from sand cultivation obtained the highest germination (100%) at a salinity of 2.9 dS m^{-1} (Figure 5A). There was a difference between the substrates under a salinity of 6.9 dS m^{-1}, and seeds from coconut fiber obtained 19 percentage points more than seeds from sand cultivation (Figure 5A).

The highest percentage of seedling emergence (87%) was obtained in the treatment with ECw 5.0 dS m^{-1}, being 17% above the control (0.5 dS m^{-1}) regardless of the substrate used (Figure 5B). Irrigation with saline water increased the electrical conductivity test of mini watermelon seeds by 3.84 and 3.19 μS m^{-1} for seeds of watermelon grown in coconut fiber and sand, respectively, for each increase of 1 dS m^{-1} in irrigation water (Figure 5C). The electrical conductivity test of mini watermelon seeds produced in a sand substrate was 8.6, 11.0, 10.3, and 20.6 μS m^{-1} higher than that of seeds produced in coconut fiber for salinities of 0.5, 2.4, 4.0, and 5.5 dS m^{-1}, respectively (Figure 5C). At a salinity of 6.9 dS m^{-1}, there was no significant difference between substrates for the electrical conductivity test of mini watermelon seeds.

The mini watermelon seeds obtained in coconut fiber cultivation after accelerated aging were not influenced by the irrigation water salinities, obtaining an average of 89% (Figure 5D). In turn, seeds produced in the sand obtained the highest germination after accelerated aging (97%) under a salinity of 3.74 dS m^{-1} (Figure 5D). There was a difference between the substrates under salinity of 4.0 dS m^{-1}, at which the seeds produced in sand obtained levels 14% above those produced with coconut fiber cultivation (Figure 5D).

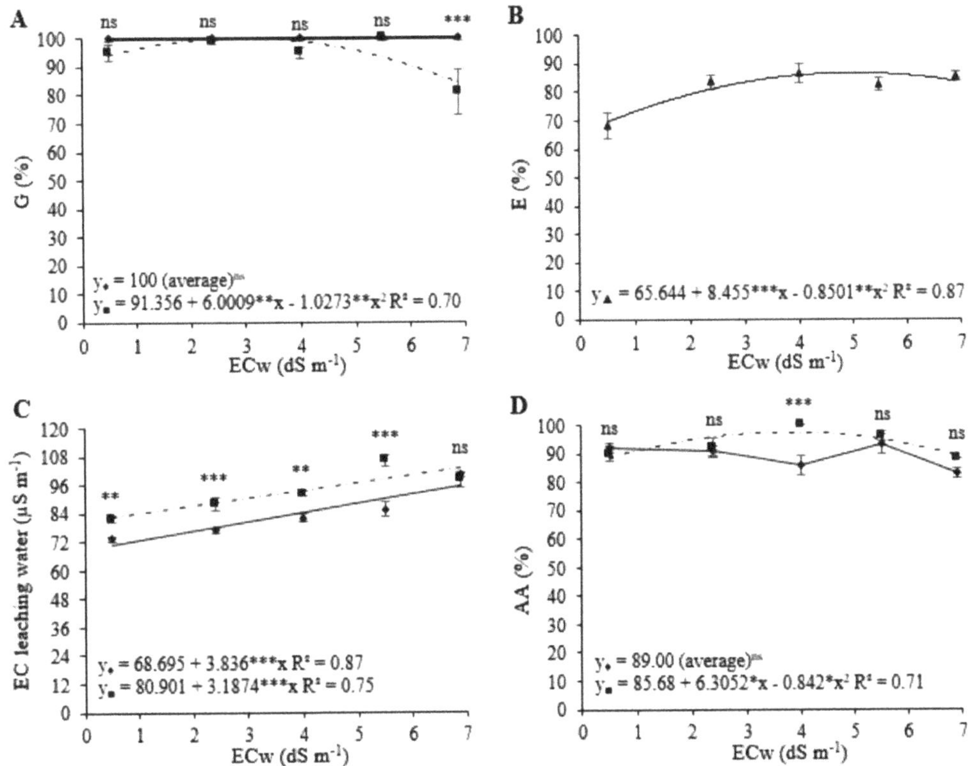

Figure 5. Regression analysis and means comparison test (Tukey, $p < 0.05$ and SD, $n = 4$) for the variables germination—G (**A**), emergence—E (**B**), electrical conductivity test—EC leaching water (**C**), and accelerated aging—AA (**D**) of seeds of mini watermelon in hydroponic cultivation using a nutrient solution prepared with reject brine and different substrates (♦ Coconut fiber; ■ Sand). ***, **, *, and ns significant at 0.001, 0.01, 0.05 probability levels, and not significant, respectively.

3. Discussion

The reuse of agro-industrial products in agriculture is essential for clean production. We produce mini watermelon fruits and seeds using brine waste from water desalination in rural communities and coconut fiber from residue from coconut water production.

The mini watermelon plant's cycle differed with the variations in the electrical conductivity (EC) of the water and cultivation substrates. The cycle of the plants grown in the coconut fiber substrate had an average duration of 81 days, whereas sand cultivation prolonged the cycle by 2.6 to 7 days. Watermelon plants cultivated in sand grew faster and reached the maximum height of the trellis (2.0 m). However, the plants needed more time to reach the top of the trellis when a nutrient solution containing water with a high concentration of salts was used. Thus, plants grown in the sand under higher salt stress presented longer cycles. Although salinity associated with sand cultivation reduced fruit maturation time, the overall length of the crop cycle did not decrease.

The physiological effects on the mini watermelon plants grown in sand and under salt stress prolonged the crop cycle. Drought induced by the reduction in osmotic potential due to the increased salinity of the water used in cultivation was the main factor that caused the mini watermelon plant's reduced growth [15,17,18]. Due to this and the sand's low water retention capacity, the plant growth rate decreased; consequently, the main branches needed more time to grow to the proper length.

The longitudinal and transversal diameters of the mini watermelons grown in the coconut fiber substrate at 45 days after anthesis (DAA) were 12.1% and 10.5% larger, respectively, compared with the diameters of the mini watermelons grown in the sand. The longitudinal and transversal diameters of the mini watermelons cultivated in coconut fiber were superior to those of the mini watermelons grown in the sand at 15 DAA for salinity treatments S1, S2, S3, S4, and S5. The longitudinal and transversal diameters resulting from coconut fiber cultivation under salinity treatments S1 and S2 were similar and, along with treatment S3, were superior to the diameters resulting from sand cultivation under treatment S1. The results from cultivation in the coconut fiber substrate under salinity treatments S4 and S5 were similar to those of sand cultivation under treatments S1 and S2, respectively. Therefore, the coconut fiber substrate promoted fruit growth and reduced the effect of salinity on the mini watermelon plants compared to sand. Coconut fiber has a greater capacity for hydration and water retention than sand without restricting aeration; as a result, coconut fiber makes more water available to the plant, even under conditions of low osmotic potential [19].

The fruits of the plants grown in the coconut fiber substrate presented greater diameters and fruit weight than those grown in sand, regardless of water salinity. The extended period required for fruit maturation and larger diameters significantly increased fruit weight. Fruit weight was 75% higher in plants grown in coconut fiber than in the sand under salinity treatment S1. For EC above 4.0 dS m^{-1}, the difference between the fruit weight of the crops grown in coconut fiber and those grown in sand decreased, which indicates that salt stress was more limiting in the coconut fiber substrate at this salinity level. Severe restrictions in fruit weight with a salinity condition of 4.0 dS·m^{-1} corroborate the reduction in the fruit's maturation period and longitudinal and transversal diameters.

Thus, decreases in fruit size and weight were not due to poor fertilization and poor fruit formation, which are capable of reducing sink strength (fruit) [20]. However, the restricted availability of water and photoassimilates due to the osmotic and ionic salinity components limits root development, leaf growth and expansion (source), and water relations and photosynthesis [2–4].

The mini watermelons grown in coconut fiber presented similar diameters and weights to those reported in the literature. However, the results we obtained for the plants grown in the sand substrate are lower than those reported by [4,15,16] for the hydroponic cultivation of the 'Smile' and 'Sugar Baby' cultivars of the mini watermelon using saline water in the nutrient solution. The authors of [21] observed mini watermelons under salinity conditions of 2.0 and 5.2 dS·m^{-1} and found that an increase in salinity resulted in a 240 g reduction in fruit weight for the non-grafted 'Tex' cultivar, 17.4 g of which corresponded to a reduction in seed weight. In the 'Sugar Baby' cultivar, fruit weight reduction between salinities of 2.4 and 5.5 dS·m^{-1} was similar to that observed in our study. However, we did not detect seed weight per fruit reduction due to the water's increased salinity, but mini watermelons grown in coconut fiber obtained more seed weight per fruit and thickness than seeds produced in sand cultivation. The higher water retention capacity of coconut fiber compared to the area favored the production of mini watermelon seeds with brackish water.

Despite the influences on the phenology and production of mini watermelon fruits and seeds grown in a hydroponic system with brackish water, seed viability in this system was acceptable. All treatments obtained germination over 90%, except for seeds from sand cultivation irrigated water of 6.9 dS·m^{-1}, which presented 84% of germination and seedling emergence between 70% and 87%. The percentages we found for germination and seedling emergence are within the appropriate range for the watermelon crop [22,23], which indicates that cultivation with brackish water did not affect the seed viability of the 'Sugar Baby' mini watermelon.

The seeds produced under higher salinity conditions presented greater EC values in the leaching water compared to the control (0.5 dS·m^{-1}), mainly when cultivated in the sand substrate. The metabolic disturbances caused by osmotic [18] and ionic [3] components during seed formation damaged the membranes, causing more significant electrolyte extravasation [7]. Electrolyte extravasation was more evident in the plants grown in the sand, in which production was affected more by salt stress than in the plants grown in coconut fiber. Although the EC values of the leaching water of the seeds rose, germination in the accelerated aging test was above 80% in all treatments, which indicates high seed vigor [23]. Although salt stress caused some damage during the production phase of 'Sugar Baby' mini watermelon fruits and seeds, the seeds showed good viability and vigor. Both [24] and [25] showed that saline stress decreases 'Sugar Baby' mini watermelon growth, fruit production, and post-harvest quality. They showed that salt stress does not affect the mini watermelon photosynthetic rate due to its high photosystem II efficiency. However, salinities from 4 dS m^{-1} significantly decrease plant growth, fruit production, and fruit post-harvest quality. They obtained marketable quality fruits only in mini watermelons grown in coconut fiber and irrigated with brackish water of up to 4.0 dS m^{-1}. They did not evaluate the salinity effect on phenology, seed yield, and seed quality. We considered it; we found that saline stress alters the phenology of mini watermelon, decreasing the cycle length and time from flowering to fruit maturation. We verified that the decrease in fruit cycle, size, and weight occurred in both substrates; however, the decrease for every 1 dS m^{-1} increase in water salinity in the mini watermelon grown on coconut fiber exceeded those obtained with sand substrate. However, we find that the seed production with good physiological quality in the coconut fiber and sand substrates occurred up to 6.9 dS m^{-1}. We found that changes in watermelon phenology by salinity decreased fruit production but did not impair seed production. The present research results reinforce the sustainable management of brackish water by the indicators such as the weight of seeds per fruit, hundred-seed weight, germination, emergence, seed electrical conductivity test, and seed accelerated aging. Therefore, our research makes it possible to identify further the potential use of reject brine in the hydroponic cultivation of mini watermelon in substrates. Thus, the seed production of mini watermelon with reject brine in the hydroponic cultivation is an alternative for regions with little available water.

4. Materials and Methods

4.1. Location and Characterization of the Environment

The study was conducted at the Federal Rural University of the Semi-Arid Region (UFERSA), Mossoró, RN, Brazil, and consisted of two phases. The first was carried out in a greenhouse (phenology and seed production) and the second in the laboratory (physiological quality of mini watermelon seeds produced in the previous phase). During the experiment in the greenhouse, the maximum and minimum values recorded in the environment were 39.2 and 20.4 °C for temperature and 86 and 22% for relative humidity, respectively.

4.2. Phase I—Phenology and Seed Production

The treatments were distributed in a split-plot scheme with a randomized block design (RBD). The plot was composed of the five electrical conductivities of water for nutrient solution preparation, ECw (S1 = 0.5, S2 = 2.4, S3 = 4.0, S4 = 5.5, and S5 = 6.9 dS m^{-1}). The subplot was composed of two substrates (coconut fiber and sand), with four replicates of two plants.

The cultivar used was 'Sugar Baby', which has a rounded shape, dark green rind, bright red flesh, and few seeds. Cultivation was performed in 6-dm^3 plastic pots filled with the growing substrates. The coconut fiber substrate has a fine texture, 95% total porosity, 507 ml L^{-1} (substrate) of water retention capacity, 0.5 dS m^{-1} (EC 1:5) of electrical conductivity, and 6.0 pH (pH 1:5). The sand substrate was sieved through a 4 mm mesh and washed with tap water. Three seeds were sown in each hole, and thinning was performed on the fifth day after sowing, leaving only one plant. Mini watermelon plants were trained in a vertical trellis with 2.0 m height, in five rows with 1.00 m spacing, with 16 plants in each row spaced 0.30 cm apart. During the growth, excess lateral shoots were eliminated up to the ninth branch by pruning, leaving the other shoots with five leaves. The apical bud was eliminated when the plants reached 2 m in height, leaving only one fruit per plant.

Pollination was carried out manually during the early morning hours, and the fruits were placed in plastic baskets.

Until the 10th day of cultivation, irrigation was performed with water from the supply network (EC = 0.54 dS m^{-1}). After this period, the nutrient solutions prepared with the different salinities began to be applied. The saline waters were obtained by mixing the public-supply water (PSW) and reject brine water from desalination (RBW), in the following proportions: S1—100% PSW, S2—85% PSW + 15% RBW, S3—70% PSW + 30% RBW; S4—55% PSW + 45% RBW; S5—40% PSW + 60% RBW. The public-supply water showed the following chemical composition: pH = 7.57; ECw = 0.5 dS m^{-1}; Ca^{2+} = 0.83 mmolc L^{-1}; Mg^{2+} = 1.20 mmolc L^{-1}; K$^+$ = 0.31 mmolc L^{-1}; Na$^+$ = 3.79 mmolc L^{-1}; Cl$^-$ = 2.40 mmolc L^{-1}; CO$_3^{2-}$ = 0.60 mmolc L^{-1}; HCO^{3-} = 3.20 mmolc L^{-1}; and SAR = 3.76 (mmolc L^{-1})$^{-0.5}$. Reject brine was collected in the Jurema Rural Settlement, Tibau, RN, Brazil, with the following chemical composition: pH = 7.10; ECw = 9.5 dS m^{-1}; Ca^{2+} = 37.8 mmolc L^{-1}; Mg^{2+} = 24.20 mmolc L^{-1}; K$^+$ = 0.83 mmolc L^{-1}; Na$^+$ = 54.13 mmolc L^{-1}; Cl$^-$ = 116.00 mmolc L^{-1}; CO$_3^{2-}$ = 0.00 mmolc L^{-1}; HCO^{3-} = 3.40 mmolc L^{-1}; and SAR = 9.70 (mmolc L^{-1})$^{-0.5}$.

The nutrient solutions were applied twice a day, in the early morning and the late afternoon, considering the volume corresponding to the actual evapotranspiration of the crop, measured by drainage lysimeters in additional plots corresponding to each treatment. A drip irrigation system applied the depth, composed of 16-mm-diameter hoses and pressure-compensating drippers with a flow rate of 1.4 L h^{-1}, connected to a self-venting Metalcorte/Eberle circulation motor pump, driven by a single-phase motor, 210 V voltage, 60 Hz frequency, installed in a reservoir with 50 L capacity.

The standard nutrient solution proposed by [26] was used for macronutrients. Micronutrients were supplied using the commercial compound Rexolin BRA, which consists of 11.68% potassium oxide (K$_2$O), 1.28% sulfur (S), 2.1% boron (B), 0.36% copper (Cu), 2.65% iron (Fe), 2.48% manganese (Mn), 0.036% molybdenum (Mo), and 3.38% zinc (Zn), following the manufacturer's recommendation (2 g L^{-1}). The nutrient solution has an electrical conductivity of 1.1 dS m^{-1}, and after preparation the solutions showed the following electrical conductivities: 1.6, 3.5, 5.1, 6.6, and 8.0 dS m^{-1}.

The experiment evaluated plants for phenology and fruit and seed production. The phenological variables considered were: days to flowering (DTF), by counting the days from sowing to the emergence of the open flower, considering the beginning of flowering as the moment when 50% of the plants in the treatment had at least one open flower per plant; days from flowering to fruit maturity (DFFTFM), by counting the days from anthesis (flower opening) to the physiological maturity of the fruit (harvest point), considering female flowers that had an ovary with a transverse diameter of 2 cm as fruits, while the harvest point of the fruits was defined based on the guidelines of [12], who consider fruits with completely dry tendril coming from the same node; and cycle length, by counting the days from sowing to fruit harvest.

The variables related to fruit production were: transverse fruit diameter (TFD) and longitudinal fruit diameter (LFD), measured with a digital caliper at 5, 10, 15, 20, 25, 30, 35, 40, and 45 days after anthesis, with results expressed in millimeters; and fruit weight (FW), determined by manual harvesting of the fruits, followed by a weighing on an analytical scale, with results expressed in grams.

After harvesting and weighing the fruits, the mini watermelon seeds were extracted manually with a spoon and a sieve, washed, and dried naturally to remove the mucilage. These seeds were then evaluated for the following variables: the number of seeds per fruit (NSF) by manually counting the seeds produced in each fruit; the weight of seeds per fruit (WSF) by weighing the fresh seeds extracted from each fruit on a precision analytical scale, with results expressed in grams; hundred-seed weight (HSW), for which eight replicates of 100 fresh seeds of each treatment were separated and subsequently weighed on a precision analytical scale, with results expressed in grams; and seed length (SL), width

(SW), and thickness (ST), measured with a digital caliper using 10 seeds per treatment, with results expressed in millimeters.

4.3. Phase II—Physiological Quality of Seeds

The viability and vigor of the mini watermelon seeds produced in the first phase were evaluated in a completely randomized delineation (CRD), with four replicates of 50 seeds. The seeds produced under five electrical conductivities of irrigation water, ECw (S1 = 0.5, S2 = 2.4, S3 = 4.0, S4 = 5.5, and S5 = 6.9 dS m^{-1}), and two substrates (coconut fiber and sand) were considered as lots. For this, the seeds were extracted from the fruit manually, washed in running water to remove the mucilage, and dried in the natural environment (30 °C) for 24 h. Then, their initial moisture content was determined using the oven method at 105 $\pm$ 3 °C, with two replicates of 4 $\pm$ 0.05 g for 24 h [27]. The results were expressed as a percentage (wet basis) (Table 1).

Table 1. The initial moisture content of mini watermelon seeds is produced with reject brine (ECw) and substrates.

ECw (dS m^{-1})	Seed Moisture Content (%)	
	Substrates	
	Sand	Coconut fiber
S1—0.5	8.4	8.4
S2—2.4	8.4	8.3
S3—4.0	8.4	7.7
S4—5.5	8.4	7.9
S5—6.9	8.7	8.6

The germination test was conducted in a Biochemical Oxygen Demand (B.O.D.)-type germination chamber, at 25 °C, with a photoperiod of eight hours and in paper towel roll substrate moistened with distilled water in an amount equivalent to 2.5 times the dry weight. Normal seedlings were counted 14 days after sowing [27].

The emergence test was performed in a greenhouse, using four replicates of 50 seeds. Sowing was carried out in polyethylene trays containing the coconut fiber substrate and irrigated with public-supply water. Emerged seedlings were counted 14 days after sowing, and the emergence percentage was calculated later.

The mass method conducted the electrical conductivity test with four replicates of 50 seeds, which were weighed, placed in plastic cups containing 75 mL of distilled water, and kept at a constant temperature of 25 °C for 24 h of incubation. After this period, the electrical conductivity of the solution was determined in a Digimed CD-21 conductivity meter, and the results were expressed in μS^{-1} cm^{-1} g^{-1} of seeds [28].

To conduct the accelerated aging test, a single layer of seeds was placed on a metal screen fixed in an acrylic box containing 40 mL of distilled water. The closed containers were kept in germination chambers (B. O. D.) at 41 °C for 48 h [29]. After this period, the germination test evaluated four subsamples of 50 seeds, computing the percentage of normal seedlings five days after sowing.

4.4. Statistical Analysis

The data were subjected to analysis of variance by the F test, and the Tukey analyzed the effects of the treatments means comparison test at a 5% probability level and polynomial regression analysis. Statistical analyses were performed using the statistical software Sisvar 5.7 [30].

5. Conclusions

High salinities for the hydroponic cultivation of the mini watermelon cultivar 'Sugar Baby' accelerated fruit maturation and crop cycle, decreasing fruit size. The greater cycle acceleration in plants grown in coconut fiber caused a marked reduction in fruit weight compared to plants grown in sand. However, at all salinities, the fruits of plants grown on coconut fiber outperform plants grown on sand. The fruits of plants grown in sand and those grown in coconut fiber with 6.9 dS m^{-1} were inferior in size and weight. However, in both substrates, the seed production of mini watermelon, seed viability, and seed vigor occurred adequately with a reject brine of 6.9 dS m^{-1} in the hydroponic nutrient solution. We found that salt stress affects the fruit production of mini watermelon, but was not harmful for seed production. The seed production of 'Sugar Baby' mini watermelons using reject brine in a hydroponic system with coconut fiber and sand substrates is viable in regions with water limitations.

Author Contributions: Conceptualization, methodology, data curation, writing—original draft preparation, T.R.C.A.; conceptualization, writing—reviewing and editing, supervision, S.B.T. and E.P.d.P.; methodology, data curation, R.R.d.O., A.L.A.F., K.T.O.P. and D.L.d.B.; conceptualization, methodology, data curation, R.R.T.O.; writing—reviewing and editing, C.Z.A., A.S.D., C.P.B., A.S.d.M. and M.F.-N.; conceptualization, writing—reviewing and editing, N.d.S.D.; conceptualization, methodology, data curation, software, validation, writing—reviewing and editing, supervision, F.V.d.S.S. All authors have read and agreed to the published version of the manuscript.

Funding: This study was financed in part by the Coordenação de Aperfeiçoamento de Pessoal de Nível Superior—Brasil (CAPES)—Finance Code 001.

Institutional Review Board Statement: Not applicable.

Informed Consent Statement: Not applicable.

Data Availability Statement: All other data are presented in the paper.

Conflicts of Interest: The authors declare no conflict of interest.

References

1. Melo, T.K.; Espínola-Sobrinho, J.; Medeiros, J.F.; Figueirêdo, V.B.; Silva, J.S.; Sá, F.V.S. Impacts of climate change scenarios in the Brazilian semiarid region on watermelon cultivars. *Rev. Caatinga* **2020**, *33*, 794–802. [CrossRef]
2. Sá, F.V.S.; Silva, I.E.; Ferreira-Neto, M.; Lima, Y.B.; Paiva, E.P.; Gheyi, H.R. Phosphorus doses alter the ionic homeostasis of cowpea irrigated with saline water. *Rev. Bras. Eng. Agrícola Ambient.* **2021**, *25*, 372–379. [CrossRef]
3. Volkov, V.; Beilby, M.J. Salinity tolerance in plants: Mechanisms and regulation of ion transport. *Front. Plant Sci.* **2017**, *8*, 17–95. [CrossRef] [PubMed]
4. Silva, S.S.; Lima, G.S.; Lima, V.L.A.; Gheyi, H.R.; Soares, L.A.A.; Lucena, R.C.L. Gas exchanges and production of watermelon plant under salinity management and nitrogen fertilization. *Pesqui. Agropecu. Trop.* **2019**, *49*, e54822. [CrossRef]
5. Silva, A.A.; Melo, S.S.; Umbelino, B.F.; Sá, F.V.S.; Dias, N.S.; Ferreira-Neto, M. Cherry tomato production and seed vigor under irrigation with saline effluent from fish farming. *Rev. Bras. Eng. Agrícola Ambient.* **2021**, *25*, 380–385. [CrossRef]
6. Paiva, E.P.; Torres, S.B.; Almeida, J.P.N.; Sá, F.V.S.; Oliveira, R.R.T. Tetrazolium test for the viability of gherkin seeds. *Cienc. Agron.* **2017**, *48*, 118–124. [CrossRef]
7. Marcos-Filho, J. *Fisiologia de Sementes de Plantas Cultivadas*, 2nd ed.; ABRATES: Londrina, Brazil, 2015; 659p.
8. Silva, R.T.; Oliveira, A.B.; Lopes, M.F.Q.; Guimarães, M.A.; Dutra, A.S. Physiological quality of sesame seeds produced from plants subjected to water stress. *Cienc. Agron.* **2016**, *47*, 643–648. [CrossRef]
9. Soares, T.M.; Silva, E.F.F.; Duarte, S.N.; Melo, R.F.; Jorge, C.A.; Bonfim-Silva, E.M. Hydroponic lettuce production using saline waters. *Irriga* **2007**, *12*, 235–248. [CrossRef]
10. Cometti, N.N.; Galon, K.; Bremenkemp, D.M. Behavior of four cultivars of lettuce in hydroponic cultivation in tropical environment. *Rev. Eixo* **2019**, *8*, 114–122. [CrossRef]
11. Casais, L.K.N.; Aviz, R.O.; Santos, N.F.A.S.; Melo, M.R.S.; Souza, V.Q.; Borges, L.S.; Lima, A.K.O.; Guerreiro, A.C. Morphophysiological indices and pepper production produced on different substrates based on organic residues in protected environment. *Agroecossistemas* **2018**, *10*, 174–190. [CrossRef]
12. Dias, N.S.; Lira, R.B.; Brito, R.F.; Souza-Neto, O.N.; Ferreira-Neto, M.; Oliveira, A.M. Melon yield in a hydroponic system with wastewater from desalination plant added in the nutrient solution. *Rev. Bras. Eng. Agrícola Ambient.* **2010**, *14*, 755–761. [CrossRef]
13. Silva, A.A.; Dias, N.S.; Jales, G.D.; Rebouças, T.C.; Fernandes, P.D.; Ferreira-Neto, M.; Morais, P.L.D.; Paiva, E.P.; Fernandes, C.S.; Sá, F.V.S. Fertigation with fish farming effluent at the adequate phenological stages improves physiological responses, production and quality of cherry tomato fruit. *Int. J. Phytoremediat.* **2022**, *24*, 283–292. [CrossRef] [PubMed]
14. Instituto Brasileiro de Geografia e Estatística—IBGE. Produção Agrícola Municipal 2020—Brasil. Available online: https://sidra.ibge.gov.br (accessed on 13 April 2022).

15. Sousa, A.B.O.; Duarte, S.N.; Sousa-Neto, O.N.; Souza, A.C.M.; Sampaio, P.R.F.; Dias, C.T.S. Production and quality of mini watermelon cv. Smile irrigated with saline water. Ver. *Bras. Eng. Agrícola Ambient.* **2016**, *20*, 897–902. [CrossRef]

16. DO Ó, L.M.G.; Cova, A.M.W.; Gheyi, H.R.; Silva, N.D.; Azevedo-Neto, A.D. Production and quality of mini watermelon under drip irrigation with brackish water. *Rev. Caatinga* **2020**, *33*, 766–774. [CrossRef]

17. Silva-Junior, E.G.; Silva, A.F.; Lima, J.S.; Silva, M.F.C.; Maia, J.M. Vegetative development and content of calcium, potassium, and sodium in watermelon under salinity stress on organic substrates. *Pesqu. Agropecu. Bras.* **2017**, *52*, 1149–1157. [CrossRef]

18. Wan, Q.; Hongbo, S.; Zhaolong, X.; Jia, L.; Dayong, Z.; Yihong, H. Salinity tolerance mechanism of osmotin and osmotin-like proteins: A promising candidate for enhancing plant salt tolerance. *Curr. Genom.* **2017**, *18*, 553–556. [CrossRef]

19. Camposeco-Montejo, N.; Robledo-Torres, V.; Ramírez-Godina, F.; Mendoza-Villarreal, R.; Pérez-Rodríguez, M.Á.; Cabrera-de la Fuente, M. Response of bell pepper to rootstock and greenhouse cultivation in coconut fiber or soil. *Agronomy* **2018**, *8*, 111. [CrossRef]

20. Bomfim, I.G.A.; Bezerra, A.D.M.; Nunes, A.C.; Freitas, B.M.; Aragão, F.A.S. Pollination requirements of seeded and seedless mini watermelon varieties cultivated under protected environment. *Pesqu. Agropecu. Bras.* **2015**, *50*, 44–53. [CrossRef]

21. Colla, G.; Roupahel, Y.; Mariateresa Cardarelli, M.; Rea, E. Effect of salinity on yield, fruit quality, leaf gas exchange, and mineral composition of grafted watermelon plants. *HortScience* **2006**, *41*, 622–627. [CrossRef]

22. Torres, S.B. Germination and seedling development as influenced by salinity. *Rev. Bras. Sementes* **2007**, *29*, 77–82. [CrossRef]

23. Radke, A.K.; Soares, V.N.; Eberhardt, P.E.R.; Martins, A.B.N.; Dias, L.W.; Xavier, F.M.; Villela, F.A. Comparison of tests for the evaluation of watermelon seed vigor. *Aust. J. Basic Appl. Sci.* **2017**, *11*, 150–156.

24. Silva, J.S.; Sá, F.V.S.; Dias, N.S.; Ferreira-Neto, M.; Jales, G.D.; Fernandes, P.D. Morphophysiology of mini watermelon in hydroponic cultivation using reject brine and substrates. *Rev. Bras. Eng. Agríc. Ambient.* **2021**, *25*, 402–408. [CrossRef]

25. Silva, J.S.; Dias, N.S.; Jales, G.D.; Reges, L.B.L.; Freitas, J.M.C.; Umbelino, B.F.; Alves, T.R.C.; Silva, A.A.; Fernandes, C.S.; Paiva, E.P.; et al. Physiological responses and production of mini-watermelon irrigated with reject brine in hydroponic cultivation with substrates. *Environ. Sci. Pollut. Res.* **2022**, *29*, 11116–11129. [CrossRef] [PubMed]

26. Marques, G.N.; Peil, R.M.N.; Lago, I.; Ferreira, L.V.; Perin, L. Phenology, water consumption, yield and quality of mini watermelon crop in hydroponics. *Rev. Fac. Agron.* **2014**, *113*, 57–65.

27. BRASIL. *Regras para Análise de Sementes*; Ministério da Agricultura, Pecuária e Abastecimento: Brasília, Brazil, 2009.

28. Vieira, R.D.; Marcos-Filho, J. *Teste de condutividade elétrica, In Vigor de Sementes: Conceitos e Testes*, 2nd ed.; Krzyzanowski, F.C., Vieira, R.D., França-Neto, J.B., Marcos-Filho, J., Eds.; Abrates: Londrina, Brazil, 2020; pp. 333–389.

29. Bhering, M.C.; Dias, D.C.F.S.; Barros, D.I.; Dias, L.A.S.; Tokuhisa, D. Vigor evaluation of watermelon seeds by accelerated aging test. *Rev. Bras. Sementes* **2003**, *25*, 1–6. [CrossRef]

30. Ferreira, D.F. Sisvar: A computer statistical analysis system to fixed effects split plot type designs. *Rev. Bras. Biom.* **2019**, *37*, 529–535. [CrossRef]

Article

Seed Priming with Glass Waste Microparticles and Red Light Irradiation Mitigates Thermal and Water Stresses in Seedlings of *Moringa oleifera*

Patrícia da Silva Costa [1], Rener Luciano de Souza Ferraz [2,*], José Dantas Neto [1], Semako Ibrahim Bonou [1], Igor Eneas Cavalcante [3], Rayanne Silva de Alencar [4], Yuri Lima Melo [4], Ivomberg Dourado Magalhães [5], Ashwell Rungano Ndhlala [6], Ricardo Schneider [7], Carlos Alberto Vieira de Azevedo [1] and Alberto Soares de Melo [4,*]

[1] Academic Unit of Agricultural Engineering, Federal University of Campina Grande, Campina Grande 58428-830, Paraíba, Brazil
[2] Academic Unit of Development Technology, Federal University of Campina Grande, Sumé 58540-000, Paraíba, Brazil
[3] Department of Plant Science and Environmental Sciences, Federal University of Paraíba, Areia 58051-900, Paraíba, Brazil
[4] Department of Biology, State University of Paraíba, Campina Grande 58429-500, Paraíba, Brazil
[5] Department of Agronomy, Federal University of Alagoas, Rio Largo 57100-000, Alagoas, Brazil
[6] Green Biotechnologies Research Centre of Excellence, University of Limpopo, Sovenga 0727, Limpopo, South Africa
[7] Department of Chemistry, Federal Technological University of Paraná, Toledo 85902-000, Paraná, Brazil
* Correspondence: ferragroestat@gmail.com (R.L.d.S.F.); alberto.melo@servidor.uepb.edu.br (A.S.d.M.)

Citation: Costa, P.d.S.; Ferraz, R.L.d.S.; Dantas Neto, J.; Bonou, S.I.; Cavalcante, I.E.; Alencar, R.S.d.; Melo, Y.L.; Magalhães, I.D.; Ndhlala, A.R.; Schneider, R.; et al. Seed Priming with Glass Waste Microparticles and Red Light Irradiation Mitigates Thermal and Water Stresses in Seedlings of *Moringa oleifera*. *Plants* **2022**, *11*, 2510. https://doi.org/10.3390/plants11192510

Academic Editor: Dayong Zhang

Received: 3 September 2022
Accepted: 21 September 2022
Published: 26 September 2022

Publisher's Note: MDPI stays neutral with regard to jurisdictional claims in published maps and institutional affiliations.

Abstract: The association between population increase and the exploitation of natural resources and climate change influences the demand for food, especially in semi-arid regions, highlighting the need for technologies that could provide cultivated species with better adaptation to agroecosystems. Additionally, developing cultivation technologies that employ waste materials is highly desirable for sustainable development. From this perspective, this study aimed to evaluate whether seed priming with glass waste microparticles used as a silicon source under red light irradiation mitigates the effects of thermal and water stress on seedlings of *Moringa oleifera*. The experimental design was set up in randomized blocks using a 2 × 2 × 2 factorial arrangement consisting of seed priming (NSP—no seed priming, and SPSi—seed priming with glass microparticles under red light irradiation), soil water replenishment (W50—50%, and W100—100% of crop evapotranspiration—ETc), and temperature change (TC30°—30 °C day/25 °C night and TC40°—40 °C day/35 °C night). Seed priming with glass microparticles under red light irradiation mitigated the effects of thermal and water stress on seedlings of *Moringa oleifera* seedlings through the homeostasis of gas exchange, leaf water status, osmotic adjustment, and the antioxidant mechanism.

Keywords: Moringaceae; abiotic stresses; gas exchange; cell membrane integrity; water status; osmotic adjustment; antioxidant mechanism

1. Introduction

The steady growth of the world population presents the agricultural sector with a challenge to increase food production and ensure food security, especially because the number of undernourished people worldwide grew from 83 to 132 million in 2020. However, the increase in production has to occur with the lowest possible impact on natural resources in order to meet one of the priority goals of the 2030 Agenda for Sustainable Development [1]. From this perspective, one alternative to increase food production and reduce the impacts of agriculture on agroecosystems is growing plants with the potential for multiple uses and genotypic and phenotypic plasticity for different cultivation environments [2].

In this scenario, *Moringa oleifera* Lamarck, a species of the family Moringaceae and native to India and Pakistan [3], could be one such alternative for cultivation given its various food, medicinal, industrial, environmental, and social purposes, and as an abundant source of essential amino acids, macronutrients, and micronutrients. This species also shows outstanding anti-fungal, analgesic, anti-inflammatory, anti-oxidant, anti-diabetic, anti-tumoral, and anti-bacterial properties, with high seed oil contents [4–8]. Furthermore, the species can also be used to purify water, generate income, and improve the quality of life of producers around the world [4–8].

However, despite the adaptive potential of *M. oleifera* to agroecosystems, its development, growth, and production can decrease when the plant is exposed to thermal and water stresses [9–12], especially in semi-arid regions with high solar radiation levels, increased air temperature, and soil water restriction, conditions that could be further aggravated by environmental climate changes [13,14]. Thermal stress reduces seed germination and vigor in seedlings of *M. oleifera* [15,16] and increases the accumulation of non-structural carbohydrates, amino acids, and phenols [12]. Water stress activates the antioxidant mechanism of this species [10], which, in turn, reduces the stomatal size, the content of photosynthetic pigments, leaf gas exchange, and root and shoot growth [9,11,17].

This scenario highlights the desire for technologies that can mitigate the effects of abiotic stresses on *M. oleifera*. In that regard, seed priming emerges as a promising alternative [2] that consists of soaking seeds in stress-mitigating solutions, e.g., silicon microparticle sources (SiMPs), with verified positive results in stress management in other crops, because the silicon is deposited in the endoplasmic reticulum, cell walls and intercellular spaces as hydrated amorphous silica, in addition to forming complexes with polyphenols to reinforce cell walls [18–21]; for example, wheat (*Triticum aestivum* L.) under cadmium (Cd) stress [22,23] and thermal stress [24], and rice (*Oryza sativa* L.) under drought [25]. Furthermore, developing accessible methodologies to obtain materials that will serve as Si sources is highly desirable. For example, the application of glass waste, a material easily found in landfill sites and used as silicon microparticle powders (SiMPs), becomes a simple approach that only requires grinding and sieving glass waste. Therefore, considering that seeds of *M. oleifera* are responsive to light [16], especially monochromatic red light [2], the irradiation of this type of light during seed priming could potentialize the stress-mitigating effect of SiMPs on plants.

From this perspective, we hypothesized that the seedlings of *M. oleifera* generated by seed priming with residual glass microparticles used as a Si source, or SiMPs, and under irradiation with monochromatic red light (RL), changed their gas exchange, osmotic adjustment, antioxidant mechanism, and dry matter accumulation as responses to overcome the effects of abiotic stresses.

Therefore, this study aimed to evaluate whether seed priming with silicon microparticles and red-light irradiation mitigate the effects of thermal and water stress on seedlings of *Moringa oleifera* cultivated in a Phytotron growth chamber.

2. Results

2.1. Principal Components and Multivariate Variance

Three principal components (PCs) with eigenvalues (λ) higher than one and percentages of variance (s^2) higher than 10% were formed by the linear combination between 13 original variables collected from seedlings of *M. oleifera* generated by seed priming (SP) with glass microparticles (SiMPs) and subjected to different combinations of soil water replenishment levels (SWR) and temperature variations (TC). The first three PCs explained 86.87% of s^2. PC_1 represented 53.67% of s^2 and was formed by combining the leaf water status (relative water content—RWC), leaf gas exchange (stomatal conductance—gs, transpiration—E, and internal CO_2 concentration—Ci), leaflet osmotic adjustment indicators (proline in leaves—PRO, total soluble proteins in leaves—TSP-L, and total soluble sugars in leaves—TSS-L) catalase activity in the leaflets (CAT-L), and total dry matter accumulation (TDM). PC_2 represented 18.72% of s^2 and was formed by the combination between the net photosynthetic rate (A), superoxide dismutase activity in the leaflets (SOD-L), and TDM accumulation. PC_3 represented 14.49% of s^2 and was formed by the osmotic adjustment indicator in the roots (total soluble proteins in roots—TSP-R) and the activity of the antioxidant mechanism of roots (CAT-R). Electrolyte leakage in the leaflets (EL), the total sugar content in roots (TSS-R), and SOD-R activity were not associated with any PC, and their s^2 values were disregarded in the principal component analysis (PCA) to be subjected to univariate analysis of variance (ANOVA). There was a significant interaction (*p*-value < 0.01) between SP x SWR x TC in the three PCs, according to the results of the multivariate analysis of variance (MANOVA) (Table 1).

Table 1. Correlation between the original variables and principal components, eigenvalues, explained and cumulative variance, and probability of significance of the hypothesis test in the interaction between the first three principal components (PCs 1, 2, and 3) and seed priming, soil water replenishment levels, and temperature change in *Moringa oleifera* seedlings.

EV—Evaluated Variables	PC—Principal Components		
	PC_1	PC_2	PC_3
EL—Electrolyte leakage	−0.30	−0.06	0.28
RWC—Relative water content	−0.94 *	−0.10	0.18
A—Net photosynthetic rate	−0.07	0.67 *	0.45
gs—Stomatal conductance	−0.80 *	0.48	0.24
E—Transpiration	−0.89 *	0.26	0.17
Ci—Internal CO_2 concentration	−0.89 *	0.30	−0.05
PRO—Proline in leaves	0.91 *	0.05	0.19
TSP-L—Total soluble proteins in leaves	0.85 *	0.34	0.05
TSP-R—Total soluble proteins in roots	−0.46	0.45	−0.71 *
TSS-L—Total soluble sugars in leaves	0.73 *	−0.43	−0.38
TSS-R—Total soluble sugars in roots	−0.33	−0.25	0.17
SOD-L—Superoxide dismutase in leaves	−0.56	−0.70 *	0.32
SOD-R—Superoxide dismutase in roots	0.36	−0.28	0.54
CAT-L—Catalase in leaves	−0.77 *	−0.46	0.23
CAT-R—Catalase in roots	0.47	−0.22	0.77 *
TDM—Total dry matter	0.66 *	0.57 *	0.36
λ—Eigenvalues	6.98	2.43	1.88
s^2 (%)—Explained variance	53.67	18.71	14.49
s^2 (%)—Cumulative variance	53.67	72.38	86.87
MANOVA—Multivariate analysis of variance	**Significance probability (*p*-value)**		
Hotelling's T-squared test for seed priming—SP	<0.01	<0.01	<0.01
Hotelling's T-squared test for soil water replenishment—SWR	<0.01	<0.01	<0.01
Hotelling's T-squared test for temperature change—TC	<0.01	<0.01	<0.01
Hotelling's T-squared test for the SP × SWR interaction	<0.01	<0.01	<0.01
Hotelling's T-squared test for the SP × TC interaction	<0.01	<0.01	<0.01
Hotelling's T-squared test for the SWR × TC interaction	<0.01	<0.01	<0.01
Hotelling's T-squared test for the SP × SWR × TC interaction	<0.01	<0.01	<0.01

Variables considered in PC formation (*).

2.2. Responses to Thermal and Water Stresses and Mitigation by Seed Priming

In the two-dimensional projection of the first four PCs (Figure 1A,D), PC_1 is seen as a process triggered by temperature variations (TC), PC_2 is triggered by the soil water replenishment levels (SWR), and PC_3 is triggered by seed priming (SP). In PC_1, temperature variation with thermal stress (40 °C day/35 °C night) increased the RWC, gs, E, Ci, and CAT-L and reduced the leaflet osmotic adjustment (PRO, TSP-L, and TSS-L) and the total dry matter accumulation (TDM) in relation to the seedlings that were not subjected to water stress (30 °C day/25 °C night). However, when seed priming (SP) with glass microparticles used as a silicon source was applied under monochromatic red light (RL), the seedlings that were subjected to thermal stress (SPSi-W100-T40°) and water stress (SPSi-W50-T30°), either in isolation or combined (SPSi-W50-T40°), showed mitigation in such stresses by reducing the RWC, gs, E, Ci, and CAT-L and increasing the PRO, TSP-L, TSS-L, and TDM. On the other hand, in the seedlings that were not subjected to the previously mentioned stresses (SPSi-W100-T30°), SP increased the RWC, gs, E, Ci, and CAT-L and decreased the PRO, TSP-L, TSS-L, and TDM in relation to those that did not receive seed priming (NSP-W100-T30°) (Figure 1A,B).

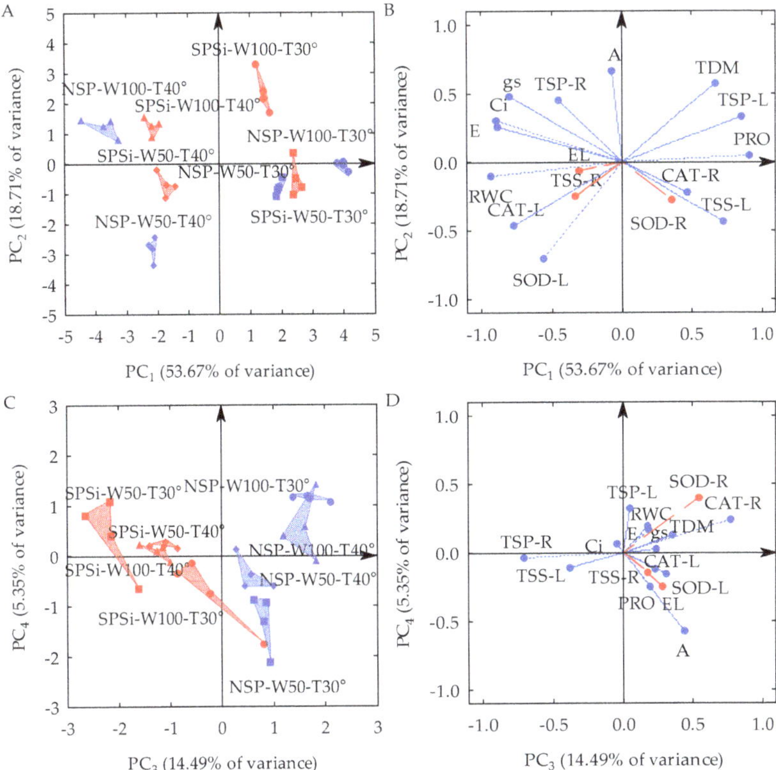

Figure 1. Two-dimensional projection of factorial scores (**A**,**C**) and variables (**B**,**D**) in the first four principal components (PCs 1, 2, 3, and 4) for the interaction between soil water replenishment levels (W50 and W100) and temperatures (T30° and T40°) in *Moringa oleifera*. PC, Principal Component; NSP, no seed priming; SPSi, seed priming with SiMPs; W100, no water stress; W50, water stress; T30°, no thermal stress; T40°, thermal stress; EL, electrolyte leakage; RWC, relative water content; TDM, total dry matter; A, net photosynthetic rate; gs, stomatal conductance; E, transpiration; Ci, internal CO_2 concentration; PRO, proline; TSP-L, total soluble proteins in leaves; TSP-R, total soluble proteins in roots; TSS-L, total soluble sugars in leaves; TSS-R, total soluble sugars in roots; SOD-L, superoxide dismutase in leaves; SOD-R, superoxide dismutase in roots; CAT-L, catalase in leaves; CAT-R, catalase in roots; •, NSP-W100-T30°; ▲, NSP-W100-T40°; ■, NSP-W50-T30°; ♦, NSP-W50-T40°; •, SPSi-W100-T30°; ▲, SPSi-W100-T40°; ■, SPSi-W50-T30°; ♦, SPSi-W50-T40°.

In PC_2, the water stress caused by 50% soil water replenishment (W50) increased the activity of the SOD-L enzyme and decreased the net photosynthetic rate (A) and TDM accumulation in the seedlings in relation to those that were not subjected to this condition (W100). SP with SiMPs under RL irradiation mitigated the combined effects of water and thermal stresses (SPSi-W50-T40°) in relation to stressed seedlings that were not generated by SP (NSP-W50-T40°) by inducing a lower SOD-L activity and higher A and TDM values.

SP did not mitigate water stress when this condition was imposed in isolation (SPSi-W50-T30°) compared to seedlings produced from seeds that did not undergo SP (NSP-W50-T30°). It is also seen that, in seedlings that were not subjected to stresses (SPSi-W100-T30°), SP reduced the activity of SOD-L and increased the A and TDM values in relation to control seedlings (NSP-W100-T30°) (Figure 1A,B).

In PC₃, SP application with SiMPs and RL irradiation increased the total soluble protein content in the roots (TSP-R) and reduced catalase activity in the roots (CAT-R) of seedlings subjected to thermal (SPSi-W100-T40°) and water stress (SPSi-W50-T30°) in isolation and also when these stresses were simultaneously imposed (SPSi-W50-T40°) in relation to the seedlings that were not generated by seed priming (NSP-W100-T40°, NSP-W50-T30°, and NSP-W50-T40°). The same was observed in the seedlings that were not subjected to the stresses (SPSi-W100-T30°) (Figure 1C,D).

Figure 2A shows that SP with SiMPs under RL irradiation increased the cell membrane integrity in seedlings of *M. oleifera* subjected to thermal (SPSi-W100-T40°) and water stresses (SPSi-W50-T30°) either in isolation or combined (SPSi-W50-T40°) by reducing electrolyte leakage (EL). This EL reduction was not observed in non-stressed seedlings (SPSi-W100-T30°). When SP was not applied, higher EL values were recorded in the seedlings subjected to thermal (NSP-W100-T40°) and water stresses (NSP-W50-T30°) in isolation.

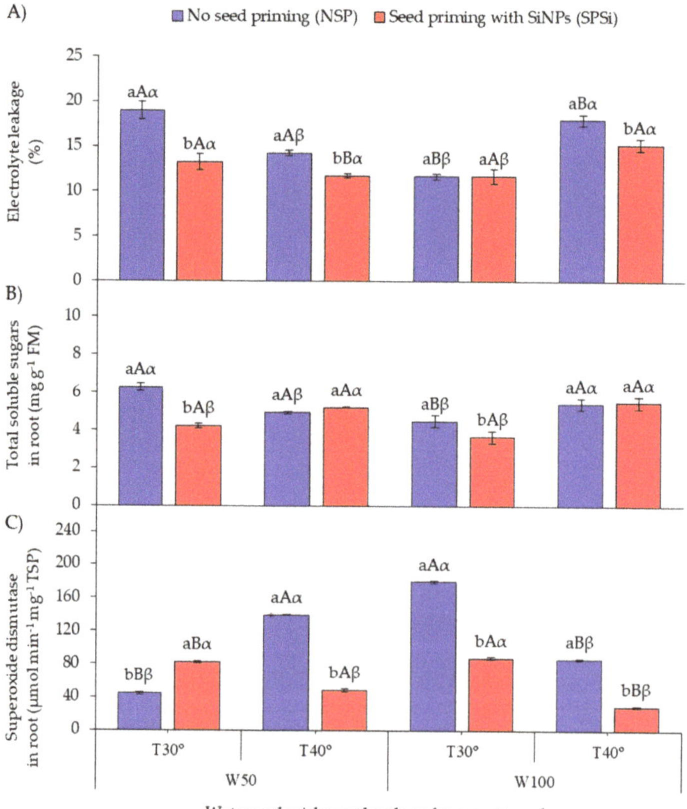

Figure 2. Electrolyte leakage (**A**), total soluble sugars in roots (**B**), and superoxide dismutase activity in roots (**C**) as a function of soil water replenishment levels and temperature change in *Moringa oleifera*. Means followed by the same lowercase letters (a, and b) for seed priming, uppercase letters (A, and B) for soil water replenishment, and Greek letters (α, and β) for temperature change do not differ ($p > 0.05$) by Student's *t*-test. FM, fresh mass; TSP, total soluble proteins.

Figure 2B shows that SP decreased the total soluble sugar content in the roots (TSS-R) of seedlings subjected to water stress (SPSi-W50-T30°) and in the roots of non-stressed seedlings (SPSi-W100-T30°) in relation to those in which SP was not applied (NSP-W50-T30° and NSP-W100-T30°). When SP was not applied, the seedlings subjected to water stress in isolation (NSP-W50-T30°) showed higher TSS-R accumulation.

Figure 2C shows that SP increased the activity of the enzyme superoxide dismutase in the roots (SOD-R) of seedlings subjected to water stress only (SPSi-W50-T30°) in relation to those that did not receive SP (NSP-W50-T30°). When SP was not applied, the seedlings subjected to thermal and water stresses combined (NSP-W50-T40°) and those that were not stressed (NSP-W100-T30°) showed high SOD-R activity. However, SP reduced the enzyme activity in these seedlings (SPSi-W50-T40° and SPSi-W100-T30°). SP also reduced the SOD-R activity in the seedlings only subjected to thermal stress (SPSi-W100-T40°).

Figure 3 shows the differences caused by thermal and water stress on the growth of *M. oleifera* at 35 days after sowing. Under thermal stress (40 °C day/35 °C night), the seedlings showed shorter petioles and smaller leaves with a dark green color. Under water stress (W50), the seedlings showed less height. However, seed priming with SiMPs under red light irradiation reversed the differences caused by the combined stresses.

Figure 3. Seedlings of *Moringa oleifera* subjected to seed priming, soil water replenishment, and temperature change. W50, water stress; W100, no water stress.

The means of the scores for each PC, the means comparison test for these scores, and the means and standard error of the original individual variables can be seen in Table 2. These results reinforce the differences caused by the interaction between temperature variations, soil water replenishment levels and seed priming, as seen in Figure 1.

Table 2. Means of factor scores for each component and means ± standard error of the original variables evaluated as a function of the interaction between seed priming, soil water replacement levels, and temperature change in *Moringa oleifera* seedlings.

Principal Components	No Seed Priming				Seed Priming with SiMPs			
	W50		W100		W50		W100	
	T30°	T40°	T30°	T40°	T30°	T40°	T30°	T40°
	Factor Score Means for Each Component							
PC_1	0.72bBα	−0.82bAβ	1.51aAα	−1.41bBβ	0.93aAα	−0.65aAβ	0.53bBα	−0.82aBβ
PC_2	−0.52aBα	−1.80bBβ	−0.04bAβ	0.80aAα	−0.31aBα	−0.45aBα	1.52aAα	0.81aAβ
PC_3	0.58aBα	0.41aBα	1.25aAα	1.17aAα	−1.56bBβ	−0.81bAα	−0.14bAα	−0.91bAβ
Variables	Means ± Standard error of the original variables							
EL (%)	19.00 ± 0.32	14.27 ± 1.02	11.71 ± 0.66	18.00 ± 0.32	13.27 ± 0.19	11.78 ± 0.92	11.73 ± 0.66	15.27 ± 0.80
RWC (%)	83.72 ± 0.79	99.75 ± 0.37	79.81 ± 0.33	111.88 ± 1.00	79.65 ± 0.33	101.98 ± 0.95	79.62 ± 0.67	95.52 ± 0.86
TDM (g)	0.56 ± 0.01	0.31 ± 0.00	1.03 ± 0.01	0.53 ± 0.01	0.52 ± 0.01	0.31 ± 0.01	0.95 ± 0.03	0.54 ± 0.02
A (μmol of CO_2 m^{-2} s^{-1})	3.09 ± 0.09	2.67 ± 0.07	2.79 ± 0.03	3.13 ± 0.06	2.53 ± 0.13	2.68 ± 0.02	3.30 ± 0.14	2.91 ± 0.03
gs (mol of H_2O m^{-2} s^{-1})	16.00 ± 0.45	20.83 ± 0.05	14.53 ± 0.34	33.97 ± 0.16	14.70 ± 0.46	19.90 ± 0.71	22.93 ± 0.61	24.93 ± 1.09
E (mmol of H_2O m^{-2} s^{-1})	0.49 ± 0.02	0.67 ± 0.01	0.43 ± 0.02	1.10 ± 0.09	0.45 ± 0.00	0.75 ± 0.02	0.57 ± 0.00	0.81 ± 0.01
Ci (μmol m^{-2} s^{-1})	86.37 ± 1.29	181.33 ± 2.68	78.53 ± 0.78	233.07 ± 3.47	119.90 ± 0.18	152.03 ± 1.42	167.17 ± 0.78	199.67 ± 1.56
PRO (μmol g^{-1} FM)	796.05 ± 2.16	60.68 ± 0.83	650.88 ± 10.7	27.25 ± 1.04	475.88 ± 2.06	44.08 ± 0.46	521.93 ± 0.77	28.15 ± 0.35
TSP-L (mg g^{-1} FM)	3.28 ± 0.04	1.08 ± 0.03	8.85 ± 0.03	2.78 ± 0.06	7.04 ± 0.03	1.61 ± 0.03	6.42 ± 0.03	2.61 ± 0.03
TSP-R (mg g^{-1} FM)	2.38 ± 0.12	1.94 ± 0.05	0.92 ± 0.05	3.50 ± 0.16	4.02 ± 0.16	4.58 ± 0.03	3.89 ± 0.14	6.58 ± 0.11
TSS-L (mg g^{-1} FM)	6.41 ± 0.26	6.17 ± 0.14	6.31 ± 0.15	3.83 ± 0.24	8.05 ± 0.13	5.30 ± 0.27	5.76 ± 0.23	4.75 ± 0.06
TSS-R (mg g^{-1} FM)	6.28 ± 0.06	4.93 ± 0.20	4.48 ± 0.30	5.39 ± 0.33	4.23 ± 0.03	5.21 ± 0.13	3.64 ± 0.32	5.47 ± 0.31
SOD-L (μmol min^{-1} mg^{-1} TSP)	69.10 ± 1.23	180.92 ± 5.13	32.55 ± 0.24	89.75 ± 1.57	22.52 ± 1.07	83.62 ± 0.48	21.76 ± 0.42	32.23 ± 0.22
SOD-R (μmol min^{-1} mg^{-1} TSP)	44.26 ± 0.60	138.98 ± 1.54	178.90 ± 0.89	86.46 ± 1.07	82.21 ± 1.57	48.22 ± 1.28	87.32 ± 1.03	29.05 ± 1.32
CAT-L (μmol H_2O_2 min^{-1} mg^{-1} TSP)	7.47 ± 0.19	16.47 ± 0.53	3.25 ± 0.15	11.15 ± 0.68	2.41 ± 0.14	7.41 ± 0.68	1.98 ± 0.13	10.68 ± 0.47
CAT-R (μmol H_2O_2 min^{-1} mg^{-1} TSP)	26.56 ± 1.42	14.81 ± 0.90	42.60 ± 1.29	18.67 ± 0.58	7.66 ± 0.39	5.91 ± 0.13	4.74 ± 0.55	4.40 ± 0.31

PC, Principal component; W50 and W100, soil water replenishment levels; T30° and T40°, temperature change; EL, electrolyte leakage; RWC, relative water content; TDM, total dry matter; A, net photosynthetic rate; gs, stomatal conductance; E, transpiration; Ci, internal CO_2 concentration; PRO, proline; TSP-L, total soluble proteins in leaf; TSP-R, total soluble proteins in root; TSS-L, total soluble sugars in leaf; TSS-R, total soluble sugars in root; SOD-L, superoxide dismutase in leaf; SOD-R, superoxide dismutase in root; CAT-L, catalase in leaf; CAT-R, catalase in root. Lowercase letters compare seed priming, uppercase letters compare soil water replenishment levels, and Greek letters compare temperature variations. Means of factor scores for each component followed by the same lowercase letters (a, and b) for seed priming, uppercase letters (A, and B) for soil water replenishment, and Greek letters (α, and β) for temperature change in the same row do not differ ($p > 0.05$) by Student's t-test.

3. Discussion

M. oleifera is a tropical tree considered tolerant to soil water restriction and high-temperature stress. Although the mechanisms that allow this tolerance are still poorly understood, one of the adaptive metabolic responses when seedlings are exposed to heat stress (35 °C day/18 °C night) is accumulating non-structural carbohydrates, proline, and phenols for osmotic adjustment and reducing water loss to the environment through gas exchange [12].

In our study, in PC_1, the seedlings that were not subjected to SP, reduction of osmotic adjustment indicators (PRO, TSP-L, and TSS-L), and increased gas exchange (gs, E, and Ci) showed an imbalance of the tolerance mechanism resulting from the severity of the thermal stress imposed (40 °C day/35 °C night). The optimal temperature for photosynthesis generally ranges from 20 to 30 °C, with limited enzyme activity and even denaturation of enzymes of the photosynthetic complex at higher temperatures (above 45 °C, for example) [26]. In fact, the increase in CAT-L activity shows that heat stress triggered photosynthetic limitations responsible for decreasing TDM accumulation, justifying the higher RWC in leaflets, which is also related to the water supply resulting from the catalysis of hydrogen peroxide molecules (H_2O_2).

Under water stress conditions, Moringa seedlings show reduced leaf water content, reduced leaf water potential, loss of turgidity, stomatal closure, inhibited photosynthesis, impaired metabolic processes, reduced cell expansion, and affected growth and development [17]. In our research, in PC_2, physiological disorders were also observed in seedlings subjected to water stress (50% of ETc) since the SOD-L activity increased, possibly to catalyze the dismutation of the superoxide anion ($O_2^{\bullet-}$) in O_2 and H_2O_2 while decreasing A and TDM accumulation due to the reduction in guard cell turgidity and the consequent reduction in the stomatal opening, resulting in lower cell expansion and seedling growth.

When SPSi was applied with RL irradiation, the effects of the isolated or combined heat and water stresses were mitigated by the increased A, PRO, TSP-L, TSP-R, and TSS-L levels under these stress conditions, in addition to TDM accumulation and reductions in RWC, gs, E, Ci, and CAT-L, CAT-R, and SOD-L activity (Figure 4). This mitigation occurs as the SiMPs cross cell wall barriers and, when in contact with plant cells, are captured by plasmodesmata pathways and translocated by apoplastic and/or symplastic pathways [19,23], reducing lipid peroxidation by decreasing the H_2O_2 levels, which restores the redox balance and mitigates oxidative stress [25]. In addition, RL could have potentiated the effect of SiMPs since RL irradiation activates the phyA, phyB, phyC, phyD, and phyE phytochromes responsible for light capture and modulation, gene expression, and the transduction of signals related to growth regulation, plant development, metabolic activities, and responses to heat and water stress [2,27].

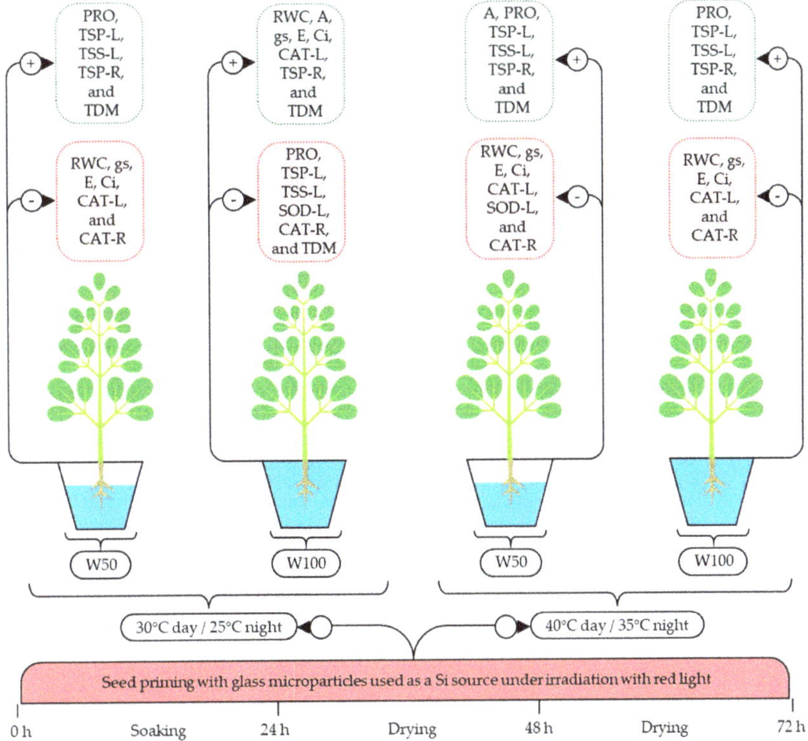

Figure 4. Effect of seed priming with SiMPs under red light irradiation on seedlings of *Moringa oleifera* subjected to thermal and water stress. RWC, relative water content; TDM, total dry matter; A, net photosynthetic rate; gs, stomatal conductance; E, transpiration; Ci, internal CO_2 concentration; PRO, proline; TSP-L, total soluble proteins in leaves; TSP-R, total soluble proteins in roots; TSS-L, total soluble sugars in leaves; SOD-L, superoxide dismutase in leaves; CAT-L, catalase in leaves; CAT-R, catalase in roots; W50, water stress; W100, no water stress.

EL increased in the leaflets of *M. oleifera* seedlings subjected to heat and water stress when SP was not applied, which was also noted by [28], when they observed that water deficit (18 days without irrigation) induced an EL increase since the accumulation of reactive oxygen species (ROS) caused lipid peroxidation and rupture of the thylakoid membrane, thus limiting photosynthesis. However, in our study, there was a reduction in EL in *M. oleifera* seedlings subjected to thermal and water stress when SP was applied, which could be due to the role of SiMPs in modulating cell wall plasticity, leaf thickness, water-plant ratio, and plant metabolism and defense [29].

The highest TSS-R accumulation in seedlings generated without SP and subjected to water stress was due to osmotic adjustment since, under water restriction conditions, plants prevent water loss through stomatal closure, growth modulation, and proline and total soluble sugar accumulation while the root growth is stimulated, characteristics associated with tolerance to water deficit [30]. However, when SP was applied, the TSS-R of *M. oleifera* seedlings decreased under water stress, indicating that SiMPs can act in the root system of the species as an osmotic adjuster in place of organic osmoregulatory since Si deposition on the cell wall, besides acting as a physical barrier against water loss, also increases the osmotic potential for maintaining the water status [31].

From this perspective, based on the increased SOD-R activity observed in seedlings generated by SP when subjected to water restriction, it can be inferred that *M. oleifera* seedlings activate the antioxidant mechanism of the root system instead of performing osmotic adjustment by organic pathways to overcome the negative effects of water stress. On the other hand, SOD-R decreased when heat stress was imposed, indicating that the tolerance induced by SiMPs in the root system is more efficient under water stress than under heat stress. A SOD-R increase was also observed by [32] in response to water stress during the initial stage of plant development, which was justified by the root being the main organ of initial perception of low soil water availability.

Therefore, as already stressed by [2], the photomorphogenesis of *Moringa oleifera* is mediated by the interaction between light, water, and phytohormones during SP, reinforcing the idea that RL has a synergistic effect on the action of SiMPs since irradiation with RL activates phyB, whose activity reduces the levels of abscisic acid (ABA), a germination repressor, and increases the levels of gibberellic acid (GA), which stimulates germination [33] and could have generated a memory effect when the seeds were dehydrated. This information is essential for sustainably managing agroecosystems cultivated with *M. oleifera*, especially in semi-arid regions where thermal and water stresses are frequent.

4. Materials and Methods

4.1. Experimental Design

The experiment was set up in a completely randomized design arranged as a $2 \times 2 \times 2$ factorial, with 5 replicates, totaling 40 experimental units. The seed priming factor (SP) consisted of a control (NSP—no seed priming, n = 20), in which the seeds were not treated, and seed priming with glass microparticles used as a Si source (SPSi, n = 20) under irradiation with monochromatic red light (RL) with an emission of 184 lumens m^{-2} and wavelengths from 600 to 680 nm. The soil water replenishment factor (SWR) consisted of two levels (W50—50% and W100—100% of crop evapotranspiration). The temperature change factor (TC) consisted of two temperature variations (TC30°—30 °C day/25 °C night and TC40°—40 °C day/35 °C night, with a 12-h photoperiod).

4.2. Seed Priming Application

Seed priming was performed at the Laboratory of Plant Physiology of the Agricultural Engineering Academic Unit (UAEA) of the Center of Technology and Natural Resources (CTRN) of the Federal University of Campina Grande (UFCG), located in the municipality of Campina Grande, Paraíba, Brazil. The location has a semi-arid climate, a mean temperature of 25 °C, and a relative air humidity ranging from 72 to 91% [34].

A biochemical oxygen demand (B.O.D.) germination chamber was adapted to provide monochromatic red light using panels with RGB (red, green, and blue) LED (light emitting diode) lamps with an emission of 184 lumens m^{-2}. Then, the seeds of *Moringa oleifera* Lam. obtained from three parent plants that were three years old located in Catolé do Rocha,

Paraíba, Brazil, were disinfected with sodium hypochlorite (1%) for three minutes [35] at 25 °C and under RL irradiation.

Next, the seeds were placed in Gerbox® boxes measuring $11 \times 11 \times 3.5$ cm in length, width, and height, respectively, for soaking in a solution containing residual glass microparticles used as a silicon source (600 mg L^{-1} of SiMPs). For that purpose, residual amber glass resulting from beverage bottles was collected at the municipal landfill of Toledo, Paraná, Brazil and contains in its composition more than 75% of SiO_2 and other elements such as Fe, S, Na, K, Ca and Al. The clean and dry bottles were manually ground and sieved through a 400 Tyler mesh fabric to obtain an upper powder particle size below 38 μm. Next, the boxes were placed in the germination chamber at 25 °C for 24 h. Subsequently, the seeds were transferred to open Gerbox® boxes with two layers of dry germitex paper and dried for 72 h at the same light and temperature conditions used during inhibition.

4.3. Seedling Formation, Temperature Variations, and Soil Water Replenishment

The control seeds (NSP—no seed priming) and those obtained by seed priming (SPSi) were sown in pots with a volumetric capacity of 0.3 dm^3 filled with a substrate composed of sandy soil and earthworm humus at a ratio of 3:1 and with moisture close to field capacity. The pots containing the seeds were then transferred to a Phytotron growth chamber (Weiss Technik, Technal) located at the Experimental Unit belonging to the State University of Paraíba (UEPB), Campina Grande, Paraíba, Brazil.

Given the availability of a single phytotron chamber and in order to meet the temperature variations, the experiment was conducted in two stages, the first with the condition of 30 °C day/25° night and an air relative humidity (RH) ranging from 50 to 60% (T30°), and the second with 40 °C day/35 °C night and an RH ranging from 40 to 50%.

Substrate moisture management was performed in a daily irrigation shift using the weighing method [36], according to which the water lost by crop evapotranspiration (ETc) is replenished. The imposition of water stress caused by the replenishment of 50% ETc (W50) was carried out 18 days after sowing (DAS), while the other plots without water stress were irrigated with 100% ETc (W100).

4.4. Variables Evaluated

The gas exchange variables were evaluated 30 days after sowing (DAS), whereas cell membrane integrity and the leaf water status, osmotic adjustment indicators, antioxidant mechanism activity, and total dry matter accumulation were evaluated 35 days after sowing. The analyses were performed at the Laboratory of Ecophysiology of Cultivated Plants (ECOLAB) of UEPB, located in the Integrated Research Complex Três Marias.

4.4.1. Exchange Evaluations

Gas exchange evaluation was performed in three fully expanded leaflets counted from the base of the seedling to its apex, from 8:00 a.m. to 9:00 a.m., using an infrared gas analyzer—IRGA (Infra-red Gas Analyzer)—GFS 3000 FL with a CO_2 concentration of 400 ppm and an artificial light source to provide a photon flux of 1000 $\mu mol\ m^{-2}\ s^{-1}$. The following variables were measured: net photosynthetic rate (A, μmol of $CO_2\ m^{-2}\ s^{-1}$), stomatal conductance (gs, mol of $H_2O\ m^{-2}\ s^{-1}$), transpiration (E, mmol of $H_2O\ m^{-2}\ s^{-1}$), and internal CO_2 concentration (Ci, $\mu mol\ m^{-2}\ s^{-1}$).

4.4.2. Cell Membrane Integrity and Leaf Water Status

The methodology described in [37] (with adaptations) was used to analyzed the intracellular electrolyte leakage (EL, %) and the relative water content (RWC, %), for which 12 leaf discs measuring 113 mm^2, 6 for EL and 6 for RWC, were collected 35 DAS using a copper pourer.

The EL was determined by washing the leaf discs three times in deionized water to remove the solutes released during cutting. Then, the material was placed in Petri dishes containing 6 mL of deionized water, after which the plates were stored at 25 °C for two hours. After incubation, the electrical conductivity in the medium (ECi) was determined using a portable conductivity meter (WATERPROOF). Then, the samples were subjected to a temperature of 80 °C for 90 min, after which the conductivity was again measured (ECf), and the leakage of electrolytes was quantified using Equation (1).

$$EL = \left(\frac{EC_i}{EC_f} \right) * 100\% ,\tag{1}$$

where EL is cell leakage, EC_i is the initial electrical conductivity of the medium (dS m^{-1}), and EC_f is the final electrical conductivity of the medium.

The RWC was determined by weighing the leaf disks to determine the fresh mass (FMD), after which the material was immediately placed in Petri dishes containing 6 mL of deionized water. Then, the dishes were placed in a B.O.D. incubator at 25 °C and 202 lumens m^{-2}. After four hours of exposure, the disks were dried with filter paper and weighed to obtain the turgid disk mass (TMD). Subsequently, the plant material was packed in paper bags and transferred to a forced-air oven at 60 °C for 48 h. Finally, the material was weighed to determine the dry disk mass (DMD), and the RWC was quantified using Equation (2).

$$RWC = \left(\frac{FMD - DMD}{TMD - DMD} \right) * 100\% ,\tag{2}$$

where RWC is the relative water content, FMD is the fresh mass, MSD is the dry mass, and DMD is the turgid disk mass.

4.4.3. Indicators of Osmotic Adjustment

The proline content (PRO, μmol g^{-1} FM—fresh matter) was determined by the colorimetric method described by [38] and modified in [39]. Initially, 250 mg of fresh leaf tissue was weighed and macerated in 5 mL of 3% sulfosalicylic acid, followed by centrifugation at 2000 rpm for 10 min. Then, the supernatant was removed and stored in tubes with a capacity of 2.5 mL for later determination of the PRO concentration.

The extraction of total soluble proteins (TSP, mg g^{-1} FM) in the leaflets (TSP-L) and roots (TSP-R) was performed using 200 mg of fresh leaflet and root mass, respectively. The plant material was macerated and then received 3.0 mL of potassium phosphate buffer (100 mM, pH 7.0 + EDTA 1 mM), after which the material was stored in Eppendorf tubes for subsequent centrifugation (5000× g) for 10 min in a refrigerated centrifuge (4 °C). After extraction, the TSP concentration was determined according to the methodology proposed by [40].

The extract used to measure the total soluble sugars (TSS, mg g^{-1} FM) in the leaflets (TSS-L) and roots (TSS-R) was obtained from 200 mg of fresh leaflet mass and 100 mg of fresh root mass. Initially, these samples were macerated in 2 mL of 80% ethanol (v/v). Then, the extract was added to Eppendorf tubes with a capacity of 2 mL and taken to a water bath (60 °C) for 30 min, after which the tubes were transferred to a centrifuge (2000× g) to obtain and collect the supernatant. After removing the supernatant, another 2 mL of ethanol (80%) was added to the same tubes for a new extraction, followed by heating in a water bath and subsequent transfer to the centrifuge. The supernatants resulting from the two washes were mixed in Falcon tubes and stored in Eppendorf tubes, totaling 4 mL of extract per sample. The concentrations of TSS-L and TSS-R were determined by the "phenol-sulfuric" method described by [41].

4.4.4. Activity of the Antioxidant Mechanism

The antioxidant mechanism activity was evaluated by measuring the enzyme activities of superoxide dismutase (SOD, μmol min^{-1} mg^{-1} TSP) and catalase (CAT, μmol H$_2$O$_2$ min^{-1} mg^{-1} TSP) in the leaflets (SOD-L and CAT-L) and roots (SOD-R and CAT-R) was determined using the enzyme extract obtained by the same procedure described TPS.

In the determination of SOD activity, the reaction mixture consisting of 0.3 mL of 130 µM methionine, 0.1 mL of 2250 µM p-nitro tetrazolium blue (NBT), 0.1 mL of 3 µM EDTA, 0.2 mL of riboflavin, 0.75 mL of deionized water, and 1.5 mL of 50 mM sodium phosphate buffer at pH 7.8 received 100 µL of the crude enzyme extract. Then, the absorbance was determined at 560 nm, which was subtracted from the absorbance reading of the reaction mixture without the enzyme extract. Under these conditions, one unit of SOD corresponded to the amount of enzyme required to inhibit the photoreduction of NBT by 50% [42].

CAT activity was determined by adding 100 µL of the crude enzymatic extract to 2.9 mL of the reaction medium consisting of 500 µL of 59 mM hydrogen peroxide (H_2O_2), 1.5 mL of 0.05 M potassium phosphate buffer at pH 7.0, and 400 µL of deionized water at 30 °C [43]. Enzyme activity was determined by the reduction in absorbance at 240 nm.

4.4.5. Total Dry Matter Accumulation

The total dry matter accumulation (TDM, g) was quantified in the seedlings by sectioning the plants into leaflets, stem, branches, and roots. Then, the plant material was packed in properly identified paper bags and dried to a constant weight in a forced-air oven at 70 °C. After drying, the plant material was weighed on an analytical balance accurate to 0.0001 g, and the TDM was determined by the sum of the dry matter of leaflets, stem, branches, and roots.

4.5. Statistical Analysis

The data on the response variables were subjected to the Shapiro–Wilk normality test [44]. Once the assumptions of normality were met, the data on each variable were standardized to obtain the Z variable with a null mean ($\overline{X}$ = 0.0) and the unit variance (s^2 = 1.0) according to Equation (3).

$$Z = \frac{X - \overline{X}}{s^2},\tag{3}$$

where: X corresponds to each observation of the data set of the variable, $\overline{X}$ is the mean, and s^2 is the variance of the data set.

The transformed data were subjected to the exploratory procedure of principal component analysis (PCA). The choice of principal components (PCs) was based on eigenvalues higher than one ($\lambda > 1.0$) according to the criterion proposed in [45], which explained a percentage of the total variance higher than 10% [2,46]. The original data referring to each PC were subjected to multivariate analysis of variance (MANOVA) by Hotelling's T-squared test.

The variables not associated with any PC were removed from the PCA and subjected to univariate analysis of variance (ANOVA) by the F-test at 95% of confidence [47]. These analyses were performed using the software Statistica v. 7.0 [48].

5. Conclusions

Seed priming with residual glass microparticles used as a silicon source under red light irradiation mitigated the effects of thermal and water stress in seedlings of Moringa oleifera through the homeostasis of gas exchange, leaf water status, osmotic adjustment, and the plant antioxidant mechanism. Additionally, it was shown that silicon can be provided through an accessible and residual glass source on a micrometer scale, which can be achieved through traditional grinding and sieving processes.

Author Contributions: Conceptualization, R.L.d.S.F.; methodology, R.L.d.S.F., P.d.S.C. and J.D.N.; validation, C.A.V.d.A., A.S.d.M. and J.D.N.; investigation, P.d.S.C., S.I.B., I.E.C., R.S.d.A. and Y.L.M.; resources, R.L.d.S.F. and A.S.d.M.; writing—original draft preparation, P.d.S.C., R.L.d.S.F. and J.D.N.; writing—review and editing, P.d.S.C., R.L.d.S.F., J.D.N., A.S.d.M., C.A.V.d.A., I.D.M., A.R.N. and R.S.;

supervision, J.D.N., C.A.V.d.A. and A.S.d.M.; funding acquisition, R.L.d.S.F. and A.S.d.M. All authors have read and agreed to the published version of the manuscript.

Funding: This research was funded by the National Council for Scientific and Technological Development of Brazil-CNPq, Research Project No. 160862/2019-1 and the Federal Management Council of the Fund for the Defense of Diffuse Rights (CFDD)—Process No. 08000.012516/2019-27.

Data Availability Statement: Not applicable.

Acknowledgments: The Coordination for the Improvement of Higher Education Personnel—Brazil (CAPES)—Finance Code 001. To the Fundação de Amparo à Pesquisa do Estado da Paraíba (FAPESQ-PB) for granting financial aid (Proc. 030/2018).

Conflicts of Interest: The authors declare no conflict of interest.

References

1. Calone, R.; Mircea, D.M.; González-Orenga, S.; Boscaiu, M.; Lambertini, C.; Barbanti, L.; Vicente, O. Recovery from Salinity and Drought Stress in the Perennial *Sarcocornia fruticosa* vs. the Annual *Salicornia europaea* and *S. veneta*. *Plants* **2022**, *11*, 1058. [CrossRef] [PubMed]

2. Costa, P.S.; Ferraz, R.L.S.; Dantas-Neto, J.; Martins, V.D.; Viégas, P.R.A.; Meira, K.S.; Ndhlala, A.R.; Azevedo, C.A.V.; Melo, A.S. Seed priming with light quality and *Cyperus rotundus* L. extract modulate the germination and initial growth of *Moringa oleifera* Lam. Seedlings. *Braz. J. Biol.* **2022**, *84*, 255836. [CrossRef] [PubMed]

3. Domenico, M.; Lina, C.; Francesca, B. Sustainable crops for food security: Moringa (*Moringa oleifera* Lam.). *Encycl. Food Secur. Sustain.* **2019**, *1*, 409–415. [CrossRef]

4. Garcia, T.B.; Soares, A.A.; Costa, J.H.; Costa, H.P.S.; Neto, J.X.S.; Rocha-Bezerra, L.C.B.; Silva, F.D.A.; Arantes, M.R.; Sousa, D.O.B.; Vasconcelos, I.M.; et al. Gene expression and spatiotemporal localization of antifungal chitinbinding proteins during *Moringa oleifera* seed development and germination. *Planta* **2019**, *249*, 1503–1519. [CrossRef]

5. Karthickeyan, V. Effect of cetane enhancer on *Moringa oleifera* biodiesel in a thermal coated direct injection diesel engine. *Fuel* **2019**, *235*, 538–550. [CrossRef]

6. Páramo-Calderón, D.E.; Aparicio-Saguilán, A.; Aguirrecruz, A.; Carrillo-Ahumada, J.; Hernández-Uribe, J.P.; Acevedo-Tello, S.; Torruco-Uco, J.G. Tortilla added with Moringa oleífera flour: Physicochemical, texture properties and antioxidant capacity. *LWT* **2019**, *100*, 409–415. [CrossRef]

7. Macário, A.P.S.; Ferraz, R.L.S.; Costa, P.S.; Brito Neto, J.F.; Melo, A.S.; Dantas Neto, J. Allometric models for estimating *Moringa oleifera* leaflets area. *Ciênc. Agrotecnol.* **2020**, *44*, 005220. [CrossRef]

8. Parveen, S.; Rasool, F.; Akram, M.N.; Khan, N.; Ullah, M.; Mahmood, S.; Rabbani, G.; Manzoor, K. Effect of *Moringa olifera* leaves on growth and gut microbiota of Nile tilapia (*Oreochromis niloticus*). *Braz. J. Biol.* **2021**, *84*, 250916. [CrossRef]

9. Vasconcelos, M.C.; Costa, J.C.; Sousa, J.P.S.; Santana, F.V.; Soares, T.F.S.N.; Oliveira Júnior, L.F.G.; Silva-Mann, R. Biometric and physiological responses to water restriction in *Moringa oleifera* seedlings. *Floresta Ambient.* **2019**, *26*, 20150165. [CrossRef]

10. Azam, A.; Nouman, W.; Rehman, U.; Ahmed, U.; Gull, T.; Shaheen, M. Adaptability of *Moringa oleifera* Lam. under different water holding capacities. *S. Afr. J. Bot.* **2020**, *129*, 299–303. [CrossRef]

11. Boumenjel, A.; Papadopoulos, A.; Ammari, Y. Growth response of *Moringa oleifera* (Lam) to water stress and to arid bioclimatic conditions. *Agrofor. Syst.* **2021**, *95*, 823–833. [CrossRef]

12. Mohammed, F.; Tesfay, S.Z.; Mabhaudhi, T.; Chimonyo, V.G.P.; Modi, A.T. Biochemical response of *Moringa oleifera* to temperature. *Acta Hortic.* **2021**, *1306*, 43–50. [CrossRef]

13. Bhadra, P.; Maitra, S.; Shankar, T.; Hossain, A.; Praharaj, S.; Aftab, T. Climate change impact on plants: Plant responses and adaptations. In *Plant Perspectives to Global Climate Changes: Developing Climate-Resilient Plants*, 1st ed.; Aftab, T., Roychoudhury, A., Eds.; Academic Press: London, UK, 2022; pp. 1–24.

14. Melo, A.S.; Melo, Y.L.; Lacerda, C.F.; Viégas, P.R.A.; Ferraz, R.L.S.; Gheyi, H.R. Water restriction in cowpea plants [*Vigna unguiculata* (L.) Walp.]: Metabolic changes and tolerance induction. *Rev. Bras. Eng. Agríc. Ambiental.* **2022**, *26*, 190–197. [CrossRef]

15. Muhl, Q.E.; du Toit, E.S.; Robbertse, P.J. Temperature effect on seed germination and seedling growth of *Moringa oleifera* Lam. *Seed Sci. Technol.* **2011**, *39*, 208–213. [CrossRef]

16. Costa, T.R.; Reis, V.C.R.; Ferreira, R.R.; Silva, L.S.; Gonzaga, A.P.D. Influência dos tratamentos pré-germinativos, térmicos e regime de luz na germinação de sementes de moringa (*Moringa oleifera* Lam.). *Divers. J.* **2021**, *6*, 3763–3778. [CrossRef]

17. Galgaye, G.G.; Beshir, H.M.; Roro, A.G. Physiological Responses of Moringa (*Moringa stenopetala* L.) Seedlings to Drought Stress under Greenhouse Conditions, Southern Ethiopia. *Asian J. Biotechnol.* **2020**, *12*, 97–107. [CrossRef]

18. Jalil, S.U.; Ansari, M.I. Nanoparticles and abiotic stress tolerance in plants: Synthesis, action, and signaling mechanisms. In *Plant Signaling Molecules: Role and Regulation under Stressful Environments*, 1st ed.; Khan, M.I.R., Ferrante, A., Reddy, P.S., Khan, N.A., Eds.; Woodhead Publishing: Cambridge, UK, 2019; pp. 549–561.

19. Singh, A.; Tiwari, S.; Pandey, J.; Lata, C.; Singh, I.K. Role of nanoparticles in crop improvement and abiotic stress management. *J. Biotechnol.* **2021**, *337*, 57–70. [CrossRef] [PubMed]

20. Tariq, M.; Choudhary, S.; Singh, H.; Siddiqui, M.A.; Kumar, H.; Amir, A.; Kapoor, N. Role of nanoparticles in abiotic stress. In *Technology in Agriculture*, 1st ed.; Ahmad, F., Sultan, M., Eds.; IntechOpen: London, UK, 2021; pp. 1–16.

21. Zellner, W.; Tubaña, B.; Rodrigues, F.A.; Datnoff, L.E. Silicon's role in plant stress reduction and why this element is not used routinely for managing plant health. *Plant Dis.* **2021**, *105*, 2033–2049. [CrossRef] [PubMed]

22. Ali, S.; Rizwan, M.; Hussain, A.; ur Rehman, M.Z.; Ali, B.; Yousaf, B.; Wijaya, L.; Alyemeni, M.N.; Ahmad, P. Silicon nanoparticles enhanced the growth and reduced the cadmium accumulation in grains of wheat (*Triticum aestivum* L.). *Plant Physiol. Biochem.* **2019**, *140*, 1–8. [CrossRef] [PubMed]

23. Hussain, A.; Rizwan, M.; Ali, Q.; Ali, S. Seed priming with silicon nanoparticles improved the biomass and yield while reduced the oxidative stress and cadmium concentration in wheat grains. *Environ. Sci. Pollut. Res. Int.* **2019**, *26*, 7579–7588. [CrossRef] [PubMed]

24. Younis, A.; Khattab, H.; Emam, M.M. Impacts of silicon and silicon nanoparticles on leaf ultrastructure and TaPIP1 and TaNIP2 gene expressions in heat stressed wheat seedlings. *Biol. Plant.* **2020**, *64*, 343–352. [CrossRef]

25. Ali, L.G.; Nulit, R.; Ibrahim, M.H.; Yien, C.Y.S. Efficacy of KNO_3, SiO_2 and SA priming for improving emergence, seedling growth and antioxidant enzymes of rice (*Oryza sativa*), under drought. *Sci. Rep.* **2021**, *11*, 3864. [CrossRef] [PubMed]

26. Yamori, W.; Hikosaka, K.; Way, D.A. Temperature response of photosynthesis in C_3, C_4, and CAM plants: Temperature acclimation and temperature adaptation. *Photosynth. Res.* **2014**, *119*, 101–117. [CrossRef] [PubMed]

27. Wang, Y.; Bian, Z.; Pan, T.; Cao, K.; Zou, Z. Improvement of tomato salt tolerance by the regulation of photosynthetic performance and antioxidant enzyme capacity under a low red to far-red light ratio. *Plant Physiol. Biochem.* **2021**, *167*, 806–815. [CrossRef]

28. Martins, A.C.; Larré, C.F.; Bortolini, F.; Borella, J.; Eichholz, R.; Delias, D.; Amarante, L. Tolerância ao déficit hídrico: Adaptação diferencial entre espécies forrageiras. *Iheringia* **2018**, *73*, 228–239. [CrossRef]

29. Mukarram, M.; Khan, M.M.A.; Corpas, F.J. Silicon nanoparticles elicit an increase in lemongrass (*Cymbopogon flexuosus* (Steud.) Wats) agronomic parameters with a higher essential oil yield. *J. Hazard. Mater.* **2021**, *412*, 125254. [CrossRef]

30. Lozano-Montaña, P.A.; Sarmiento, F.; Mejía-Sequera, L.M.; Álvarez-Flórez, F.; Melgarejo, L.M. Physiological, biochemical and transcriptional responses of *Passiflora edulis* Sims f. *edulis* under progressive drought stress. *Sci. Hortic.* **2021**, *275*, 109655. [CrossRef]

31. Sarkar, M.M.; Mathur, P.; Roy, S. Silicon and nano-silicon: New frontiers of biostimulants for plant growth and stress amelioration. In *Silicon and Nano-Silicon in Environmental Stress Management and Crop Quality Improvement: Progress and Prospects*, 1st ed.; Etesami, H., El-Ramady, H., Pessarakli, M., Al Saeedi, A.H., Fujita, M., Hossain, M.A., Eds.; Academic Press: London, UK, 2022; pp. 17–36.

32. Deus, K.E.; Lanna, A.C.; Abreu, F.R.M.; Silveira, R.D.D.; Pereira, W.J.; Brondani, C.; Vianello, R.P. Molecular and biochemical characterization of superoxide dismutase (SOD) in upland rice under drought. *Aust. J. Crop Sci.* **2015**, *9*, 744–753.

33. Neff, M.M. Light-mediated seed germination: Connecting phytochrome B to gibberellic acid. *Dev. Cell* **2012**, *22*, 687–688. [CrossRef]

34. Ferraz, R.L.S.; Costa, P.S.; Magalhães, I.D.; Viégas, P.R.A.; Dantas Neto, J.; Melo, A.S. Physiological adjustments, yield increase and fiber quality of 'BRS Rubi' naturally colored cotton under silicon doses. *Rev. Caatinga* **2022**, *35*, 371–381. [CrossRef]

35. Carvalho, D.B.; Carvalho, R.I.N. Qualidade fisiológica de sementes de guanxuma em influência do envelhecimento acelerado e da luz. *Acta Sci. Agron.* **2009**, *31*, 489–494. [CrossRef]

36. Silva, A.E.; Ferraz, R.L.S.; Silva, J.P.; Costa, P.S.; Viégas, P.R.A.; Brito Neto, J.F.; Melo, A.S.; Meira, K.S.; Soares, C.S.; Magalhães, I.D.; et al. Microclimate changes, photomorphogenesis, and water consumption by Moringa oleifera cuttings under light spectrum variations and exogenous phytohormones concentrations. *Aust. J. Crop Sci.* **2020**, *14*, 751–760. [CrossRef]

37. Brito, G.G.; Sofiatti, V.; Lima, M.M.A.; Carvalho, L.P.; Silva Filho, J.L. Physiological traits for drought phenotyping in cotton. *Acta Sci. Agron.* **2011**, *33*, 117–125. [CrossRef]

38. Bates, L.S.; Waldren, R.P.; Teare, I.D. Rapid determination of free proline for water-stress studies. *Plant Soil* **1973**, *39*, 205–207. [CrossRef]

39. Bezerra Neto, E.; Barreto, L.P.N. *Análises Químicas e Bioquímicas em Plantas*, 1st ed.; UFRPE: Recife, Brazil, 2011; p. 267.

40. Bradford, M.M. A rapid and sensitive method for the quantitation of microgram quantities of protein utilizing the principle of protein-dye binding. *Anal. Biochem.* **1976**, *72*, 248–254. [CrossRef]

41. Dubois, M.; Gilles, K.A.; Hamilton, J.K.; Rebers, P.A.; Smith, F. Colorimetric method for determination of sugars and related substances. *Anal. Chem.* **1956**, *28*, 350–356. [CrossRef]

42. Giannopolitis, C.N.; Ries, S.K. Superoxide dismutases: I. occurrence in higher plants. *Plant Physiol.* **1977**, *59*, 309–314. [CrossRef]

43. Havir, E.A.; McHale, N.A. Biochemical and developmental characterization of multiple forms of catalase in tobacco leaves. *Plant Physiol.* **1987**, *84*, 450–455. [CrossRef]

44. Shapiro, S.S.; Wilk, M.B. An analysis of variance test for normality (complete samples). *Biometrika* **1965**, *52*, 591–609. [CrossRef]

45. Kaiser, H.F. The Varimax Criterion for Analytic Rotation in Factor Analysis. *Psychometrika* **1958**, *23*, 187–200. [CrossRef]

46. Hair, J.F., Jr.; Black, W.C.; Babin, B.J.; Anderson, R.E.; Tatham, R.L. *Análise Multivariada de Dados*, 6th ed.; Bookman: Porto Alegre, Brazil, 2009; p. 688.

47. Barbosa, J.C.; Maldonado Junior, W. *AgroEstat: Sistema para Análises Estatísticas de Ensaios Agronômicos*, 1st ed.; FCAV/UNESP: Jaboticabal, Brazil, 2015; p. 396.

48. STATSOFT Inc. *Statistica: Data Analysis Software System*, version 7; STATSOFT Inc.: Tulsa, OK, USA, 2004.

Article

Exogenously Applied GA₃ Enhances Morphological Parameters of Tolerant and Sensitive *Cyclamen persicum* Genotypes under Ambient Temperature and Heat Stress Conditions

Mihaiela Cornea-Cipcigan [1], Mirela Irina Cordea [1,*], Rodica Mărgăoan [2,*] and Doru Pamfil [3]

[1] Department of Horticulture and Landscaping, Faculty of Horticulture,
    University of Agricultural Sciences and Veterinary Medicine, 400372 Cluj-Napoca, Romania;
    mihaiela.cornea@usamvcluj.ro
[2] Laboratory of Cell Analysis and Spectrometry, Advanced Horticultural Research Institute of Transylvania,
    University of Agricultural Sciences and Veterinary Medicine of Cluj-Napoca, 400372 Cluj-Napoca, Romania
[3] Research Centre for Biotechnology in Agriculture Affiliated to Romanian Academy,
    University of Agricultural Sciences and Veterinary Medicine, 400372 Cluj-Napoca, Romania;
    presedinte.arfc@academia-cj.ro
*   Correspondence: mcordea@usamvcluj.ro (M.I.C.); rodica.margaoan@usamvcluj.ro (R.M.)

Abstract: *Cyclamen* genus is part of the Primulaceae family consisting of 24 species widely cultivated as ornamental and medicinal plants. They also possess high plasticity in terms of adaptability to alternating environmental conditions. In this regard, the present study investigates the germination and morphological parameters of heat-tolerant and heat-sensitive *Cyclamen persicum* accessions in the presence of different GA₃ solutions (0, 30, 70 and 90 mg/L) under ambient temperature and heat stress conditions. Heat-tolerant genotypes, mainly C3-Smartiz Victoria (6.42%), C15-Merengue magenta (6.47%) and C16-Metis silverleaf (5.12%) had the highest germination rate with 90 mg/L GA₃ treatment compared with control. Regarding heat-sensitive genotypes, C11-Verano (5.11%) and C13-Metis Origami (4.28%) had the lowest values in mean germination time, along with the Petticoat genotypes C1 (73.3%) and C2 (80.0%) with a high germination percentage. Heat-tolerant genotypes positively responded to GA₃ (70 and 90 mg/L) even under heat stress conditions, by their higher values in plant height, an ascending trend also seen in heat-sensitive genotypes under GA₃ treatment (70 and 90 mg/L). According to the hierarchical clustering, several heat-tolerant genotypes showed peculiar behavior under heat stress conditions, namely C3 (Smartiz Victoria), C7 (Halios falbala) and C8 (Latinia pipoca) which proved to be susceptible to heat stress even under GA₃ application, compared with the other genotypes which showed tolerance to higher temperatures. In the case of heat-sensitive genotypes, C4 (Smartiz violet fonce), C6 (Metis blank pur), C11 (Verano) and C13 (Metis origami) possessed higher positive or negative values compared with the other heat-sensitive genotypes with increased doses of GA₃. These genotypes were shown to be less affected by heat stress, suggesting their positive response to hormone treatment. In conclusion, the above-mentioned genotypes, particularly heat-tolerant C15 and heat-sensitive C2 with the highest germination capacity and development can be selected as heat-resistant genotypes to be deposited in gene banks and used in further amelioration programs under biotic and/or abiotic stresses to develop resistant genotypes.

Keywords: crop tolerance; germination; heat mapping; heat stress; plant development

Citation: Cornea-Cipcigan, M.; Cordea, M.I.; Mărgăoan, R.; Pamfil, D. Exogenously Applied GA₃ Enhances Morphological Parameters of Tolerant and Sensitive *Cyclamen persicum* Genotypes under Ambient Temperature and Heat Stress Conditions. *Plants* **2022**, *11*, 1868. https://doi.org/10.3390/plants11141868

Academic Editors: Alberto Soares De Melo and Hans Raj Gheyi

Received: 28 June 2022
Accepted: 14 July 2022
Published: 18 July 2022

Publisher's Note: MDPI stays neutral with regard to jurisdictional claims in published maps and institutional affiliations.

1. Introduction

Abiotic stress has become a key area of concern in crop production as a result of global warming. Substantial research is being carried out to develop approaches that deal with abiotic stresses, to develop heat-, sun-, and drought-tolerant cultivars, adjust crop calendars, and resource management approaches, among other things [1]. As these practices are costly and time-consuming, recent studies denote that the introduction of high-yield phenotyping techniques, along with other non-destructive methods, has increasingly taken over the plant phenomics area, where novel technologies such as non-invasive imaging, spectroscopy and high computation performance are combined to assess phenotypic performances at high resolution, flow and accuracy. This will aid breeders and plant scientists to conduct phenotypic experiments with large plant populations in different environments to non-destructively monitor the performance of plants over time [2].

Cyclamen is widely distributed in the Mediterranean regions of Cyprus, Rhodes and (Eastern) Crete, as well as coastal areas of the Eastern Aegean and from southern Turkey to northern Israel [3]. It presents adaptation to habitat modifications and environmental changes [4], but its exploitation has had negative effects on native populations. Special attention is also regarded to their antioxidant and strong anticancer activity [5–8], but as they mostly germinate from seeds and have a slow development, they are hard to use extensively. Moreover, being sensitive to inbreeding depression, germination capacity of pollen grains is employed for efficient artificial hybridization [9]. Although it is highly acclimated to the Mediterranean environment, an increase in temperature, alongside water shortage and other biotic and abiotic factors, might affect both the quality and quantity of the crop, traits that are affected by both temperature and water accessibility [10]. The Schoneveld Breeding Company recently evaluated the effect of high levels of light exposure on leaf temperature in *Cyclamen*. They demonstrated that on sunny days if the light intensity is high, the leaves' temperature will gradually rise above the air temperature. In this case, the plant cannot maintain enough evaporation to cool and will therefore close the stomata to prevent wilting. This results in an inactive plant for up to 5 h/day due to no water/nutrients uptake because of closed stomata. The optimal temperature in the first period of growth is around 18–20 °C and between 15–20 °C in the flowering period. Therefore, to promote cooler temperatures, up to 50% shading is applied in the summer months, together with lateral and/or roof ventilation [11]. Recent studies demonstrated the correlation between GA application and light exposure to cardinal temperatures and thermal times required for seed germination [4,12].

Molecules that protect plants against several adverse climate conditions are starting to gain interest among plant researchers. Exogenous application of growth regulators like gibberellic acids (GAs) have been demonstrated to be effective in ameliorating abiotic stresses, including in heat-stress induced plants [13,14]. Moreover, gibberellins are key plant regulators that control the developmental stage of plants [4]. They have been extensively used in economically important crops and ornamentals due to their positive effects in early seed germination [15,16], reduced juvenile stage [17], leaf and root development and extension [16,18], early flowering and fruit formation [19,20]. It has been shown that mutant plants deficient in GA have a diminutive phenotype and blossom late. A growing demand for patterned foliage and prolonged vase life in *Cyclamen* cut-flowers has led researchers to design new amelioration strategies to meet market demands [21]. GA_3 foliar applications increased flower stem length in cut flowers cultivated in open-fields and delayed leaf browning and senescence [22].

Seed-priming is one of the most low-cost and low-risk methods of improving germination, seedling development and yield [23]. Exogenously applied GA_3 enhances seed germination and dormancy release, thus highlighting the importance, fast effectiveness and low cost of seed-priming in multiple crops [24]. The long-term effect of seed priming with GA_3 was assessed in multiple crops to determine the plant growth and production [15], which has been demonstrated to improve germination and growth parameters of shoot and root length, and seedling fresh and dry weights [15,25]. Furthermore, seed priming has been shown to enhance resistance to abiotic stress via various pathways involved in different metabolic processes, with best results in early and uniform germination [26]. Thus, seed simulation modeling, a topic of interest in quantifying germination and growth dynamics of plants, may assist researchers to forecast seed germination and final germination potential at the end of the experiment. Thus, it is critical to assess the genetic diversity of planted germplasm for heat-stress resistance and to choose genotypes with higher levels of heat tolerance. In this regard, the treatment estimate provides a framework for evaluating the genetic potential of germination and quality-related plant characteristics under optimum and heat-stress circumstances. The aims of the study were: to non-destructively evaluate the germination capacity (germination capacity and mean germination time) and plant development (seedling vigor index, leaf area, plant, petiole and root lengths) of heat-tolerant and heat-sensitive *Cyclamen* genotypes subjected to ambient temperature (AT) or heat stress (HS) conditions; to select the most advantageous GA_3 dosage applied; to evaluate the cultivated germplasm's genetic diversity for stress tolerance; and to select genotypes which possess increased tolerance to heat.

2. Results

2.1. Variation in Germination Parameters of Heat-Resistant and Heat-Sensitive Cyclamen Genotypes under AT and HS

Heat-resistant and heat-sensitive *Cyclamen* responded variably to HS conditions and GA_3 treatments (Table 1). Under HS heat-resistant genotypes responded positively to GA_3 treatment concentrations in a dose-dependent and specie-specific manner. Thus, under control, relatively low germination percentages were noticed especially in C10 (53.4%) and C14 (60%), and the highest were found in both C8 and C15 with 86.7%. Following treatment with 30 mg/L GA_3, the highest germination was observed in C10 (93.3%) and the lowest in C7 (60.0%). Increased GA_3 concentrations (70 and 90 mg/L) had a positive effect in all genotypes except C3 which presented a slightly lower germination between 80 and 83.3%. Regarding the heat-sensitive genotypes, significant differences were noticed between temperature conditions and hormone treatment. Under HS, the heat-sensitive genotypes presented lower germination percentages even among those subjected to GA_3 treatments, except C1, C2 and C4. The highest germination and resistance to HT was noticed in C2 in all GA_3 concentrations, followed by C4 which presented the highest germination with 30 and 70 mg/L GA_3, whereas C1 had the best germination performance under 70 and 90 mg/L GA_3 concentrations with 73.3 and 80.0% germination percentages, respectively. In heat-resistant genotypes under AT, the greatest germination percentage was observed under 70 mg/L GA_3 in C3, C7, C10 and C16. Conversely, relatively low germination was noticed in C14 and C15 under all treatment concentrations. Treatment with 30 mg/L GA_3 concentrations revealed relatively close values with control in all genotypes. Compared with the other concentrations, the highest 90 mg/L GA_3 concentration presented the best results in all genotypes with the highest germination percentage in C9. Under AT, the highest germination percentage was noticed using 30 and 70 mg/L GA_3 with best results in C4, C6 and C13. Conversely, under the 90 mg/L GA_3, the highest germination was noticed in C2 and C11.

Table 1. Effects of GA_3 concentrations (mg/L) on germination parameters of the selected heat-resistant and heat-sensitive *Cyclamen* genotypes under ambient temperatures and heat stress.

Parameters	Genotype	GA₃ Concentrations Treatment under AT				GA₃ Concentrations Treatment under HS			
		0 mg/L GA₃	30 mg/L GA₃	70 mg/L GA₃	90 mg/L GA₃	0 mg/L GA₃	30 mg/L GA₃	70 mg/L GA₃	90 mg/L GA₃
		Heat-resistant genotypes							
Germination percentage (%)	C3	80.0 ± 8.0 [b]	93.3 ± 9.1 [a]	100.0 ± 10.0 [a]	93.3 ± 9.4 [a]	73.4 ± 3.3 [c]	86.7 ± 2.1 [b]	80.0 ± 2.0 [c]	83.3 ± 3.4 [c]
	C7	73.3 ± 9.1 [b]	93.3 ± 9.5 [ab]	100.0 ± 10.2 [a]	86.7 ± 8.8 [ab]	73.3 ± 5.1 [c]	60.0 ± 1.8 [d]	86.7 ± 3.1 [de]	90.0 ± 1.8 [b]
	C8	86.7 ± 8.3 [a]	100.0 ± 8.9 [a]	86.7 ± 8.0 [a]	93.3 ± 8.8 [a]	86.7 ± 4.5 [a]	86.7 ± 5.7 [b]	93.3 ± 6.0 [ab]	95.0 ± 3.4 [ab]
	C9	60.0 ± 8.0 [b]	73.3 ± 8.8 [ab]	93.3 ± 8.3 [ab]	100.0 ± 9.7 [a]	73.3 ± 4.6 [c]	86.7 ± 2.9 [b]	100.0 ± 3.3 [a]	93.3 ± 1.7 [b]
	C10	86.7 ± 7.9 [b]	93.3 ± 9.0 [ab]	100.0 ± 8.9 [a]	93.3 ± 9.5 [ab]	53.4 ± 6.7 [d]	93.3 ± 6.6 [a]	86.7 ± 2.8 [d]	93.3 ± 2.5 [b]
	C14	53.3 ± 5.3 [c]	60.0 ± 6.0 [b]	86.7 ± 7.8 [a]	86.7 ± 7.8 [a]	60.0 ± 5.5 [d]	73.3 ± 2.0 [c]	86.7 ± 5.8 [de]	93.3 ± 3.3 [ab]
	C15	40.0 ± 4.1 [c]	73.3 ± 7.3 [b]	80.0 ± 7.9 [b]	93.3 ± 9.2 [a]	86.7 ± 4.5 [a]	86.7 ± 6.0 [ab]	93.3 ± 5.2 [ab]	100.0 ± 4.0 [a]
	C16	90.0 ± 8.4 [b]	95.0 ± 9.5 [ab]	100.0 ± 9.7 [a]	95.0 ± 9.1 [ab]	73.3 ± 4.4 [c]	73.3 ± 3.5 [c]	86.7 ± 2.7 [d]	95.0 ± 1.1 [b]
		Heat-sensitive genotypes							
	C1	80.0 ± 7.9 [a]	93.3 ± 8.6 [a]	93.3 ± 8.8 [a]	93.3 ± 9.4 [a]	40.0 ± 3.8 [e]	40.0 ± 2.5 [f]	73.3 ± 0.8 [f]	80.0 ± 2.4 [c]
	C2	80.0 ± 7.6 [b]	86.7 ± 8.8 [ab]	93.3 ± 9.3 [a]	100.0 ± 9.5 [a]	80.0 ± 0.6 [b]	86.7 ± 2.6 [b]	93.3 ± 1.5 [b]	93.3 ± 2.1 [b]
	C4	73.3 ± 7.4 [b]	86.7 ± 8.3 [b]	100.0 ± 9.6 [a]	80.0 ± 7.8 [b]	53.4 ± 2.4 [d]	73.3 ± 5.3 [c]	86.7 ± 1.6 [d]	60.0 ± 2.8 [e]
	C5	80.0 ± 7.6 [b]	80.0 ± 7.3 [b]	93.3 ± 9.7 [a]	86.7 ± 8.9 [ab]	40.0 ± 5.6 [e]	53.3 ± 2.3 [e]	53.3 ± 1.9 [h]	73.3 ± 3.9 [d]
	C6	93.3 ± 9.0 [a]	93.3 ± 8.7 [a]	100.0 ± 9.9 [a]	93.3 ± 8.7 [a]	33.3 ± 3.2 [ef]	60.0 ± 3.3 [d]	73.3 ± 2.9 [f]	40.0 ± 1.7 [g]
	C11	73.3 ± 6.8 [b]	86.7 ± 8.0 [ab]	93.3 ± 8.9 [a]	100.0 ± 9.9 [a]	33.3 ± 5.8 [ef]	40.0 ± 3.3 [f]	40.0 ± 1.6 [i]	73.3 ± 7.1 [d]
	C12	86.7 ± 8.2 [a]	80.0 ± 7.9 [a]	93.3 ± 8.9 [a]	93.3 ± 9.0 [a]	40.0 ± 7.7 [e]	60.0 ± 4.9 [d]	73.3 ± 3.1 [f]	60.0 ± 8.0 [ef]
	C13	86.7 ± 9.2 [b]	100.0 ± 9.9 [a]	100.0 ± 10.4 [a]	86.7 ± 8.4 [b]	33.3 ± 7.6 [f]	40.0 ± 5.6 [f]	60.0 ± 3.4 [g]	60.0 ± 5.6 [ef]
		Heat-resistant genotypes							
MGT	C3	13.90 ± 5.7 [a]	17.64 ± 7.2 [a]	3.24 ± 2.0 [b]	10.49 ± 6.0 [a]	10.31 ± 3.49 [e]	7.50 ± 5.40 [g]	5.088 ± 2.34 [ef]	6.42 ± 3.84 [e]
	C7	4.02 ± 2.1 [ab]	8.70 ± 6.1 [a]	3.03 ± 1.5 [b]	8.98 ± 5.9 [a]	8.48 ± 0.91 [f]	6.13 ± 3.06 [gh]	5.35 ± 2.33 [e]	10.34 ± 6.56 [d]
	C8	6.18 ± 3.3 [a]	4.25 ± 2.2 [a]	3.90 ± 2.0 [a]	8.66 ± 7.7 [a]	16.91 ± 2.66 [d]	14.33 ± 2.86 [cd]	4.62 ± 2.08 [f]	13.85 ± 1.87 [b]
	C9	11.44 ± 5.0 [a]	3.71 ± 2.0 [b]	3.39 ± 1.9 [b]	4.32 ± 2.2 [b]	19.35 ± 5.48 [c]	26.24 ± 1.90 [a]	4.68 ± 2.49 [f]	12.98 ± 3.71 [c]
	C10	11.57 ± 6.1 [a]	2.34 ± 0.9 [b]	3.70 ± 2.0 [ab]	9.00 ± 7.9 [a]	18.72 ± 4.46 [cd]	12.86 ± 7.89 [de]	12.28 ± 1.69 [c]	11.89 ± 3.94 [d]
	C14	7.01 ± 3.1 [a]	12.19 ± 7.3 [a]	9.89 ± 6.2 [a]	2.65 ± 0.3 [b]	8.42 ± 3.53 [f]	16.77 ± 3.52 [b]	23.51 ± 3.16 [a]	12.69 ± 2.90 [c]
	C15	19.95 ± 6.2 [a]	4.82 ±1.9 [b]	4.25 ± 1.6 [b]	3.72 ± 1.6 [b]	17.10 ± 2.08 [d]	14.30 ± 1.09 [cd]	4.54 ± 1.99 [g]	6.47 ± 2.66 [e]
	C16	4.26 ± 2.1 [a]	5.14 ± 2.6 [a]	2.01 ± 0.7 [b]	1.76 ± 0.6 [b]	19.39 ± 3.95 [c]	6.47 ± 3.00 [gh]	6.05 ± 2.64 [e]	5.12 ± 2.14 [f]
		Heat-sensitive genotypes							
	C1	13.43 ± 7.4 [a]	2.49 ± 0.7 [a]	10.10 ± 6.0 [a]	9.06 ± 4.8 [a]	44.89 ± 4.94 [a]	5.21 ± 1.81 [h]	4.31 ± 3.01 [g]	12.98 ± 5.87 [c]
	C2	4.77 ± 2.1 [a]	10.43 ± 4.3 [a]	10.91 ± 4.6 [a]	3.74 ± 1.6 [a]	7.98 ± 4.88 [g]	13.069 ± 3.81 [de]	9.944 ± 3.30 [d]	14.61 ± 2.48 [a]
	C4	11.55 ± 6.4 [a]	3.69 ± 1.5 [b]	1.64 ± 0.4 [c]	4.54 ± 1.7 [b]	17.05 ± 2.62 [d]	12.33 ± 2.67 [e]	12.63 ± 4.72 [c]	11.80 ± 1.70 [d]
	C5	10.37 ± 4.8 [ab]	22.73 ± 12.5 [a]	9.63 ± 4.4 [b]	21.09 ± 12.7 [a]	6.01 ± 3.46 [h]	14.95 ± 1.87 [c]	11.35 ± 3.74 [d]	10.82 ± 3.57 [d]
	C6	3.86 ± 2.3 [ab]	11.01 ± 8.0 [a]	2.90 ± 1.9 [b]	9.01 ± 5.7 [a]	6.27 ± 3.02 [h]	11.34 ± 4.68 [e]	4.33 ± 2.19 [g]	12.46 ± 1.67 [c]
	C11	20.54 ± 12.1 [a]	3.94 ± 1.6 [b]	3.18 ± 0.6 [b]	8.78 ± 3.1 [b]	6.08 ± 3.18 [h]	14.02 ± 3.11 [d]	16.23 ± 1.13 [b]	5.11 ± 2.13 [f]
	C12	4.61 ± 1.6 [a]	7.27 ± 3.1 [a]	3.36 ± 1.5 [ab]	3.18 ± 1.5 [b]	32.35 ± 1.11 [b]	5.88 ± 3.19 [h]	5.32 ± 2.49 [e]	14.45 ± 6.95 [a]
	C13	17.87 ± 9.1 [a]	7.60 ± 3.6 [b]	4.18 ± 3.4 [a]	22.78 ± 13.8 [a]	5.84 ± 3.19 [h]	10.60 ± 4.98 [ef]	4.95 ± 2.49 [ef]	4.28 ± 1.96 [f]

Values represent the mean ± standard deviations of three independent determinations. Different letters (a,b,c,d,e,f,g,h,i) within a column denote significant differences ($p < 0.05$).

The period required for germination under HS mainly corresponded to the germination percentage (Table 1). Thus, genotypes under control presented a delayed germination as noticed by the increased mean germination time (MGT) values (8.42–19.39%). Following GA$_3$ treatment with 30 mg/L, a lower germination rate was noticed in C9 and C14 genotypes with an MGT of 26.24% and 16.77%, respectively. Conversely, genotypes C3, C7 and C16 presented lower MGT values of 7.50%, 6.13% and 6.47%, respectively, corresponding to an earlier germination rate. Treatment with 70 mg/L GA$_3$ presented a higher germination rate as seen by the lower values in MGT, except C10 and C14 which had a lower germination rate compared with control and the other treatments. As seen in the previous GA$_3$ treatments, genotypes C3, C15 and C16 had the earliest germination with 90 mg/L GA$_3$ compared with control and the other genotypes subjected to different hormone concentration treatments. In heat-resistant genotypes, the lowest MGT values (3.03–9.89%) were noticed under 70 mg/L GA$_3$ which presented an earlier germination compared with control and the other treatments. The lowest MGT values correspond to the highest germination percentages (i.e., shorter period required for germination) compared with control which presented higher MGT values in most genotypes, except in C7 and C16. Under 30 mg/L GA$_3$ treatment, all genotypes presented earlier germination, except C14 which showed a higher MGT (9.89%). Lastly, C14–C16 genotypes showed earlier germination rates under the highest GA$_3$ level (90 mg/L). In heat-resistant genotypes, under HS the highest MGT values (i.e., delayed germination) were recorded under control as also seen by the lowest germination percentages, except C7 which had a slightly lower MGT of 8.42%. Treatment with 30 and 70 mg/L GA$_3$ showed significantly higher MGT values in almost all genotypes, except C1 and C12 which at both concentrations had the lowest MGT values compared with control presenting an increased germination rate. Finally, using 90 mg/L GA$_3$, the lowest MGT values were noticed in C11 (Verano) and C13 (Metis Origami) making them favorable to be used in future amelioration programs. Significant differences were noticed in the heat-sensitive genotypes which had the highest germination rate under AT as shown by the relatively low MGT, especially in the case of 70 mg/L GA$_3$. Under control, the lowest MGTs were noticed in C2, C6 and C12 with values between 3.86 and 4.77 which also had a relatively high germination percentage (80.0–93.3%). Under treatment with 30 mg/L GA$_3$, the genotypes had a slightly higher germination rate (lower MGT) compared with control, except C5 which had a delayed germination but a higher germination percentage. As seen by the higher germination percentage values (93.3–100%), the heat-sensitive genotypes had the earliest germination rate under the 70 mg/L GA$_3$ treatment (2.01–9.89) compared with control and other treatments. The final GA$_3$ presented the lowest MGT values in C14 and C16. Overall, genotypes C3 (Smartiz Victoria), C15 (Merengue magenta) and C16 (Metis silverleaf) had the highest germination rates compared with control and GA$_3$ treatments, which makes them suitable to be selected as heat-resistant genotypes and used further in several amelioration programs under biotic and/or abiotic stresses, especially drought stress to develop resistant genotypes. Overall, heat-sensitive C11 and C13, which had the lowest MGT values, along with C1 and C2 with a high germination percentage can be selected as genotypes which might be used in future studies to assess their resistance under heat and/or drought stress, but also subjected to different growth hormone regulators. Finally, all heat-resistant genotypes presented high germination rate and percentages making them economically important and suitable to be cultivated in arid regions with elevated temperatures.

2.2. Variation in Plant Development of Heat-Resistant and Heat-Sensitive Cyclamen Genotypes under AT and HS

The plants' height was measured from the beginning of the tuber to the highest point of the leaf (Table 2). The development of heat-resistant and heat-sensitive genotypes was significantly influenced by GA$_3$ treatment concentrations. Under HS, significantly lower values were noticed in the heat-resistant genotypes under control, except C14 which had an increased height (6.27 cm) compared with the other genotypes. Under 70 mg/L GA$_3$, the genotypes that presented the highest development were C9, C10 and C14 with values between 12.79 and 15.43 cm. The last treatment significantly increased the plants' height, except genotypes C3 and C7 with values between 5.41 and 7.01 cm. Under HS, significantly lower values were observed in heat-sensitive genotypes under control, which persisted even in those subjected to 30 mg/L GA$_3$ treatment. As it can be foreseen, in the subsequent treatments of 70 and 90 mg/L GA$_3$, the plants developed in a dose-dependent manner. The most sensitive to heat were C1, C2 and C12 which presented relatively lower development compared with the other genotypes. Under AT, the heat-resistant genotypes under control presented significantly lower development compared with genotypes subjected to hormone treatment. Under 30 mg/L GA$_3$, the C8, C10 and C14 positively responded to hormone treatment with values between 11.15 cm and 13.25 cm compared with the other genotypes which had relatively close values with control. The subsequent treatment concentrations (70 and 90 mg/L GA$_3$) had a positive influence in all tested genotypes in a dose-dependent manner. In the case of heat-sensitive genotypes, a significantly lower development was noticed in control plants under AT. Similar development was also noticed with 30 mg/L GA$_3$, with increased plant height noticed solely in C2. The ensuing 70 mg/L had a positive influence regarding the plant's height in all genotypes with regard to C4 (10.08 cm). The same case was noticed with increased hormone treatment of 90 mg/L in all genotypes. Overall, in the case of heat-resistant genotypes, the plants had the best development under the 90 mg/L GA$_3$, as also seen by the high levels in leaf area (Figure 1), except C3 and C7 which might be resistant to moderate heat.

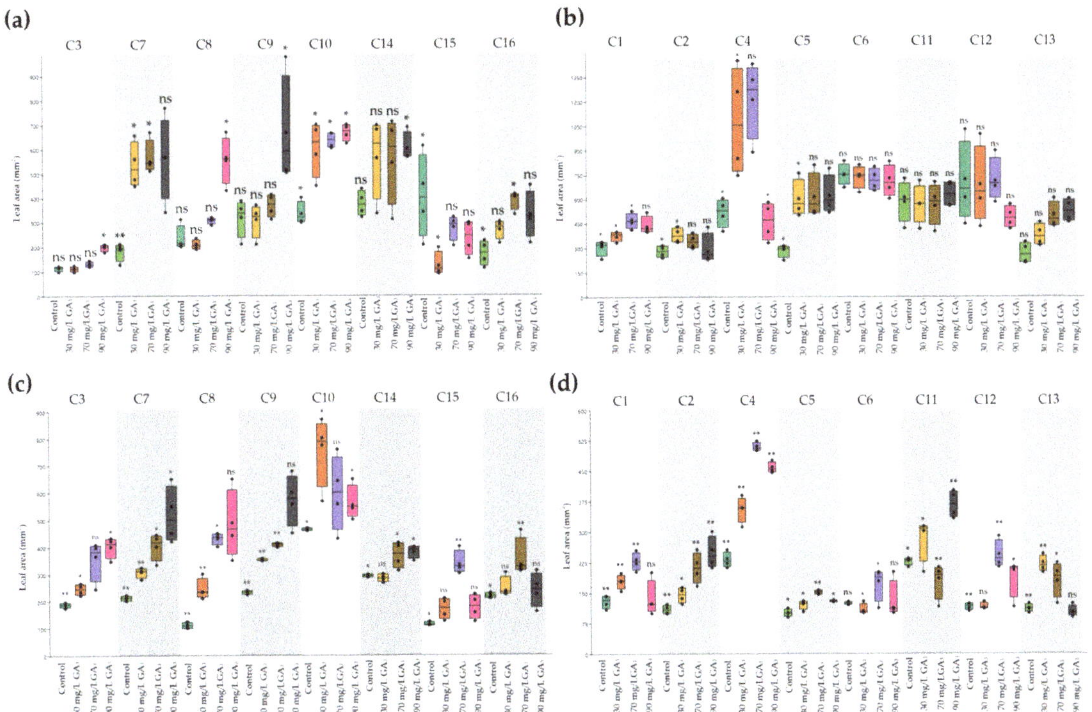

Figure 1. The effects of ambient temperature and high temperature stress and exogenously applied GA$_3$ on leaf area of heat-tolerant (**a,c**) and heat-sensitive (**b,d**) genotypes. * Significant at 5% probability level; ** significant at 1% probability level; ns, no significant difference.

Table 2. Effects of GA_3 concentrations (mg/L) on plant development of the selected heat-resistant and heat-sensitive *Cyclamen* genotypes under ambient temperatures and heat stress.

Parameters	Genotype	GA₃ Concentrations Treatment under AT				GA₃ Concentrations Treatment under HS			
		0 mg/L GA₃	30 mg/L GA₃	70 mg/L GA₃	90 mg/L GA₃	0 mg/L GA₃	30 mg/L GA₃	70 mg/L GA₃	90 mg/L GA₃
		Heat-resistant genotypes							
Plant height (cm)	C3	7.39 ± 0.2 [b]	8.10 ± 0.5 [ab]	10.74 ± 1.1 [a]	10.17 ± 0.5 [a]	2.31 ± 0.0 [ef]	6.46 ± 1.1 [cd]	8.78 ± 1.2 [d]	5.41 ± 0.2 [f]
	C7	7.04 ± 0.7 [c]	6.83 ± 0.9 [c]	12.11 ± 0.9 [a]	11.44 + 0.7 [b]	1.64 ± 0.1 [g]	2.85 ± 0.5 [h]	4.41 ± 0.1 [g]	7.01 + 0.3 [de]
	C8	7.39 ± 0.4 [b]	11.15 ± 1.0 [a]	12.12 ± 0.7 [a]	12.99 + 1.1 [a]	1.83 ± 0.0 [f]	6.17 ± 0.9 [de]	6.89 ± 0.2 [e]	10.22 + 1.2 [c]
	C9	5.24 ± 0.1 [c]	7.75 ± 0.6 [b]	12.15 ± 0.8 [a]	11.82 + 0.1 [a]	3.94 ± 0.3 [c]	7.04 ± 0.4 [c]	13.54 ± 1.7 [ab]	14.89 + 0.4 [b]
	C10	7.93 ± 0.2 [c]	13.25 ± 0.5 [b]	14.06 ± 0.2 [b]	16.07 ± 0.6 [a]	5.52 ± 0.0 [b]	10.77 ± 0.7 [a]	15.43 ± 1.1 [a]	17.98 ± 0.7 [a]
	C14	6.08 ± 0.8 [c]	12.63 ± 0.1 [b]	11.02 ± 0.4 [d]	12.76 ± 0.3 [a]	6.27 ± 0.4 [a]	10.59 ± 0.4 [a]	12.79 ± 0.4 [b]	14.29 ± 0.8 [b]
	C15	5.74 ± 0.3 [d]	8.88 ± 0.5 [c]	9.94 ± 0.7 [b]	12.07 ± 1.2 [a]	2.94 ± 0.2 [d]	7.14 ± 0.6 [c]	10.55 ± 1.5 [c]	16.17 ± 1.8 [a]
	C16	5.72 ± 0.5 [c]	7.82 ± 0.3 [b]	9.94 ± 0.2 [a]	10.06 ± 1.2 [a]	4.12 ± 0.1 [c]	6.64 ± 0.3 [cd]	8.96 ± 0.1 [d]	12.35 ± 1.1 [bc]
		Heat-sensitive genotypes							
	C1	3.24 ± 0.4 [d]	3.98 ± 0.8 [c]	9.76 ± 1.5 [a]	6.73 ± 0.8 [b]	1.56 ± 0.2 [g]	3.29 ± 0.3 [fg]	7.19 ± 1.1 [e]	5.90 ± 0.7 [f]
	C2	2.51 ± 0.1 [c]	7.92 ± 0.4 [b]	8.47 ± 0.8 [a]	9.43 ± 1.0 [a]	1.31 ± 0.3 [h]	3.45 ± 0.4 [fg]	5.43 ± 0.8 [f]	6.60 ± 3.5 [e]
	C4	4.10 ± 0.3 [c]	6.70 ± 2.1 [b]	10.08 ± 0.9 [a]	10.50 + 1.1 [a]	2.27 ± 0.2 [f]	3.81 ± 0.8 [f]	8.85 ± 0.6 [d]	7.10 + 0.7 [de]
	C5	6.96 ± 0.2 [c]	7.18 ± 0.3 [a]	6.72 ± 0.4 [d]	7.09 + 0.5 [b]	6.24 ± 0.4 [a]	7.72 ± 0.3 [b]	6.68 ± 0.5 [e]	7.19 + 0.5 [d]
	C6	5.72 ± 0.4 [d]	6.55 ± 0.3 [c]	8.47 ± 0.9 [a]	7.58 + 0.6 [b]	1.23 ± 0.2 [h]	3.62 ± 0.3 [f]	8.22 ± 0.3 [d]	6.24 + 2.4 [e]
	C11	5.60 ± 0.2 [a]	6.31 ± 0.7 [a]	6.54 ± 0.4 [a]	7.02 ± 0.8 [a]	5.74 ± 0.2 [b]	5.76 ± 0.2 [e]	9.44 ± 0.9 [c]	10.80 ± 0.3 [c]
	C12	5.73 ± 0.4 [c]	6.60 ± 0.5 [b]	8.89 ± 1.0 [a]	7.50 ± 0.7 [a]	2.56 ± 0.1 [d]	3.19 ± 0.3 [gh]	6.37 ± 1.1 [ef]	4.44 ± 0.2 [g]
	C13	5.59 ± 0.6 [d]	6.75 ± 0.5 [c]	8.24 ± 0.4 [b]	9.51 ± 0.8 [a]	2.22 ± 0.3 [f]	6.19 ± 0.5 [d]	7.59 ± 0.8 [de]	8.19 ± 0.9 [d]
		Heat-resistant genotypes							
Seedling Vigor Index (SVI)	C3	1095.9 ± 1.4 [d]	1476.2 ± 1.3 [c]	1936.5 ± 1.2 [a]	1844.4 ± 0.9 [b]	167.17 ± 9.8 [e]	489.88 ± 10.1 [c]	647.3 ± 15.7 [cd]	359.02 ± 8.4 [h]
	C7	965.3 ± 2.3 [d]	1228.2 ± 0.9 [c]	1795.5 ± 2.1 [a]	1683.7 ± 3.1 [b]	65.25 ± 9.8 [hi]	78.66 ± 5.5 [i]	249.26 ± 8.7 [j]	422.64 ± 10.3 [g]
	C8	797.7 ± 1.2 [d]	1102.3 ± 3.1 [c]	1499.8 ± 0.9 [b]	2229.2 ± 1.6 [a]	60.26 ± 9.9 [i]	342.16 ± 15.6 [e]	432.91 ± 6.5 [f]	673.11 ± 9.8 [d]
	C9	849.6 ± 1.3 [d]	1133.1 ± 1.8 [c]	1331.4 ± 0.9 [b]	1562.1 ± 1.2 [a]	215.45 ± 11.9 [d]	375.10 ± 11.0 [e]	1136.54 ± 17.8 [a]	1190.01 ± 13.8 [b]
	C10	803.4 ± 1.3 [d]	1916.0 ± 0.9 [c]	1997.4 ± 1.5 [b]	2168.3 ± 2.6 [a]	263.29 ± 12.1 [c]	856.54 ± 16.7 [a]	858.50 ± 9.12 [b]	1295.57 ± 7.8 [a]
	C14	504.7 ± 1.2 [d]	1044.8 ± 1.0 [c]	1326.2 ± 0.5 [b]	1635.0 ± 0.3 [a]	335.21 ± 12.4 [a]	671.95 ± 12.1 [b]	867.20 ± 10.9 [b]	1216.39 ± 15.9 [a]
	C15	317.3 ± 0.3 [d]	1097.1 ± 0.5 [c]	1505.7 ± 0.7 [b]	1727.8 ± 1.2 [a]	116.93 ± 3.6 [f]	409.38 ± 13.0 [d]	844.30 ± 9.7 [b]	1184.10 ± 20.7 [b]
	C16	601.0 ± 0.5 [d]	883.7 ± 0.7 [ac]	1976.1 ± 0.0 [a]	1763.0 ± 1.2 [b]	303.88 ± 7.9 [b]	361.43 ± 10.0 [e]	676.57 ± 6.3 [c]	1068.72 ± 23.4 [c]
		Heat-sensitive genotypes							
	C1	333.8 ± 1.7 [d]	563.3 ± 3.4 [c]	1261.5 ± 5.3 [b]	1354.7 ± 0.3 [a]	79.4 ± 8.3 [h]	113.9 ± 3.3 [h]	439.5 ± 14.5 [f]	449.1 ± 7.8 [fg]
	C2	501.2 ± 0.4 [d]	1146.0 ± 0.8 [c]	1306.4 ± 1.2 [b]	1829.0 ± 2.5 [a]	132.42 ± 11.1 [f]	258.1 ± 14.5 [f]	408.24 ± 14.2 [g]	509.28 ± 8.5 [f]
	C4	492.8 ± 1.5 [d]	687.5 ± 0.9 [c]	1245.6 ± 0.9 [b]	1319.5 ± 2.5 [a]	67.496 ± 13.4 [hi]	198.46 ± 5.6 [g]	591.72 ± 20.9 [d]	371.26 ± 9.9 [h]
	C5	976.1 ± 1.3 [d]	1041.7 ± 0.9 [c]	1524.1 ± 0.8 [a]	1358.4 ± 1.3 [b]	146.22 ± 10.9 [ef]	274.20 ± 7.6 [f]	327.88 ± 12.4 [h]	533.99 ± 5.6 [e]
	C6	1108.6 ± 2.4 [d]	1240.3 ± 3.1 [c]	1657.5 ± 0.8 [a]	1401.9 ± 1.2 [b]	16.71 ± 4.6 [j]	84.61 ± 6.9 [i]	480.35 ± 10.9 [e]	249.41 ± 9.7 [i]
	C11	385.9 ± 2.4 [d]	466.5 ± 2.1 [c]	830.3 ± 3.1 [b]	1276.2 ± 0.7 [a]	107.30 ± 9.9 [g]	173.79 ± 14.5 [gh]	279.35 ± 11.3 [i]	565.46 ± 5.6 [e]
	C12	746.0 ± 0.9 [d]	830.2 ± 2.1 [c]	1614.7 ± 3.1 [a]	1284.2 ± 2.6 [b]	94.12 ± 12.7 [g]	160.85 ± 18.8 [gh]	304.41 ± 12.8 [h]	71.94 ± 6.2 [j]
	C13	543.5 ± 0.6 [d]	855.2 ± 2.3 [c]	1379.1 ± 0.4 [a]	1358.0 ± 0.8 [b]	74.90 ± 13.5 [h]	200.03 ± 11.8 [g]	440.02 ± 13.7 [f]	534.29 ± 8.8 [e]

Values represent the mean ± standard deviations of three independent determinations. Different letters (ab,c,d,e,f,g,h,i,j) within a column denote significant differences ($p < 0.05$).

Leaf area was significantly influenced by GA_3 treatment concentrations in heat-resistant genotypes in a dose-dependent manner, except C14 for which treatment at 90 mg/L reduced the leaf area (Figure 1c). Regarding the heat-sensitive genotypes, significantly lower leaf areas were noticed compared with heat-resistant genotypes. Although the leaf area increased with GA_3 concentrations, reduced levels were noticed under 90 mg/L in almost all genotypes, except C1 (Figure 1d).

Under HS, increased values of petiole length were noticed in heat-resistant genotypes with increased concentrations of GA_3, with the highest values under 90 mg/L hormone treatment (Figure 2c). Conversely, heat-sensitive genotypes exhibited relatively low levels in petiole height compared with heat-resistant genotypes. As also seen in heat-resistant genotypes, although higher values were noticed with increased application of GA_3, reduced levels were recorded at 90 mg/L treatment (Figure 2d).

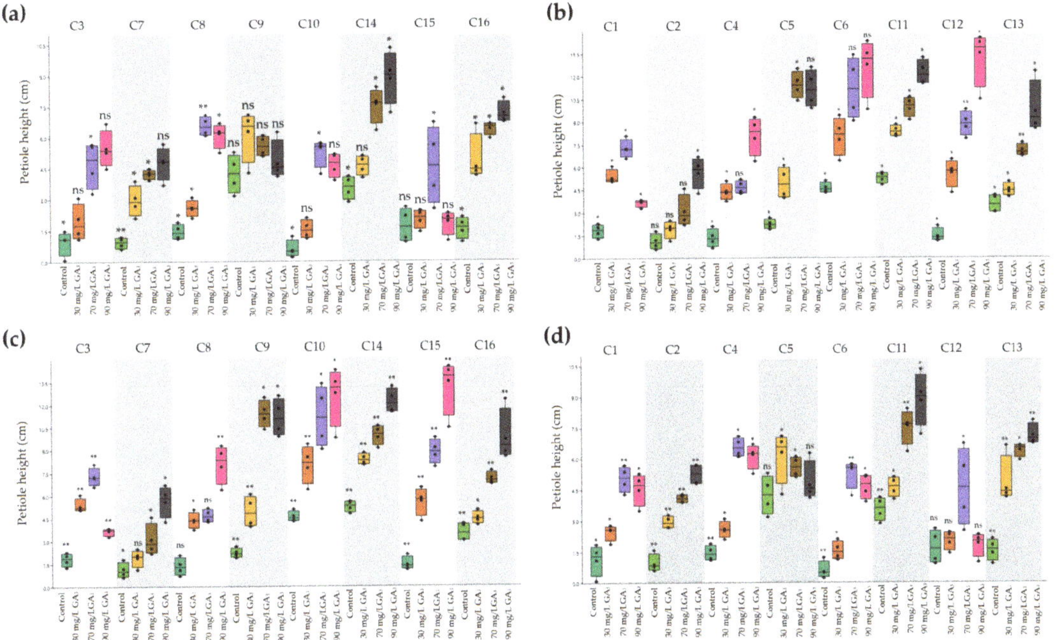

Figure 2. The effects of ambient temperature and high temperature stress and exogenously applied GA_3 on petiole height of heat-tolerant (**a,c**) and heat-sensitive (**b,d**) genotypes. * Significant at 5% probability level; ** significant at 1% probability level; ns, no significant difference.

Regarding root length, significant differences were noticed in heat-resistant genotypes which exhibited longer roots with increased GA_3 application, except C9 with higher values using 70 mg/L of growth regulator (Figure 3c). In heat-sensitive genotypes, higher values in root length were noticed under the application of 70 mg/L GA_3.

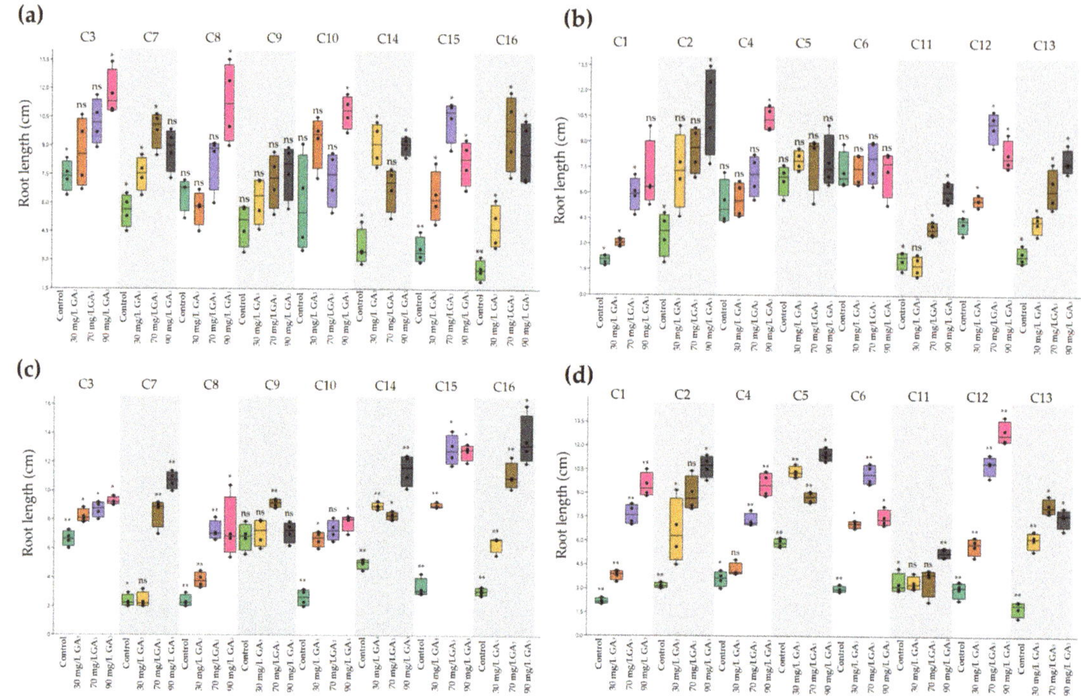

Figure 3. The effects of ambient temperature and high temperature stress and exogenously applied GA$_3$ on root length of heat-tolerant (**a,c**) and heat-sensitive (**b,d**) genotypes. * Significant at 5% probability level; ** significant at 1% probability level; ns, no significant difference.

The SVI was significantly different among genotypes and GA$_3$ treatments. Under HS, significantly lower SVI values were noticed under control, especially in C7 and C8 with values of 65.25 and 60.26, respectively. With 30 mg/L GA$_3$, higher SVI levels were noticed in C10, followed by C14 and C3. The increasing dose of hormone concentrations led to a higher SVI especially in C9. Following the highest GA$_3$ dose (90 mg/L), an increase in SVI was noticed in almost all genotypes except C3 and C7 with values of 59.02 and 422.64, respectively. The highest SVI were noticed in C2 and C13 which had the best development among all genotypes. HS negatively affected the heat-sensitive plant's development as seen by the relatively low SVI values under control, even with GA$_3$ treatment. The highest doses of hormone treatment led to higher SVI in C4 under 70 mg/L, and in C5, C11 and C13 with 90 mg/L GA$_3$ treatment. Under AT, the heat-resistant genotypes under control presented relatively low SVI which gradually increased by GA$_3$ treatment in a dose-dependent manner (Table 2). Thus, the use of 30 mg/L GA$_3$ significantly increased the SVI compared with control, with the highest values in C10, followed by C3. This ascending trend was noticed with the increased dose of GA$_3$, which had the strongest effect in C10, C16 and C3 under the 70 mg/L GA$_3$, whereas the highest value of SVI with 90 mg/L GA$_3$ was noticed in C8 and C10. The heat-sensitive genotypes under control presented significantly different SVI values, with the highest in C5 and C6 and the lowest in C1 and C11, respectively. The use of GA$_3$ influenced the plant development in a positive way. Thus, the SVI presented a positive tendency under 30 mg/L GA$_3$, which maintained its trend with increased dose of GA$_3$ (70 and 90 mg/L). The variation in SVI could be attributed to an enhanced germination rate, which strongly depends on the seedlings' root and shoot development, along with the stimulation of enzymatic activities (Figure 4).

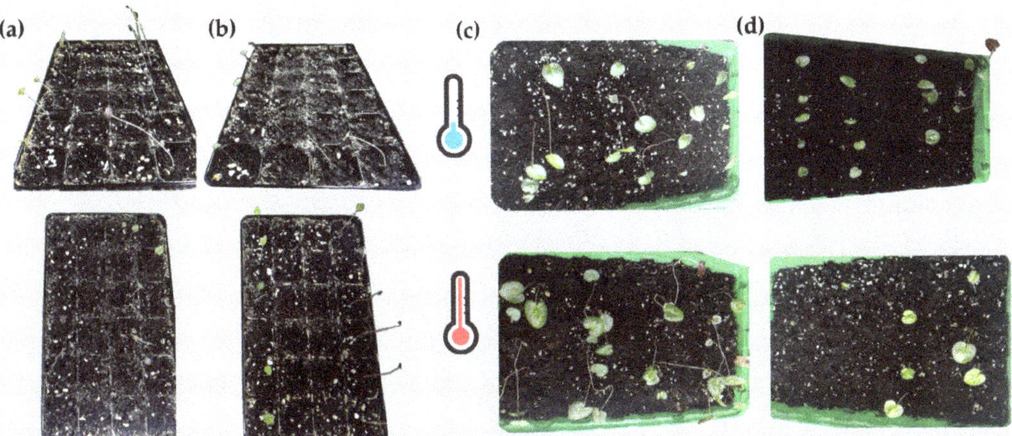

Figure 4. Germination capacity (**a,b**) and plant development (**c,d**) of heat-resistant C15 (**upper left-a**) and heat-sensitive C2 (**upper right-b**) under ambient temperature or heat stress.

2.3. Hierarchical Clustering and Heat Mapping of Seed Germination and Plant Development Parameters of Heat-Tolerant and Heat-Sensitive Genotypes under Ambient Temperature and Heat Stress Conditions

Hierarchical clustering and heat mapping were used to visualize similarities and differences between AT, HS and GA$_3$ applications. Several heat-resistant genotypes exhibited unusual behavior in HS conditions, namely C3, C7 and C8, as these genotypes showed negative correlations in terms of germination parameters and plant development (plant height) under control compared with GA$_3$ application. The other genotypes were found to be less affected by heat stress as compared with the above-mentioned genotypes as shown by the positive correlation between SVI, germination percentage and root length under 70 and 90 mg/L GA$_3$. Conversely, a negative correlation was shown in these genotypes under control in the majority of the traits which were drastically influenced by heat stress. Thus, these were referred to as susceptible genotypes as vindicated by their performance trend in Figure 5. As shown in the heatmap, it is clear that GA$_3$ treatment elevated the negative effect of HS by inducing seed germination and plant development with increased dose application. In HS under control, a negative correlation in GP, SVI and plant height was noticed as all genotypes were strongly affected by higher temperatures. Following the importance score and the increased GA$_3$ application, a positive correlation is shown in terms of plant height and root length, demonstrating the effectiveness of hormone treatment under HS conditions. Except for C3, C7 and C8 which proved to be susceptible to HS even under GA$_3$ application, the other genotypes showed tolerance to higher temperatures.

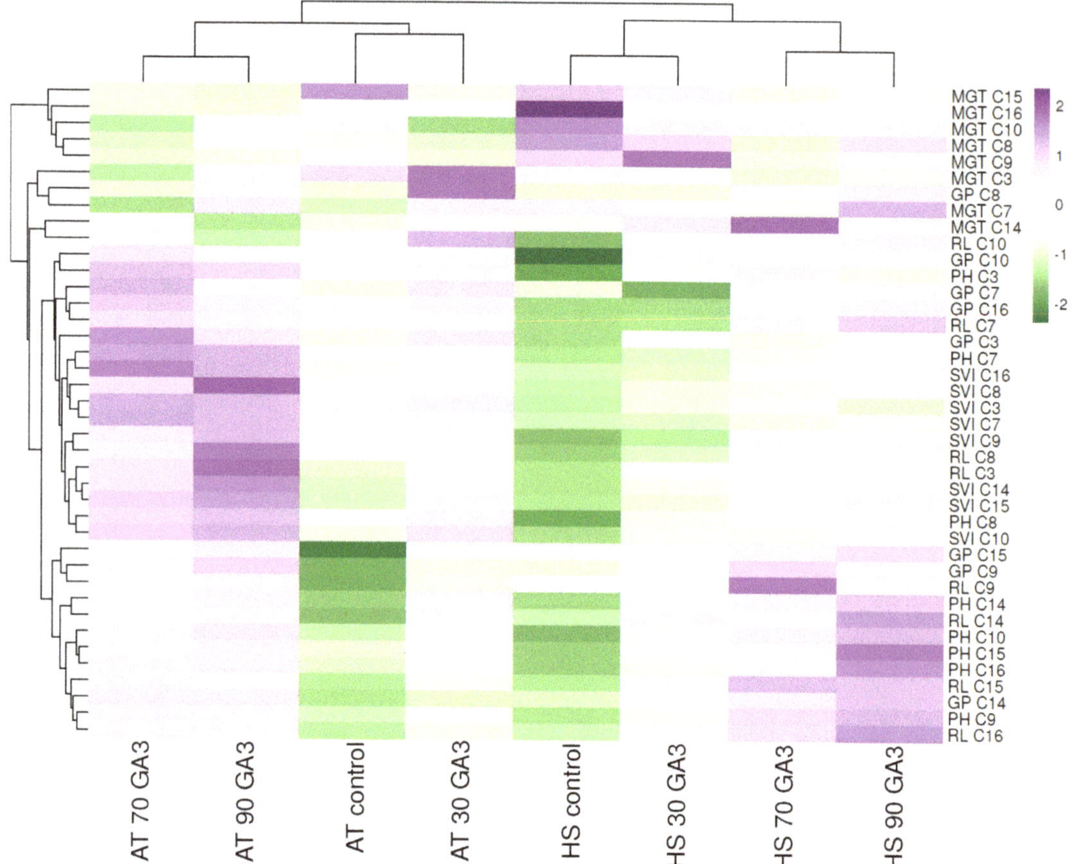

Figure 5. Hierarchical clustering and heatmap visualization of heat-resistant genotypes under ambient temperature and heat stress conditions. Columns indicate the control and GA₃ treatments, and rows indicate morphological parameters. Cells are colored based on values of seed and plant development, where purple represents a strong positive correlation and green a strongly negative correlation.

In the case of heat-sensitive genotypes, significant differences were observed under ambient and increased temperatures as observed by the importance scores (Figure 6). Under AT, the genotypes presented increased germination rate and plant development with increased GA₃ application as shown by their overall positive trend in the red colored gradient strips especially with 70 and 90 mg/L hormone treatment. Under HS, these genotypes showed rather negative values under control conditions, emphasizing their low susceptibility to increased temperatures. Following GA₃ application, genotypes C4, C6, C11 and C13 possessed higher positive or negative values compared with the other heat-sensitive genotypes. These genotypes were shown to be less affected by heat stress, suggesting their positive response to hormone treatment. As shown in the heatmap, these genotypes presented negative values in terms of germination parameters and positive values in root length compared with the other heat-sensitive genotypes, emphasizing their tolerance to heat stress under GA₃ application. In the present conditions, both heat-resistant and heat-sensitive genotypes were found to have higher petiole and root lengths and leaf area values as compared with control. Overall, heat-resistant C9, C10, C11, C14 and C16 genotypes are highly resistant to heat stress especially under hormone treatment, whereas the heat-sensitive genotypes, C4, C6, C11 and C13 proved to be less susceptible to HS as illustrated in Figure 6. These genotypes can be further employed in breeding programs by using multiple stress designs and hormone treatments to achieve maximum heat tolerance in crop generations, which may contribute to enhanced productivity.

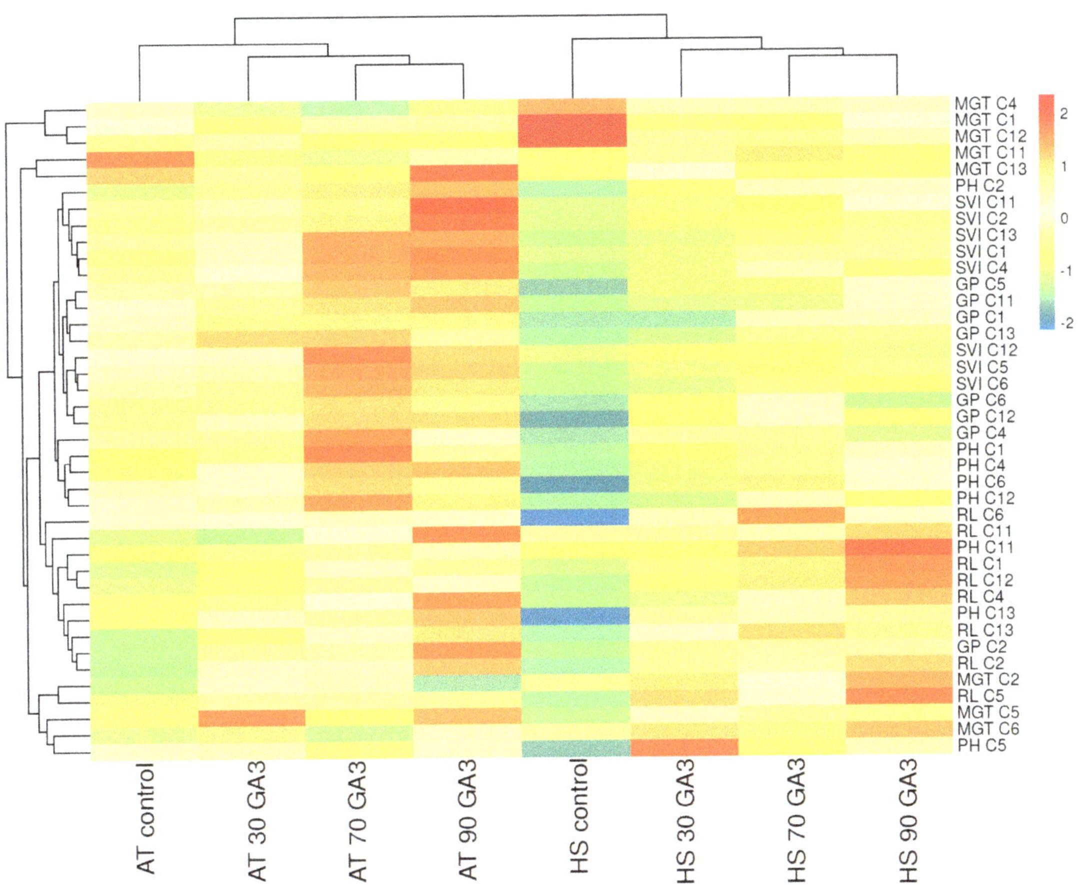

Figure 6. Hierarchical clustering and heatmap visualization of heat-sensitive genotypes under ambient temperature and heat stress conditions. Columns indicate the control and GA$_3$ treatments, and rows indicate morphological parameters. Cells are colored based on values of seed and plant development, where red represents a strong positive correlation and blue a strongly negative correlation.

3. Discussion

A prevalent concern due to climate change urges researchers to develop heat-tolerant genotypes to cope with the decline in water sources and rising temperatures so as to diminish their damaging effect on crop development. Utilization of low-cost hormone treatment along with tolerant genotypes is useful in the long term in genetic breeding programs. Moreover, identification of efficient screening techniques to detect heat-resistant plants proves to be significant approach for future studies. At seedling stage, phenotypic screening can be evaluated using destructive methods (fresh and dry weight) and non-destructive ones (image processing, canopy temperature). Moreover, root development was shown to be a reliable predictor of the plant's response to drought and heat stresses through its direct connection with soil, water relation and nutrient absorption under unfavorable environmental conditions [27]. It was demonstrated that heat stress strongly affects seed germination and vigor causing thermal injury or seed death. Various physiological parameters in plants are also affected by heat stress, resulting mainly in earlier leaf senescence, shoot and root growth inhibition, reduction in flower number and fruit development, all of which eventually lead to loss of crop yield [28,29]. Recently, Siddiqui et al. (2015) demonstrated that growth parameters of faba bean were significantly reduced under heat stress [30]. Seed priming with GA$_3$ overcomes seed dormancy mainly due to deterioration of the endosperm layer and activation of embryo development [31]. Pre-sowing treatments with GA$_3$ increased seed yield and qualitative characteristics under both drought stress and non-stress conditions compared with genotypes under control (no treatment) [23]. Germination rate of *T. terscheckii* seeds was strongly influenced under light regimes, heat stress and GA$_3$ treatments. Increased germination proportion was noticed under ambient temperatures compared with higher temperatures with increased GA$_3$ concentrations under both conditions. Conversely, significantly low germination was noticed under darkness only with increased dose of GA$_3$ (1000 mg/L) [32]. GA$_3$ (60 mg/L) was demonstrated to be a quick and efficient treatment in breaking rice seed dormancy compared with lower concentrations [33]. It was demonstrated that application of GA$_3$ between 50 and 100 mg/L proves to be efficient in breaking seed dormancy of *Cyclamen* species under high light exposure treatment. Thus, the highest germination rate in *C. africanum* and *C. hederifolium* was noticed under 50 mg/L GA$_3$ treatment, whereas *C. cyprium* presented the highest germination percentage with application of 100 mg/L GA$_3$ [4]. Early emergence, high germination percentage and normal seedlings development with least mortality were noticed in seeds of Hevea brasiliensis (rubber) with 100 mg/L GA$_3$ treatment [34]. Conversely, the germination rate of Lavandula angustifolia 'Codreanca' and 'Sevtopolis' were favorably influenced with application of GA$_3$ at doses of 200 and 300 mg/L [35]. Industrial hemp (*Cannabis sativa* L.) seed pre-treatments with high levels of GA$_3$ (500 and 1000 mg/L) were associated with a decreasing trend in germination, but a positive effect on early growth responses was observed [36]. Priming treatment of *Brassica napus* L. seeds with 300 mg/L GA$_3$ showed a significantly increased drought tolerance index compared with control and improved seedling tolerance to drought stress [37]. In the present study, heat stress significantly influenced the germination rate and plant development in heat-tolerant and heat-sensitive genotypes. Both heat-tolerant and heat-sensitive genotypes presented increased germination rate with GA$_3$, especially with 70 and 90 mg/L under both AT and HS conditions. Heat-sensitive genotypes positively responded to GA$_3$ especially in the case of C1 and C2 which might be selected as resistant genotypes as seen by their high germination rate (73.33–93.33%).

Plant response to increased temperatures depends on several factors in which plant regulators are considered significant and are involved in the mechanisms of susceptibility or tolerance of plants. Under abiotic stress, several proteins (DELLAs) are produced by the induction of different plant hormone levels. Seed priming with optimal concentrations of GA$_3$ was proven advantageous to increase early seedling growth under abiotic stress conditions [38]. Under HS, treatment with GA$_3$ (288.7 µM) increased the emergence percentage and emergence rate of sweet sorghum (*Sorghum bicolor* L. Moench) under heat stress compared with control. Lower levels of shoot and higher levels in root length and number were noticed under GA$_3$ treament under HS [39]. Combined treatment of silicon (Si) and GA$_3$ significantly influenced the development of date palm (*Phoenix dactylifera* L.) resulting in alleviation of adverse effect of HS and greatest shoot length (31.87 cm) and root length (11.56 cm) compared with solely hormone treatment and control. By comparison, relatively close results were noticed under sole treatment with GA$_3$ emphasizing its usage, alone or in combination with other growth regulators, as an efficient treatment under HS conditions [40]. Exogenous GA$_3$ application elevated the adverse effects of HS in Arabidopsis [41]. Heat-tolerant and heat-sensitive perennial ryegrass (*Lolium perenne*) accessions were both affected by HS resulting in decreased plant height and leaf water content, but with delayed negative effect in the tolerant genotype [42]. According to previous reports, the optimal temperature for germination in *Cyclamen* is 15 °C, whereas temperatures above 20 °C lead to inhibition of germination [43]. Supra-optimal temperatures (25–30 °C) inhibit germination with longer exposure (4 weeks) [44]. Moreover, temperatures higher than 25 °C or hypoxia inhibit tuber formation and lead to very elongated tubers which affects the plant's development [45]. In the present study, heat-resistant genotypes positively responded to GA_3 treatment in a dose-dependent manner under HS conditions. Elevated levels in SVI were noticed with 90 mg/L GA_3 with regard to Merengue genotypes, mainly C10, C14 and C15, which presented earlier and higher seedling development under HS conditions. Conversely, heat-sensitive genotypes were negatively affected by HS, whereas GA_3 acted in a protective way by enhancing seedling development and reducing the adverse effects of higher temperature exposure. Regarding the plant development, increased petiole and plant height were noticed in heat-tolerant genotypes under HS conditions with increased doses of GA_3, especially with 70 and 90 mg/L treatment regimes.

Regarding the root development under HS conditions, relatively few reports have evaluated their responses with growth hormone treatments. The effect of GA_3 application on the growth and development of roots was evaluated in paper flower (*Bougainvillea glabra*), jungle geranium (*Ixora coccinea*) and Chinese rose (*Rosa chinensis*) finding the highest levels with the dose of 100 mg/L GA_3 in all above-mentioned genotypes [46]. In a different study, GA_3 concentrations between 0.05 and 50 mg/L were applied to several *Pelargonium* (geranium) cultivars to assess their influence on root development. Their results showed that with increased GA_3 concentration, an increased growth rate and decreased shoot:root ratio were noticed [47]. Exposure of cucumber seedlings to a relatively low root-zone temperature of 16 °C led to significantly lower root growth and development which was reversed by exogenous GA_3 application [48]. Conversely, an increase in ambient temperature was reported to stimulate GA production, reduce DELLA levels and promote stem elongation in Arabidopsis [49]. Root lengths at seedling stage were severely reduced by HS with significant variations among wheat genotypes. Heat-tolerant genotypes at seedling stage showed less root length decrease compared with heat-susceptible ones [50]. In the present study, significant differences were noticed between root development of heat-sensitive and heat-tolerant genotypes with increased GA_3 doses. Under HS, heat-resistant genotypes revealed significantly higher root lengths with 70 and 90 mg/L GA_3 treatment. Conversely, significant differences between AT and HS conditions were noticed only in Petticoat genotypes (C2, and C12) and Metis genotypes (C5 and C6), respectively. Overall, heat-resistant C15 and heat-sensitive C2 might be selected for future studies under different abiotic and/or biotic stresses, according to their highest germination capacity (86.67–100% and 80.0–93.33%, respectively) and plant vigor (116.93–1184.1 and 132.42–509.28, respectively).

4. Materials and Methods

4.1. Plant Materials

The experiments were carried out at the Advanced Horticultural Research Institute of Transylvania (AHRIT), University of Agricultural Sciences and Veterinary Medicine Cluj-Napoca, Romania, using 16 *Cyclamen* accessions selected for their foliar and ornamental quality. A dual experiment was carried out: plants were maintained under control conditions of optimum temperature and heat stress. The 16 genotypes were divided into two agronomic groups according to their provenance: group 1 (Morel company breeder, Rue de Montourey, Fréjus, France) and group 2 (Schoneveld Breeding, Zeewolde, The Netherlands) (Supplementary Table S1). The plants were sown in May 2021 so that they would reach the juvenile vegetative stage for the heat treatment approximately in June–July 2021.

4.2. Growth Regulator Treatment

Seeds of *Cyclamen* genotypes were subjected to hormone seed priming using different concentrations of GA_3 and using distilled water for control for 24 h according to [38]. Afterward, the seeds were washed three times with distilled water, surface dried and transferred to growth chambers. Seeds were germinated on two-layer filter paper in Petri dishes (14.5 cm diameter) with 10 mL of distilled water and different levels of GA$_3$ (0, 30, 70 and 90 mg/L) in altering temperature conditions of either ambient temperature (11–17 °C) or heat stress (23–36 °C) with 60% relative humidity (Figure 7).

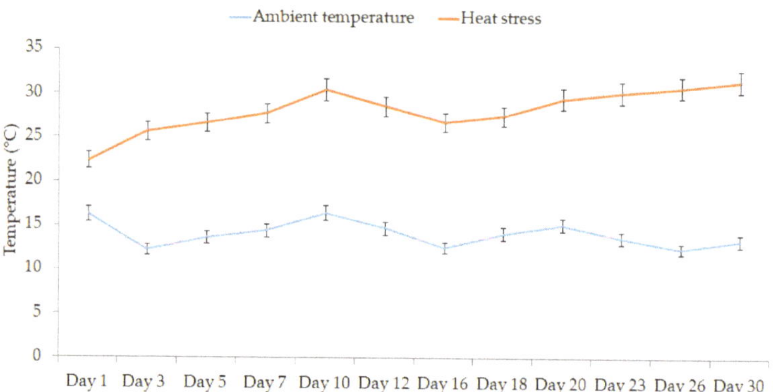

Figure 7. Average of the growth chambers recordings during the experiment.

4.3. Pot Experiment

After germination, the seedlings were transplanted to pots (30 cm length and 14 cm width) and watered according to the plants' requirements (at 3 days' interval) with distilled water (control) and GA$_3$ solutions in different concentrations. The growing substrate (60/20/20 *v/v*) was a mixture of sowing and propagation soil (pH = 6.0) with a content of NPK 0.1:0.01:0.03 m/m%, *Cyclamen* substrate (pH = 6.2) with a content of NPK (1.0:0.1:0.3 m/m%), 70% organic substances and perlite. Seeds germinated in approximately 3 weeks. Afterward, the influence of exposure to ambient temperatures (AT) and heat stress conditions (HS) under the influence of different GA$_3$ concentrations on morphological parameters was evaluated. Non-destructive methods were used by scanning all seedlings with the ImageJ Programme 1.52a, Wayne Rasband, National Institutes of Health Bethesda, Maryland, USA for image processing. Plant height, root and petiole lengths (cm), and leaf area were determined for heat-resistant and heat-sensitive genotypes under control and heat stress conditions. The germination percentage [51] was calculated accordingly:

$$GP = \frac{\sum Ns}{Fs} \times 100\%$$ (1)

where Ns corresponds to the number of seeds at the establishment of the experiment and Fs represents the germinated seeds at the end of the experiment. Germination rate [51,52] was assessed by estimating the mean germination time (MGT) using the following equation:

$$MGT = \frac{\sum DNs}{\sum Ns}$$ (2)

where D is the day at the start of the germination test, and Ns is the number of recently germinated seeds on day D [53]. The Seedling Vigor Index (SVI) [52] was evaluated using the equation described below:

$$SVI = GP \times (Mrl + Mhl)$$ (3)

GP represents the germination percentage, MrL the average root length (mm) and Mhl the average seedling length (mm) [54].

4.4. Statistical Analysis

For each randomized block (16 blocks in total), 5 plants per treatment were used (a total of 20 plants per exposure treatment and genotype). The aggregated dataset comprises 16 genotypes of 320 plants (80 control and 240 stress) together with germination parameters and morphological characteristics. Data collected were analyzed with the analysis of variance technique (ANOVA) using HSD Tukey's test ($p \leq 0.05$) and presented as the interaction of ambient and high temperature stress and GA_3 treatment using GraphPad Prism 8.2.1.441. Results were given as average $\pm$ standard deviation. Heatmap and dendrograms were generated using the Euclidean distance based on Ward's algorithm for clustering [55].

5. Conclusions

The present study highlights the priming effects with GA_3 on seed germination and plant development and the most advantageous GA_3 dosage applied to *Cyclamen* accessions under AT and HS conditions. Moreover, the cultivated germplasm's genetic diversity for HS tolerance was assessed to select genotypes which possess increased tolerance to heat. Heat-tolerant genotypes, mainly C3 (6.42%), C15 (6.47%) and C16 (5.12%) had the lowest MGT (highest germination rate) with 90 mg/L GA_3 treatment compared with control. Regarding heat-sensitive genotypes, C11 (5.11%) and C13 (4.28%) which had the lowest MGT values, along with C1 (73.3%) and C2 (80.0%) with a high germination percentage, can be selected as genotypes which might be used in future studies to assess their resistance under heat and/or drought stress, but also subjected to different growth hormone regulators. Under HS in heat-tolerant genotypes with 70 mg/L GA_3, the genotypes that presented the highest plant development were C9, C10 and C14 with values between 12.79 and 15.43 cm. The above genotypes which presented the highest values in germination and plant development might be selected as heat-resistant genotypes to be deposited in germplasm banks and used in further amelioration programs under biotic and/or abiotic stresses to develop resistant genotypes.

Supplementary Materials: The following supporting information can be downloaded at: https://www.mdpi.com/article/10.3390/plants11141868/s1, Table S1: Overview of 16 *Cyclamen* genotypes used in this study.

Author Contributions: Conceptualization, M.C.-C. and R.M.; methodology, M.C.-C.; formal analysis, M.C.-C.; investigation, M.C.-C.; writing—original draft preparation, M.C.-C., R.M., D.P. and M.I.C.; writing—review and editing, M.C.-C., R.M., D.P. and M.I.C.; supervision, D.P.; funding acquisition, M.I.C. All authors have read and agreed to the published version of the manuscript.

Funding: The publication was supported by funds from the National Research Development Projects to finance excellence (PFE)-14/2022–2024 granted by the Romanian Ministry of Research and Innovation.

Institutional Review Board Statement: Not applicable.

Informed Consent Statement: Not applicable.

Data Availability Statement: Not applicable.

Acknowledgments: The authors thank Céline Pesteil and Hein Desmet from S.A.S. Morel Diffusion (rue de Montourey, Fréjus, France, www.cyclamen.com (accessed on 16 June 2022)) and Dinja Lankhorst from Schoneveld Breeding (Sluinerweg 15, The Netherlands, https://schoneveld-breeding.com (accessed on 16 June 2022)) for providing the seeds of *Cyclamen* accessions.

Conflicts of Interest: The authors declare no conflict of interest.

References

1. Venkateswarlu, B.; Shanker, A.K. Climate change and agriculture: Adaptation and mitigation stategies. *Indian J. Agron.* **2009**, *54*, 226–230.
2. Eberius, M.; Lima-Guerra, J. High-Throughput Plant Phenotyping—Data Acquisition, Transformation, and Analysis. In *Bioinformatics: Tools and Applications*; Edwards, D., Stajich, J., Hansen, D., Eds.; Springer: New York, NY, USA, 2009; pp. 259–278.
3. Wilson, G. *Cyclamen (A Guide for Gardeners, Horticulturists and Botanists)*; Timber Press: London, UK, 2015.
4. Cornea-Cipcigan, M.; Pamfil, D.; Sisea, C.R.; Mărgăoan, R. Gibberellic Acid Can Improve Seed Germination and Ornamental Quality of Selected Cyclamen Species Grown Under Short and Long Days. *Agronomy* **2020**, *10*, 516. [CrossRef]

5. Karagur, E.R.; Ozay, C.; Mammadov, R.; Akca, H. Anti-invasive effect of Cyclamen pseudibericum extract on A549 non-small cell lung carcinoma cells via inhibition of ZEB1 mediated by miR-200c. *J. Nat. Med.* **2018**, *72*, 686–693. [CrossRef] [PubMed]

6. El Hosry, L.; Di Giorgio, C.; Birer, C.; Habib, J.; Tueni, M.; Bun, S.-S.; Herbette, G.; De Meo, M.; Ollivier, E.; Elias, R. In vitro cytotoxic and anticlastogenic activities of saxifragifolin B and cyclamin isolated from Cyclamen persicum and Cyclamen libanoticum. *Pharm. Biol.* **2014**, *52*, 1134–1140. [CrossRef]

7. Zengin, G.; Mahomoodally, M.F.; Sinan, K.I.; Picot-Allain, M.C.N.; Yildiztugay, E.; Cziáky, Z.; Jekő, J.; Saleem, H.; Ahemad, N. Chemical characterization, antioxidant, enzyme inhibitory and cytotoxic properties of two geophytes: Crocus pallasii and Cyclamen cilicium. *Food Res. Int.* **2020**, *133*, 109129. [CrossRef]

8. Cornea-Cipcigan, M.; Bunea, A.; Bouari, C.M.; Pamfil, D.; Páll, E.; Urcan, A.C.; Mărgăoan, R. Anthocyanins and Carotenoids Characterization in Flowers and Leaves of Cyclamen Genotypes Linked with Bioactivities Using Multivariate Analysis Techniques. *Antioxidants* **2022**, *11*, 1126. [CrossRef]

9. Cordea, M.I.; Tiriplică, A. Influence of Pollen Germination Capacity on a Successful Artificial Hybridization in *Cyclamen* sp. *Int. J. Innov. Appr. Agric. Res.* **2019**, *3*, 53. [CrossRef]

10. Cornea-Cipcigan, M.; Pamfil, D.; Sisea, C.R.; Gavriş, C.P.; da Graça Ribeiro Campos, M.; Margaoan, R. A review on Cyclamen species: Ttranscription factors vs. pharmacological effects. *Acta Pol. Pharm.-Drug Res.* **2019**, *76*, 919–938. [CrossRef]

11. Breeding, S. *Growing Cyclamen under High Light Levels*. Available online: https://schoneveld-breeding.com/en/growing-cyclamen-under-high-light-levels/ (accessed on 16 June 2022).

12. Savaedi, Z.; Parmoon, G.; Moosavi, S.A.; Bakhshande, A. The role of light and Gibberellic Acid on cardinal temperatures and thermal time required for germination of Charnushka (*Nigella sativa*) seed. *Ind. Crops Prod.* **2019**, *132*, 140–149. [CrossRef]

13. Jha, U.C.; Nayyar, H.; Siddique, K.H.M. Role of Phytohormones in Regulating Heat Stress Acclimation in Agricultural Crops. *J. Plant Growth Regul.* **2022**, *41*, 1041–1064. [CrossRef]

14. Wang, H.-Q.; Liu, P.; Zhang, J.-W.; Zhao, B.; Ren, B.-Z. Endogenous Hormones Inhibit Differentiation of Young Ears in Maize (*Zea mays* L.) Under Heat Stress. *Front. Plant Sci.* **2020**, *11*, 533046. [CrossRef]

15. Ma, H.-Y.; Zhao, D.-D.; Ning, Q.-R.; Wei, J.-P.; Li, Y.; Wang, M.-M.; Liu, X.-L.; Jiang, C.-J.; Liang, Z.-W. A Multi-year Beneficial Effect of Seed Priming with Gibberellic Acid-3 (GA3) on Plant Growth and Production in a Perennial Grass, Leymus chinensis. *Sci. Rep.* **2018**, *8*, 13214. [CrossRef] [PubMed]

16. Tsegay, B.A.; Andargie, M. Seed Priming with Gibberellic Acid (GA3) Alleviates Salinity Induced Inhibition of Germination and Seedling Growth of *Zea mays* L., Pisum sativum Var. abyssinicum A. *Braun* and *Lathyrus sativus* L. *J. Crop. Sci. Biotechnol.* **2018**, *21*, 261–267. [CrossRef]

17. Miceli, A.; Moncada, A.; Sabatino, L.; Vetrano, F. Effect of Gibberellic Acid on Growth, Yield, and Quality of Leaf Lettuce and Rocket Grown in a Floating System. *Agronomy* **2019**, *9*, 382. [CrossRef]

18. Muniandi, S.K.M.; Hossain, M.A.; Abdullah, M.P.; Ab Shukor, N.A. Gibberellic acid (GA3) affects growth and development of some selected kenaf (*Hibiscus cannabinus* L.) cultivars. *Ind. Crops Prod.* **2018**, *118*, 180–187. [CrossRef]

19. Garmendia, A.; Beltrán, R.; Zornoza, C.; García-Breijo, F.J.; Reig, J.; Merle, H. Gibberellic acid in *Citrus* spp. flowering and fruiting: A systematic review. *PLoS ONE* **2019**, *14*, e0223147. [CrossRef]

20. Dong, B.; Deng, Y.; Wang, H.; Gao, R.; Stephen, G.u.K.; Chen, S.; Jiang, J.; Chen, F. Gibberellic Acid Signaling Is Required to Induce Flowering of Chrysanthemums Grown under Both Short and Long Days. *Int. J. Mol. Sci.* **2017**, *18*, 1259. [CrossRef] [PubMed]

21. Bergmann, B.A.; Dole, J.M.; McCall, I. Gibberellic Acid Shows Promise for Promoting Flower Stem Length in Four Field-grown Cut Flowers. *HortTechnology Hortte* **2016**, *26*, 287–292. [CrossRef]

22. Toscano, S.; Trivellini, A.; Ferrante, A.; Romano, D. Physiological mechanisms for delaying the leaf yellowing of potted geranium plants. *Sci. Hortic.* **2018**, *242*, 146–154. [CrossRef]

23. Khan, M.N.; Khan, Z.; Luo, T.; Liu, J.; Rizwan, M.; Zhang, J.; Xu, Z.; Wu, H.; Hu, L. Seed priming with gibberellic acid and melatonin in rapeseed: Consequences for improving yield and seed quality under drought and non-stress conditions. *Ind. Crops Prod.* **2020**, *156*, 112850. [CrossRef]

24. Lutts, S.; Benincasa, P.; Wojtyla, L.; Kubala, S.; Pace, R.; Lechowska, K.; Quinet, M.; Garnczarska, M. *Seed Priming: New Comprehensive Approaches for an Old Empirical Technique*; InTech: London, UK, 2016.

25. Heydariyan, M.; Basirani, N.; Sharifi-Rad, M.; Khmmari, I.; Poor, S.R. Effect of seed priming on germination and seedling growth of the caper (*Capparis spinosa*) under drought stress. *Int. J. Adv. Biol. Biomed. Res.* **2014**, *2*, 2381–2389.

26. Jisha, K.C.; Vijayakumari, K.; Puthur, J.T. Seed priming for abiotic stress tolerance: An overview. *Acta Physiol. Plant.* **2013**, *35*, 1381–1396. [CrossRef]

27. Lozano, Y.M.; Aguilar-Trigueros, C.A.; Flaig, I.C.; Rillig, M.C. Root trait responses to drought are more heterogeneous than leaf trait responses. *Funct. Ecol.* **2020**, *34*, 2224–2235. [CrossRef]

28. Teixeira, E.I.; Fischer, G.; van Velthuizen, H.; Walter, C.; Ewert, F. Global hot-spots of heat stress on agricultural crops due to climate change. *Agric. For. Meteorol.* **2013**, *170*, 206–215. [CrossRef]

29. Hemantaranjan, A.; Bhanu, A.N.; Singh, M.N.; Yadav, D.K.; Patel, P.K.; Singh, R.; Katiyar, D. Heat stress responses and thermotolerance. *Adv. Plants Agric. Res.* **2014**, *1*, 00012. [CrossRef]

30. Siddiqui, M.H.; Al-Khaishany, M.Y.; Al-Qutami, M.A.; Al-Whaibi, M.H.; Grover, A.; Ali, H.M.; Al-Wahibi, M.S. Morphological and physiological characterization of different genotypes of faba bean under heat stress. *Saudi J. Biol. Sci.* **2015**, *22*, 656–663. [CrossRef]

31. Pallaoro, D.S.; Avelino, A.C.D.; Camili, E.C.; Guimarães, S.C.; Albuquerque, M.C.D.F. Priming corn seeds with plant growth regulator. *J. Seed Sci.* **2016**, *38*, 227–232. [CrossRef]

32. Ortega-Baes, P.; Rojas-Aréchiga, M. Seed germination of Trichocereus terscheckii (Cactaceae): Light, temperature and gibberellic acid effects. *J. Arid. Environ.* **2007**, *69*, 169–176. [CrossRef]

33. Vieira, A.R.; Vieira, M.D.G.G.C.; Fraga, A.C.; Oliveira, J.A.; Santos, C.D.D. Action of gibberellic acid (GA3) on dormancy and activity of alpha-amylase in rice seeds. *Rev. Bras. Sementes* **2002**, *24*, 43–48.

34. Sappalani, G.I.; Cabahug, L.M.; Valleser, V.C. Impact of gibberellic acid and organic growth media on seed germination and seedling development of rubber (*Hevea brasiliensis*). *Int. J. Hortic. Sci. Technol.* **2021**, *8*, 165–174.

35. Szekely-Varga, Z.; Kentelky, E.; Cantor, M. Effect of Gibberellic Acid on the Seed Germination of *Lavandula angustifolia* Mill. *RJH* **2021**, *2*, 169–176. [CrossRef]

36. Islam, M.M.; Rengel, Z.; Storer, P.; Siddique, K.H.M.; Solaiman, Z.M. Industrial Hemp (*Cannabis sativa* L.) Varieties and Seed Pre-Treatments Affect Seed Germination and Early Growth of Seedlings. *Agronomy* **2022**, *12*, 6. [CrossRef]

37. Li, Z.; Lu, G.Y.; Zhang, X.K.; Zou, C.S.; Cheng, Y.; Zheng, P.Y. Improving drought tolerance of germinating seeds by exogenous application of gibberellic acid (GA3) in rapeseed (*Brassica napus* L.). *Seed Sci. Technol.* **2010**, *38*, 432–440. [CrossRef]

38. Rhaman, M.S.; Rauf, F.; Tania, S.S.; Khatun, M. Seed priming methods: Application in field crops and future perspectives. *Asian J. Res. Crop. Sci.* **2020**, *5*, 8–19. [CrossRef]

39. Nimir, N.E.A.; Lu, S.; Zhou, G.; Guo, W.; Ma, B.; Wang, Y. Comparative effects of gibberellic acid, kinetin and salicylic acid on emergence, seedling growth and the antioxidant defence system of sweet sorghum (*Sorghum bicolor*) under salinity and temperature stresses. *Crop Pasture Sci.* **2015**, *66*, 145–157. [CrossRef]

40. Khan, A.; Bilal, S.; Khan, A.L.; Imran, M.; Shahzad, R.; Al-Harrasi, A.; Al-Rawahi, A.; Al-Azhri, M.; Mohanta, T.K.; Lee, I.-J. Silicon and Gibberellins: Synergistic Function in Harnessing ABA Signaling and Heat Stress Tolerance in Date Palm (*Phoenix dactylifera* L.). *Plants* **2020**, *9*, 620. [CrossRef]

41. Alonso-Ramírez, A.; Rodríguez, D.; Reyes, D.; Jiménez, J.S.A.; Nicolás, G.; López-Climent, M.A.; Gómez-Cadenas, A.; Nicolás, C. Evidence for a Role of Gibberellins in Salicylic Acid-Modulated Early Plant Responses to Abiotic Stress in Arabidopsis Seeds. *Plant Physiol.* **2009**, *150*, 1335–1344. [CrossRef]

42. Li, M.; Jannasch, A.H.; Jiang, Y. Growth and Hormone Alterations in Response to Heat Stress in Perennial Ryegrass Accessions Differing in Heat Tolerance. *J. Plant Growth Regul.* **2020**, *39*, 1022–1029. [CrossRef]

43. Neveur, N.; Corbineau, F.; Côme, D. Some characteristics of *Cyclamen persicum* L. seed germination. *J. Hortic. Sci.* **1986**, *61*, 379–387. [CrossRef]

44. Heydecker, W.; Wainwright, H. More rapid and uniform germination of *Cyclamen persicum* L. *Sci. Hortic.* **1976**, *5*, 183–189. [CrossRef]

45. Corbineau, F.; Neveur, N.; Côme, D. Seed Germination and Seedling Development in *Cyclamen persicum*. *Ann. Bot.* **1989**, *63*, 87–96. [CrossRef]

46. Fagge, A.A.; Manga, A.A. Effect of sowing media and gibberellic acid on the growth and seedling establishment of Bougainvillea glabra, Ixora coccinea and Rosa chinensis. 2. Root Characters. *Bayero J. Pure Appl. Sci.* **2011**, *4*, 155–159. [CrossRef]

47. Arteca, R.N.; Schlagnhaufer, C.D.; Arteca, J.M. Root applications of gibberellic acid enhance growth of seven *Pelargonium cultivars*. *HortScience* **1991**, *26*, 555–556. [CrossRef]

48. Bai, L.; Deng, H.; Zhang, X.; Yu, X.; Li, Y. Gibberellin Is Involved in Inhibition of Cucumber Growth and Nitrogen Uptake at Suboptimal Root-Zone Temperatures. *PLoS ONE* **2016**, *11*, e0156188. [CrossRef] [PubMed]

49. Stavang, J.A.; Gallego-Bartolomé, J.; Gómez, M.D.; Yoshida, S.; Asami, T.; Olsen, J.E.; García-Martínez, J.L.; Alabadí, D.; Blázquez, M.A. Hormonal regulation of temperature-induced growth in Arabidopsis. *Plant J.* **2009**, *60*, 589–601. [CrossRef]

50. Lu, L.; Liu, H.; Wu, Y.; Yan, G. Wheat genotypes tolerant to heat at seedling stage tend to be also tolerant at adult stage: The possibility of early selection for heat tolerance breeding. *Crop J.* **2022**. [CrossRef]

51. Kader, M.A. A comparison of seed germination calculation formulae and the associated interpretation of resulting data. *J. Proc. R. Soc. N. S. W.* **2005**, *138*, 65–75.

52. Al-Ansari, F.; Ksiksi, T. A Quantitative Assessment of Germination Parameters: The Case of Crotalaria Persica and Tephrosia apollinea. *Open Ecol. J.* **2016**, *9*, 13–21. [CrossRef]

53. Ellis, R.H.; Roberts, E.H. *Towards a Rational Basis for Testing Seed Quality*; Butterworths: London, UK, 1980.

54. Abdul-Baki, A.A.; Anderson, J.D. Relationship Between Decarboxylation of Glutamic Acid and Vigor in Soybean Seed. *Crop Sci.* **1973**, *13*, 227–232. [CrossRef]

55. Ward, J.H. Hierarchical Grouping to Optimize an Objective Function. *J. Am. Stat. Assoc.* **1963**, *58*, 236–244. [CrossRef]

Article

Photosynthetic Responses, Growth, Production, and Tolerance of Traditional Varieties of Cowpea under Salt Stress

Saulo Samuel Carneiro Praxedes [1], Miguel Ferreira Neto [1,*], Aline Torquato Loiola [1], Fernanda Jessica Queiroz Santos [1], Bianca Fernandes Umbelino [1], Luderlândio de Andrade Silva [2], Rômulo Carantino Lucena Moreira [2], Alberto Soares de Melo [3], Claudivan Feitosa de Lacerda [4], Pedro Dantas Fernandes [2], Nildo da Silva Dias [1] and Francisco Vanies da Silva Sá [1,*]

[1] Department of Agronomic and Forest Sciences, Federal Rural University of the Semi-Arid—UFERSA, Mossoró 59625-900, Brazil; saulosamuel@rn.gov.br (S.S.C.P.); ninator4@gmail.com (A.T.L.); fernanda.santos76421@alunos.ufersa.edu.br (F.J.Q.S.); bianca.umbelino@alunos.ufersa.edu.br (B.F.U.); nildo@ufersa.edu.br (N.d.S.D.)

[2] Center of Technology and Natural Resources, Federal University of Campina Grande—UFCG, Campina Grande 58428-830, Brazil; luderlandioandrade@gmail.com (L.d.A.S.); romulocarantino@gmail.com (R.C.L.M.); pedrodantasfernandes@gmail.com (P.D.F.)

[3] Department of Biology, Universidade Estadual da Paraíba, Campina Grande 58429-500, Brazil; alberto@uepb.edu.br

[4] Department of Agricultural Engineering, Federal University of Ceará, Fortaleza 60440-593, Brazil; cfeitosa@ufc.br

* Correspondence: miguel@ufersa.edu.br (M.F.N.); vanies@ufersa.edu.br (F.V.d.S.S.); Tel.: +55-(83)9-9861-9267 (M.F.N.)

Citation: Praxedes, S.S.C.; Ferreira Neto, M.; Loiola, A.T.; Santos, F.J.Q.; Umbelino, B.F.; Silva, L.d.A.; Moreira, R.C.L.; Melo, A.S.d.; Lacerda, C.F.d.; Fernandes, P.D.; et al. Photosynthetic Responses, Growth, Production, and Tolerance of Traditional Varieties of Cowpea under Salt Stress. *Plants* **2022**, *11*, 1863. https://doi.org/10.3390/plants11141863

Academic Editor: Pedro Diaz-Vivancos

Received: 24 June 2022
Accepted: 6 July 2022
Published: 18 July 2022

Publisher's Note: MDPI stays neutral with regard to jurisdictional claims in published maps and institutional affiliations.

Highlights:

- Salinity tolerance among traditional varieties of cowpea is variable.
- High net photosynthesis is observed in cowpea-tolerant varieties under salt stress.
- The decrease in photosynthesis is caused by stomatal restriction and low photochemical efficiency.
- Cowpea varieties are similar in photochemical efficiency and photochemical quenching.
- The decrease in pod number and seeds per pod are the main effects of salinity on cowpea production.

Abstract: Cowpea is the main subsistence crop—protein source—for the Brazilian semi-arid region. The use of salt-stress-tolerant varieties can improve crop yields. We evaluated the effect of irrigation with brackish water on the growth, photosynthetic responses, production, and tolerance of fifteen traditional varieties of cowpea. The experiment was conducted in randomized blocks, in a 15×2 factorial scheme, composed of 15 traditional varieties of cowpea and two salinity levels of irrigation water (0.5 and 4.5 dS m^{-1}), with five replicates. Plants were grown in pots containing 10 dm^3 of soil for 80 days. The reduction in the photosynthetic rate of cowpea varieties occurs mainly due to the decrease in stomatal conductance caused by salt stress. Salt stress increased the electron transport rate and photochemical quenching of cowpea varieties, but stress-tolerant varieties increased the CO$_2$ assimilation rate and instantaneous carboxylation efficiency. The Ceará, Costela de Vaca, Pingo de Ouro, Ovo de Peru, and Sempre Verde varieties are tolerant to salt stress. Salt stress decreases 26% of the production of tolerant varieties to salt stress and 54% of susceptible varieties. The present findings show the existence of variability for saline stress tolerance in traditional varieties of cowpea and that Ceará, Costela de Vaca, Pingo de Ouro, and Ovo de Peru varieties are more suitable for crops irrigated with saline water.

Keywords: *Vigna unguiculata* (L.) Walp.; salinity; gas exchange; photochemical efficiency; photochemical quenching; yield

1. Introduction

Cowpea beans (*Vigna unguiculata* (L.) Walp.), also known as rope or Macassar beans, are a significant source of protein in the north and northeastern regions of Brazil [1]. Family and small farms are the leading producers, using mostly native or traditional seeds [2]. These farmers have little technological apparatus and use family labor and seeds from selection made by the farmer himself, with well-defined and recognized phenotypic characteristics that characterize them as native or traditional seeds. Traditional or local varieties are highly adapted to the places where they are conserved and managed, and they are part of family autonomy, constituting the main factor in people's food security.

Farms in the northeastern region, especially those in Mossoró/Açu, RN, Brazil, require a substantial amount of water, which has driven the use of saline water up to 4.5 dS m^{-1}, e.g., groundwater from the Calcário Jandaíra aquifer [3]. Cowpea beans are moderately tolerant to salinity, up to 3.3 dS m^{-1} in irrigation water and 4.8 dS m^{-1} in the soil [4]. However, it is necessary to use saline water due to the scarcity of good quality water and the increase in water consumption to meet population growth and irrigated agriculture.

Water with a high sodium adsorption ratio (SAR) can modify the physicochemical conditions of the soil, and excess soluble salts result in lower osmotic potential, water deficit, stomatal closure, limited CO_2 assimilation/water usage, and alterations to the photochemical process [5,6]. Osmotic restrictions combined with ionic restrictions and nutritional imbalance limit gas exchange and biomass accumulation and production [5–8]. The decrease in productivity of cowpea under saline stress occurs due to the decrease in the water potential, survival rate, plant's initial vigor, growth, and photosynthetic activity and excessive accumulation of Cl and Na ions and reactive oxygen species (ROS) [5–11]. Therefore, studies that evaluate gas exchange and photosynthetic efficiency are important for the identification of salinity-tolerant plants.

Identifying and selecting salinity-tolerant cowpea bean varieties can facilitate the use of saline water without affecting cowpea production. While there are early growth stage studies that designate salinity-tolerant cowpea bean genotypes [2,12,13], studies investigating the complete production cycle are lacking. In this study, we hypothesized that the genetic diversity of cowpea beans grown in the semi-arid region would create salinity-tolerant cowpea bean varieties that may be cultivated in farms with poor water quality. Thus, the goals of this study were to assess the effect of irrigation with brackish water on the growth, gas exchange, and production of 15 traditional varieties of cowpea and determine the most saltwater-tolerant varieties.

2. Results

2.1. Soil Salinity

There was a significant interaction between varieties and irrigation-water salinity levels ($p < 0.05$) (Table 1). Irrigation with saline water of 4.5 dS m^{-1} increased the ECse of the soil contained in the pots cultivated with the plants of cowpea varieties compared to pots that received the water of 0.5 dS m^{-1} (control). The ECse of the soil contained in the pots irrigated with high-salinity water varied between 6.3 and 9.7 dS m^{-1}, that is, between 1.4 and 2.15 times the salinity of the irrigation water. When irrigation was performed with high-salinity water in pots cultivated with the varieties Canário and Roxão, the highest values of ECse in the soil were recorded, 9.7 and 8.4 dS m^{-1}, respectively, with no difference for salinity in pots cultivated with the other varieties (Table 1).

Table 1. F-test and means test for electrical conductivity of saturation extract (ECse, dS m^{-1}) of soils cultivated with traditional varieties of cowpea subjected to two levels of irrigation-water salinity.

F-Test (*p*-Value)	
Sources of Variation	ECse
Block	0.061
Salinity	0.000
Varieties	0.011
Salinity × Varieties	0.018

Means comparison test (Standard Deviation, *n* = 5)		
Varieties	ECse	
	0.5 dS m^{-1}	4.5 dS m^{-1}
V1—Boquinha	1.3 ± 0.19 Ba	7.8 ± 0.35 Ac
V2—Ceará	1.1 ± 0.11 Ba	7.3 ± 0.72 Ac
V3—Costela de Vaca	1.4 ± 0.23 Ba	6.7 ± 0.42 Ac
V4—Lisão	1.5 ± 0.20 Ba	6.5 ± 0.69 Ac
V5—Canário	1.4 ± 0.20 Ba	9.7 ± 0.79 Aa
V6—Pingo de Ouro	1.4 ± 0.22 Ba	7.8 ± 0.40 Ac
V7—Roxão	1.7 ± 0.17 Ba	8.4 ± 0.34Ab
V8—Feijão Branco	1.4 ± 0.08 Ba	7.8 ± 0.41 Ac
V9—Canapu Branco	1.5 ± 0.17 Ba	7.4 ± 0.48 Ac
V10—Canapu Miúdo	1.3 ± 0.17 Ba	6.7 ± 0.64 Ac
V11—Ovo de Peru	1.3 ± 0.21 Ba	7.9 ± 1.07 Ac
V12—Baeta	1.9 ± 0.13 Ba	7.2 ± 0.41 Ac
V13—Coruja	1.3 ± 0.05 Ba	7.0 ± 0.73 Ac
V14—Paulistinha	1.4 ± 0.10 Ba	6.3 ± 0.47 Ac
V15—Sempre Verde	1.4 ± 0.19 Ba	7.5 ± 0.38 Ac

Equal uppercase letters in the rows and lowercase letters in the column do not differ by Student's *t*-test and Scott–Knott test at 5% probability level, respectively.

2.2. Gas Exchange

The interaction between varieties and salinity levels was significant for the net CO_2 assimilation rate ($p < 0.01$), internal CO_2 concentration ($p < 0.01$), and instantaneous carboxylation efficiency ($p < 0.01$) (Table 2).

The A_N of the varieties Boquinha, Roxão, Feijão Branco, Canapu Miúdo, Coruja, and Paulistinha was reduced under salt stress (4.5 dS m^{-1}), varying, on average, between 13.19 and 74.21% compared to the control (0.5 dS m^{-1}). The varieties Ceará, Canário, Pingo de Ouro, Canapu Branco, Baeta, and Sempre Verde had increased A_N in the saline treatment (4.5 dS m^{-1}) with an average variation between 11.95 and 56.71% compared to the control (0.5 dS m^{-1}). Salt stress did not influence ($p < 0.05$) the A_N of the varieties Costela de Vaca, Lisão, and Ovo de Peru (Table 2).

There was the formation of three clusters regarding photosynthetic performance under the salt-stress condition, with the varieties Sempre Verde, Costela de Vaca, Pingo de Ouro, and Baeta in the cluster of higher A_N, Ceará, Lisão, Canário, Canapu Branco, Boquinha, Roxão, and Paulistinha in the cluster of intermediate A_N, and Ovo de Peru, Feijão Branco, Coruja, and Canapu Miúdo in the cluster of low A_N, in this sequence (Table 2).

The application of 4.5 dS m^{-1} water reduced the *Ci* in the varieties Canapu Branco, Ovo de Peru, and Baeta compared to the control (0.5 dS m^{-1}), with reductions ranging between 11.80 and 23.05%. A different behavior was observed in the varieties Ceará, Canário, Pingo de Ouro, Canapu Miúdo, Coruja, and Sempre Verde, which had an average increment ranging between 11.82 and 42.68% in *Ci* under salt stress (4.5 dS m^{-1}) compared to the control (0.5 dS m^{-1}). The varieties Boquinha, Costela de Vaca, Lisão, Roxão, Feijão Branco, and Paulistinha did not have their *Ci* influenced by salt stress (Table 2).

Table 2. F-test and means test for net CO_2 assimilation rate (A_N, μmol (CO_2) m^{-2} s^{-1}), internal CO_2 concentration (Ci, mol (CO_2) m^{-2} s^{-1}), and instantaneous carboxylation efficiency (CEi, (μmol (CO_2) m^{-2} s^{-1}) (mol (CO_2) m^{-2} s^{-1})$^{-1}$) of traditional varieties of cowpea subjected to salinity levels of irrigation water.

	F-Test (*p*-Value)		
Sources of Variation	A_N	Ci	CEi
Block	0.543	0.000	0.0284
Salinity	0.483	0.000	0.006
Varieties	0.000	0.000	0.000
Salinity × Varieties	0.000	0.000	0.000

	Means comparison test (Standard Deviation, *n* = 5)					
	A_N		Ci		CEi	
Varieties	0.5 dS m^{-1}	4.5 dS m^{-1}	0.5 dS m^{-1}	4.5 dS m^{-1}	0.5 dS m^{-1}	4.5 dS m^{-1}
V1—Boquinha	16.92 ± 0.37 Aa	10.13 ± 0.82 Be	187 ± 6.67 Ac	198 ± 13.50 Ab	0.091 ± 0.003 Aa	0.053 ± 0.007 Bd
V2—Ceará	8.15 ± 0.08 Be	11.56 ± 0.90 Ad	198 ± 3.27 Bc	239 ± 8.21 Aa	0.041 ± 0,001 Ad	0.049 ± 0.005 Ad
V3—Costela de Vaca	15.04 ± 0.34 Ab	15.11 ± 0.38 Ab	155 ± 5.76 Ad	164 ± 3.67 Ad	0.098 ± 0.006 Aa	0.092 ± 0.004 Aa
V4—Lisão	11.77 ± 0.03 Ac	12.27 ± 0.49 Ad	195 ± 2.03 Ac	200 ± 4,52 Ab	0.060 ± 0.001 Ac	0.062 ± 0.004 Ac
V5—Canário	7.39 ± 0.54 Be	11.12 ± 0.16 Ad	137 ± 11.98 Be	174 ± 4.06 Ad	0.057 ± 0.009 Ac	0.064 ± 0.002 Ac
V6—Pingo de Ouro	8.34 ± 0.61 Be	13.07 ± 0.27 Ac	165 ± 6.27 Bd	189 ± 4.23 Ac	0.051 ± 0.005 Bc	0.069 ± 0.003 Ab
V7—Roxão	11.58 ± 0.12 Ac	9.81 ± 0.82 Be	190 ± 6.16 Ac	192 ± 8.08 Ac	0.061 ± 0.002 Ac	0.052 ± 0.007 Ad
V8—Feijão Branco	10.78 ± 0.12 Ad	7.98 ± 0.06 Bf	215 ± 3.85 Ab	208 ± 7.00 Ab	0.050 ± 0.001 Ac	0.039 ± 0.001 Be
V9—Canapu Branco	9.96 ± 0.42 Bd	11.15 ± 0.30 Ad	243 ± 2.94 Aa	187 ± 7.19 Bc	0.041 ± 0.002 Bd	0.060 ± 0.004 Ac
V10—Canapu Miúdo	10.47 ± 0.40 Ad	2.70 ± 0.19 Bg	164 ± 6.03 Bd	234 ± 10.72 Aa	0.065 ± 0.005 Ac	0.011 ± 0.001 Bf
V11—Ovo de Peru	8.35 ± 0.14 Ae	8.03 ± 0.27 Af	231 ± 1.88 Aa	203 ± 3.25 Bb	0.036 ± 0.001 Ad	0.040 ± 0.002 Ae
V12—Baeta	9.27 ± 0.24 Be	13.61 ± 0.16 Ac	178 ± 9.71 Ad	157 ± 5.34 Bd	0.053 ± 0.002 Bc	0.087 ± 0.003 Aa
V13—Coruja	11.98 ± 0.25 Ac	7.65 ± 0.28 Bf	159 ± 4.23 Bd	210 ± 3.61 Ab	0.076 ± 0.003 Ab	0.036 ± 0.001 Be
V14—Paulistinha	11.07 ± 0.42 Ac	9.61 ± 0.19 Be	199 ± 3.45 Ac	213 ± 4.97 Ab	0.056 ± 0.002 Ac	0.045 ± 0.002 Bd
V15—Sempre Verde	11.30 ± 0.06 Bc	17.06 ± 0.20 Aa	203 ± 4.72 Bc	227 ± 2.22 Aa	0.056 ± 0.001 Bc	0.075 ± 0.001 Ab

Equal uppercase letters in the rows and lowercase letters in the column do not differ by Student's *t*-test and Scott–Knott test at 5% probability level, respectively.

The CEi of the varieties Pingo de Ouro, Canapu Branco, Baeta, and Sempre Verde were increased under the salt-stress conditions compared to the control, with a variation between 33.93 and 64.15%. The varieties Boquinha, Feijão Branco, Canapu Miúdo, Coruja, and Paulistinha had reduced CEi, between 19.64 and 83.08%, on average, under the salt-stress conditions compared to the control (Table 2). The varieties Ceará, Costela de Vaca, Lisão, Canário, Roxão, and Ovo de Peru did not have their CEi influenced by salt stress (Table 2). The best CEi values under the salt-stress conditions were observed in the varieties Costela de Vaca, Pingo de Ouro, Baeta, and Sempre Verde. The worst CEi values under the salt-stress conditions were observed in the varieties Feijão Branco, Canapu Miúdo, Ovo de Peru, and Coruja (Table 2).

The interaction between varieties and salinity levels was significant for stomatal conductance ($p < 0.01$), transpiration ($p < 0.01$), and instantaneous water use efficiency ($p < 0.01$) (Table 3).

When comparing the conditions of salt stress (4.5 dS m^{-1}) and control (0.5 dS m^{-1}) for *gs* (Table 3), the Boquinha, Roxão, Feijão Branco, Canapu Branco, Canapu Miúdo, Ovo de Peru, Coruja, and Paulistinha varieties had an average reduction ranging between 7.69 and 70.00%. The Ceará, Canário, Pingo de Ouro, and Sempre Verde varieties showed an average increase ranging between 62.50 and 100.00%. On the other hand, the stomatal conductance values of the Costela de Vaca, Lisão, and Baeta varieties were not influenced by salt stress. Under this condition, the lowest values of *gs* were verified in the Canapu Miúdo, Ovo de Peru, Roxão, Feijão Branco, and Coruja varieties, while the highest values were recorded in the varieties Sempre Verde and Ceará.

Table 3. F-test and means test for stomatal conductance (gs) (mol (H_2O) m^{-2} s^{-1}), transpiration (E) (mmol (H_2O) m^{-2} s^{-1}), and instantaneous water use efficiency ($WUEi$) [(μmol (CO_2) m^{-2} s^{-1}) (mmol (H_2O) m^{-2} s^{-1})$^{-1}$] of traditional varieties of cowpea subjected to salinity levels of irrigation water.

F-Test			
Sources of Variation	gs	E	$WUEi$
Block	0.000	0.000	0.000
Salinity	0.031	0.000	0.000
Varieties	0.000	0.000	0.000
Salinity × Varieties	0.000	0.000	0.000

Means comparison test (Standard Deviation, $n = 5$)						
Varieties	gs		E		$WUEi$	
	0.5 dS m^{-1}	4.5 dS m^{-1}	0.5 dS m^{-1}	4.5 dS m^{-1}	0.5 dS m^{-1}	4.5 dS m^{-1}
V1—Boquinha	0.21 ± 0.016 Aa	0.11 ± 0.004 Be	4.32 ± 0.19 Aa	1.98 ± 0.07 Be	3.94 ± 0.11 Bb	5.12 ± 0.38 Af
V2—Ceará	0.09 ± 0.001 Be	0.18 ± 0.006 Ab	2.47 ± 0.03 Bf	3.49 ± 0.11 Ab	3.30 ± 0.07 Ac	3.29 ± 0.19 Ah
V3—Costela de Vaca	0.14 ± 0.001 Ac	0.13 ± 0.004 Ac	3.48 ± 0.02 Ac	1.84 ± 0.04 Be	4.33 ± 0.12 Ba	8.22 ± 0.15 Ab
V4—Lisão	0.14 ± 0.002 Ac	0.14 ± 0.005 Ac	3.77 ± 0.03 Ab	2.57 ± 0.05 Bc	3.12 ± 0.03 Bc	4.78 ± 0.14 Af
V5—Canário	0.06 ± 0.002 Bg	0.11 ± 0.002 Ae	1.69 ± 0.07 Bh	2.47 ± 0.01 Ad	4.37 ± 0.23 Aa	4.51 ± 0.08 Ag
V6—Pingo de Ouro	0.08 ± 0.007 Bf	0.13 ± 0.001 Ac	2.28 ± 0.14 Ag	1.74 ± 0.01 Be	3.66 ± 0.12 Bb	7.50 ± 0.14 Ac
V7—Roxão	0.13 ± 0.007 Ac	0.09 ± 0.006 Bf	3.22 ± 0.12 Ad	1.64 ± 0.06 Bf	3.61 ± 0.11 Bb	5.96 ± 0.29 Ae
V8—Feijão Branco	0.15 ± 0.006 Ab	0.09 ± 0.004 Bf	3.83 ± 0.11 Ab	2.40 ± 0.09 Bd	2.82 ± 0.07 Bd	3.35 ± 0.14 Ah
V9—Canapu Branco	0.16 ± 0.006 Ab	0.11 ± 0.002 Be	3.63 ± 0.10 Ac	1.64 ± 0.02 Bf	2.74 ± 0.06 Bd	6.83 ± 0.28 Ad
V10—Canapu Miúdo	0.10 ± 0.002 Ae	0.03 ± 0.004 Bg	2.59 ± 0.02 Af	0.89 ± 0.09 Bg	4.05 ± 0.13 Aa	3.11 ± 0.20 Bh
V11—Ovo de Peru	0.12 ± 0.002 Ad	0.08 ± 0.002 Bf	2.83 ± 0.04 Ae	1.80 ± 0.02 Be	2.95 ± 0.04 Bd	4.45 ± 0.13 Ag
V12—Baeta	0.10 ± 0.007 Ae	0.11 ± 0.002 Ae	2.86 ± 0.17 Ae	1.44 ± 0.02 Bf	3.28 ± 0.15 Bc	9.44 ± 0.20 Aa
V13—Coruja	0.11 ± 0.002 Ad	0.09 ± 0.004 Bf	2.87 ± 0.05 Ae	2.42 ± 0.11 Bd	4.18 ± 0.08 Aa	3.17 ± 0.07 Bh
V14—Paulistinha	0.13 ± 0.007 Ac	0.12 ± 0.002 Bd	3.32 ± 0.13 Ad	2.69 ± 0.04 Bc	3.34 ± 0.06 Ac	3.58 ± 0.12 Ah
V15—Sempre Verde	0.14 ± 0.006 Bc	0.26 ± 0.005 Aa	3.37 ± 0.08 Bd	4.02 ± 0.05 Aa	3.36 ± 0.08 Bc	4.24 ± 0.04 Ag

Equal uppercase letters in the rows and lowercase letters in the column do not differ by Student's t-test and Scott–Knott test at 5% probability level, respectively.

Except for the Ceará, Canário, and Sempre Verde varieties, which had an increase in E, the others showed a reduction in E, ranging from 15.68 to 65.64% on average, under the condition of salt stress (4.5 dS m^{-1}) compared to the control (0.5 dS m^{-1}). Under the salt-stress condition, the lowest E values were verified in the Canapu Miúdo, Baeta, Canapu Branco, and Roxão varieties (Table 3).

The $WUEi$ of the Canapu Miúdo and Coruja varieties decreased on average by 23.69% under the condition of salt stress (4.5 dS m^{-1}) compared to the control (0.5 dS m^{-1}) (Table 3). Under the condition of salt stress, there was an average increase between 18.79 and 187.80% in the $WUEi$ of the Boquinha, Costela de Vaca, Lisão, Pingo de Ouro, Roxão, Feijão Branco, Canapu Branco, Ovo de Peru, Baeta, and Sempre Verde varieties compared to the control treatment. The Ceará, Canário, and Paulistinha varieties did not have their $WUEi$ influenced by salt stress (Table 3). Under this condition, the highest values of $WUEi$ were verified in the Baeta, Costela de Vaca, Pingo de Ouro, and Canapu Branco varieties, while the lowest values were found in the Canapu Miúdo, Coruja, Ceará, Feijão Branco, and Paulistinha varieties (Table 3).

2.3. Chlorophyll Fluorescence

There were significant effects of water salinity levels on initial fluorescence ($p < 0.05$), variable fluorescence (Fv) ($p < 0.05$), the maximum quantum efficiency of photosystem II ($p < 0.01$), the quantum efficiency of photosystem II ($p < 0.01$), electron transport rate ($p < 0.01$), and quantum yield of regulated photochemical quenching ($p < 0.01$) (Table 4).

Table 4. F-test and means test for initial fluorescence (*Fo*) (μmol (photons) m^{-2} s^{-1}), maximum fluorescence (*Fm*) (μmol (photons) m^{-2} s^{-1}), variable fluorescence (*Fv*) (μmol (photons) m^{-2} s^{-1}), the maximum quantum efficiency of photosystem II (*Fv/Fm*), the quantum efficiency of photosystem II (*Y(II)*), electron transport rate (*ETR*) (μmol (photons) m^{-2} s^{-1}), minimum fluorescence of the illuminated plant tissue (Fo') (μmol (photons) m^{-2} s^{-1}), photochemical quenching coefficient (*qL*), the quantum yield of regulated photochemical quenching (*Y(NPQ)*), and the quantum yield of non-regulated photochemical quenching (*Y(NO)*) of traditional varieties of cowpea subjected to salinity levels of irrigation water.

F-Test (*p*-Value)				
Sources of Variation	*Fo*	*Fm*	*Fv*	*Fv/Fm*	*Y(II)*
Block	0.000	0.000	0.000	0.442	0.445
Salinity	0.037	0.235	0.046	0.003	0.010
Varieties	0.314	0.485	0.475	0.313	0.881
Salinity $\times$ Varieties	0.616	0.963	0.983	0.668	0.222

Means comparison test (Standard Deviation, *n* = 75)				
Salinity (dS m^{-1})	*Fo*	*Fm*	*Fv*	*Fv/Fm*	*Y(II)*
0.5	777.94 $\pm$ 11.21 A	3128.91 $\pm$ 35.94 A	2351.0 $\pm$ 33.99 B	0.75 $\pm$ 0.004 B	0.68 $\pm$ 0.008 A
4.5	749.48 $\pm$ 9.16 B	3178.00 $\pm$ 31.37 A	2428.5 $\pm$ 26.05 A	0.76 $\pm$ 0.002 A	0.65 $\pm$ 0.011 B

F-test (*p*-value)				
Sources of Variation	*ETR*[1]	*Fo'*[1]	*qL*[1]	*Y(NPQ)*[1]	*Y(NO)*[1]
Block	0.000	0.2068	0.110	0.397	0.378
Salinity	0.000	0.9214	0.140	0.005	0.736
Varieties	0.308	0.5083	0.840	0.812	0.833
Salinity $\times$ Varieties	0.635	0.3448	0.626	0.203	0.393

Means comparison test (Standard Deviation, *n* = 75)				
Salinity (dS m^{-1})	*ETR*	*Fo'*	*qL*	*Y(NPQ)*	*Y(NO)*
0.5	29.66 $\pm$ 2.47 B	2.85 $\pm$ 0.102 A	0.014 $\pm$ 0.0005 A	0.27 $\pm$ 0.007 B	0.05 $\pm$ 0.0010 A
4.5	53.10 $\pm$ 4.07 A	2.82 $\pm$ 0.082 A	0.013 $\pm$ 0.0005 A	0.30 $\pm$ 0.010 A	0.05 $\pm$ 0.0012 A

[1] Data transformed to square root. Equal uppercase letters in columns do not differ by Student's *t*-test at a 5% probability level.

The cowpea varieties irrigated with high-salinity water (4.5 dS m^{-1}) showed increments in the values of *Fv*, *Fv/Fm*, *ETR*, and *Y (NPQ)* of 3.29, 1.33, 79.02, and 11.11% compared to the control treatment (0.5 dS m^{-1}), respectively (Table 4). However, irrigation with high-salinity water (4.5 dS m^{-1}) reduced *Fo* and *Y(II)* by 3.65 and 4.41% compared to the control treatment (0.5 dS m^{-1}), respectively (Table 4).

2.4. Growth and Biomass Accumulation

There were simple effects of salinity levels and varieties for stem diameter and number of leaves (Table 5). Irrigation with high-salinity water decreased on average by 14.44 and 50.0% the stem diameter (SD) and the number of leaves (NL), respectively, of cowpea plants compared to the control (Table 5). The varieties Boquinha, Ceará, Canapu Miúdo, and Ovo de Peru had the highest SD, while Ceará showed the highest NL regardless of the water salinity level (Table 5).

Table 5. F-test and means test for stem diameter (SD, mm) and the number of leaves (NL) of traditional varieties of cowpea subjected to salinity levels of irrigation water.

F-Test (*p*-Value)		
Sources of Variation	SD	NL
Block	0.871	0.035
Salinity	0.000	0.000
Varieties	0.000	0.000
Salinity × Varieties	0.159	0.329
Means comparison test (Standard Deviation, *n* = 10)		
Varieties	SD	NL
V1—Boquinha	10.1 ± 1.04 a	15.5 ± 3.36 c
V2—Ceará	9.6 ± 0.65 a	22.8 ± 3.77 a
V3—Costela de Vaca	7.9 ± 1.01 b	14.0 ± 3.19 c
V4—Lisão	8.2 ± 0.61 b	17.7 ± 3.45 b
V5—Canário	8.1 ± 0.41 b	16.9 ± 3.26 b
V6—Pingo de Ouro	8.1 ± 0.32 b	13.6 ± 2.35 c
V7—Roxão	7.9 ± 0.29 b	14.7 ± 2.11 c
V8—Feijão Branco	7.6 ± 0.54 b	11.4 ± 3.19 c
V9—Canapu Branco	8.0 ± 0.41 b	12.9 ± 2.10 c
V10—Canapu Miúdo	9.1 ± 0.58 a	16.7 ± 3.21 b
V11—Ovo de Peru	9.2 ± 0.44 a	18.9 ± 4.21 b
V12—Baeta	7.9 ± 0.29 b	11.4 ± 1.88 c
V13—Coruja	7.9 ± 0.46 b	17.7 ± 2.81 b
V14—Paulistinha	7.7 ± 0.57 b	13.5 ± 2.80 c
V15—Sempre Verde	8.1 ± 0.43 b	16.1 ± 3.86 b
Means comparison test (Standard Deviation, *n* = 75)		
Salinity	SD (mm)	NL
0.5 dS m^{-1}	9.0 ± 0.15 A	20.8 ± 0.74 A
4.5 dS m^{-1}	7.7 ± 0.14 B	10.4 ± 0.34 B

Equal uppercase letters in the rows and lowercase letters in the column do not differ by Student's *t*-test and Scott–Knott test at 5% probability level, respectively.

There was a significant interaction between varieties and salinity levels for main branch length ($p < 0.01$) and shoot dry mass ($p < 0.05$) (Table 6). Main branch length (*MBL*) was reduced, with an average variation between 26.52 and 63.73% in the varieties Costela de Vaca, Pingo de Ouro, Roxão, Canapu Branco, Canapu Miúdo, and Baeta, when irrigated with water of 4.5 dS m^{-1} compared to the control (Table 6, Figure 1). Among the varieties irrigated with saline water, Ceará, Canário, Roxão, and Ovo de Peru had the highest *MBL* values (Table 6).

Table 6. F-test and means test for main branch length (MBL, cm) and shoot dry mass (SDM, g) of traditional varieties of cowpea subjected to salinity levels of irrigation water.

	F-Test (*p*-Value)			
Sources of Variation	MBL		SDM	
Block	0.277		0.040	
Salinity	0.000		0.000	
Varieties	0.000		0.002	
Salinity x Varieties	0.006		0.015	
	Means comparison test (Standard Deviation, *n* = 5)			
	MBL		SDM	
Varieties	0.5 dS m^{-1}	4.5 dS m^{-1}	0.5 dS m^{-1}	4.5 dS m^{-1}
V1—Boquinha	202.4 ± 19.95 Ab	169.0 ± 16.79 Ab	26.9 ± 2.36 Ab	20.4 ± 2.00 Ba
V2—Ceará	253.8 ± 4.02 Aa	200.4 ± 9.21 Aa	32.8 ± 1.41 Aa	13.8 ± 1.67 Bb
V3—Costela de Vaca	266.4 ± 10.61 Aa	178.6 ± 26.79 Bb	31.7 ± 4.09 Aa	13.4 ± 2.50 Bb
V4—Lisão	193.4 ± 16.58 Ab	167.0 ± 27.82 Ab	23.3 ± 1.46 Ab	11.9 ± 1.70 Bb
V5—Canário	222.4 ± 22.46 Ab	222.2 ± 22.44 Aa	30.5 ± 0.91 Aa	17.4 ± 2.05 Ba
V6—Pingo de Ouro	242.8 ± 11.62 Aa	162.4 ± 18.52 Bb	24.5 ± 2.23 Ab	16.6 ± 2.59 Ba
V7—Roxão	276.0 ± 16.90 Aa	202.8 ± 29.67 Ba	25.8 ± 1.43 Ab	21.3 ± 1.80 Aa
V8—Feijão Branco	186.6 ± 14.18 Ab	167.0 ± 15.01 Ab	26.6 ± 1.28 Ab	18.3 ± 1.12 Ba
V9—Canapu Branco	297.4 ± 28.01 Aa	165.0 ± 19.02 Bb	23.2 ± 2.38 Ab	15.9 ± 2.10 Bb
V10—Canapu Miúdo	236.6 ± 14.94 Aa	148.0 ± 25.96 Bb	26.8 ± 1.85 Ab	13.4 ± 1.45 Bb
V11—Ovo de Peru	250.6 ± 14.32 Aa	247.8 ± 22.97 Aa	31.0 ± 4.26 Aa	21.3 ± 1.22 Ba
V12—Baeta	228.3 ± 8.22 Ab	82.8 ± 13.33 Bc	29.1 ± 1.28 Aa	17.9 ± 1.20 Ba
V13—Coruja	212.2 ± 11.62 Ab	168.0 ± 25.91 Ab	27.1 ± 1.73 Ab	16.1 ± 1.23 Bb
V14—Paulistinha	213.2 ± 19.62 Ab	164.6 ± 15.05 Ab	21.8 ± 1.43 Ab	11.6 ± 1.15 Bb
V15—Sempre Verde	238.3 ± 10.00 Aa	184.0 ± 36.33 Ab	32.9 ± 5.45 Aa	13.7 ± 3.01 Bb

Equal uppercase letters in the rows and lowercase letters in the column do not differ by Student's *t*-test and Scott–Knott test at 5% probability level, respectively.

Figure 1. Traditional varieties of cowpea subjected to two levels of irrigation-water salinity.

Irrigation with water of 4.5 dS m^{-1} reduced shoot dry mass (*SDM*) in all cowpea varieties, between 24.16 and 58.36%, compared to those irrigated with water of 0.5 dS m^{-1} (Table 6, Figure 1). Under the condition of irrigation with water of 4.5 dS m^{-1}, the highest accumulations of shoot dry mass (Table 6) were recorded in the varieties Boquinha, Canário, Pingo de Ouro, Roxão, Feijão Branco, Ovo de Peru, and Baeta.

2.5. Grain Production

The interaction between salinity levels and cowpea varieties was significant (*p* < 0.01) for the number of pods per plant (NPP), the number of seeds per pod (NSPo), the number of seeds per plant (NSPl), and production per plant (PP) (Table 7).

The number of pods per plant (NPP), the number of seeds per plant (NSPl), and production per plant (PP) were reduced by up to 68.89, 71.96, and 61.26% on average, respectively, under irrigation with water of 4.5 dS m^{-1} compared to water of 0.5 dS m^{-1}, in all cowpea varieties except for Ceará, Costela de Vaca, Pingo de Ouro, Ovo de Peru, and Sempre Verde, whose production per plant (PP) was not influenced by salinity (Table 7). However, there was no difference in PP between cowpea varieties when irrigated with high-salinity water (Table 7).

The number of seeds per pod (NSPo) of the varieties Canário, Roxão, and Coruja was reduced by up to 30.07% by irrigation with saline water, but the NSPo of the Boquinha variety was increased by 31.90% under salt stress compared to the control (Table 7).

2.6. Salinity Tolerance

In the cluster analysis, based on the Euclidean distance of 0.90 in the formation of five clusters of combinations between salinity levels (S) and cowpea varieties (V) (Figure 2), the first three clusters (I) are characterized by the 15 varieties of cowpea irrigated with low-salinity water (0.5 dS m^{-1}). Those irrigated with high-salinity water (4.5 dS m^{-1}) were grouped in clusters IV and V. Cluster two (IV) comprises the varieties V2 (Ceará), V3 (Costela de Vaca), V4 (Lisão), V5 (Canário), V9 (Canapu Branco), V10 (Canapu Miúdo), V14 (Paulistinha), and V15 (Sempre Verde). The third cluster (V) contains the varieties V1 (Boquinha), V6 (Pingo de Ouro), V7 (Roxão), V8 (Feijão Branco), V11 (Ovo de Peru), V12 (Baeta), and V13 (Coruja) (Figure 2).

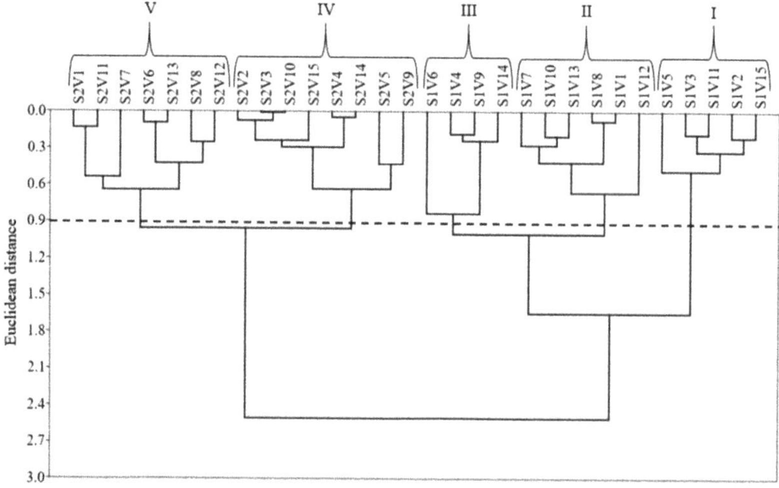

Figure 2. Dissimilarity dendrogram of the clusters formed by the combination of salinity levels (S) and traditional varieties of cowpea (V). S1—low salinity (0.5 dS m^{-1}). S2—high salinity (4.5 dS m^{-1}). V1—Boquinha, V2—Ceará, V3—Costela de Vaca, V4—Lisão, V5—Canário, V6—Pingo de Ouro, V7—Roxão, V8—Feijão Branco, V9—Canapu Branco, V10—Canapu Miúdo, V11—Ovo de Peru, V12—Baeta, V13—Coruja, V14—Paulistinha, and V15—Sempre Verde.

Table 7. F-test and means test for the number of pods per plant (*NPP*), the number of seeds per pod (*NSPo*), the number of seeds per plant (*NSPl*), and production per plant (*PP*, g) of traditional varieties of cowpea subjected to salinity levels of irrigation water.

F-Test (*p*-Value)		
Sources of Variation	NPP	NSPo
Block	0.409	0.037
Salinity	0.000	0.222
Varieties	0.000	0.000
Salinity × Varieties	0.005	0.009

Means comparison test (Standard Deviation, *n* = 5)			
	NPP		NSPo	
Varieties	0.5 dS m^{-1}	4.5 dS m^{-1}	0.5 dS m^{-1}	4.5 dS m^{-1}
V1—Boquinha	6.8 ± 0.97 Ab	2.8 ± 0.58 Ba	11.6 ± 1.15 Bb	15.3 ± 0.48 Aa
V2—Ceará	4.8 ± 0.37 Ac	2.6 ± 0.68 Ba	13.4 ± 1.20 Aa	15.0 ± 0.61 Aa
V3—Costela de Vaca	4.2 ± 0.86 Ad	3.0 ± 0.55 Aa	9.9 ± 1.74 Ab	9.3 ± 0.79 Ab
V4—Lisão	4.8 ± 0.58 Ac	2.2 ± 0.58 Ba	11.9 ± 0.58 Ab	13.6 ± 0.84 Aa
V5—Canário	6.0 ± 0.55 Ac	2.4 ± 0.24 Ba	14.3 ± 1.34 Aa	10.0 ± 0.57 Bb
V6—Pingo de Ouro	5.2 ± 0.73 Ac	2.4 ± 0.24 Ba	13.9 ± 1.32 Aa	14.1 ± 0.62 Aa
V7—Roxão	6.2 ± 0.80 Ac	2.2 ± 0.49 Ba	13.5 ± 0.92 Aa	10.5 ± 1.50 Bb
V8—Feijão Branco	9.2 ± 1.46 Aa	3.6 ± 0.24 Ba	10.5 ± 1.03 Ab	10.1 ± 0.76 Ab
V9—Canapu Branco	6.0 ± 1.26 Ac	3.0 ± 1.76 Ba	13.9 ± 1.42 Aa	14.0 ± 1.66 Aa
V10—Canapu Miúdo	6.6 ± 0.51 Ab	3.0 ± 0.71 Ba	14.7 ± 0.94 Aa	12.6 ± 0.54 Aa
V11—Ovo de Peru	2.8 ± 0.37 Ad	2.4 ± 0.24 Aa	11.6 ± 1.13 Ab	9.9 ± 0.76 Ab
V12—Baeta	8.6 ± 0.24 Aa	3.6 ± 0.40 Ba	13.3 ± 1.10 Aa	12.9 ± 0.95 Aa
V13—Coruja	8.0 ± 0.84 Aa	4.0 ± 0.71 Ba	15.1 ± 0.85 Aa	12.1 ± 0.91 Ba
V14—Paulistinha	9.0 ± 0.01 Aa	2.8 ± 0.37 Ba	14.1 ± 1.10 Aa	13.0 ± 1.11 Aa
V15—Sempre Verde	3.6 ± 1.03 Ad	2.0 ± 0.32 Aa	11.8 ± 0.58 Ab	13.9 ± 1.08 Aa

F-test (*p*-value)		
Sources of Variation	NSPl	PP (g)
Block	0.273	0.351
Salinity	0.000	0.000
Varieties	0.000	0.000
Salinity x Varieties	0.000	0.003

Means comparison test (Standard Deviation, *n* = 5)			
	NSPl		PP	
Varieties	0.5 dS m^{-1}	4.5 dS m^{-1}	0.5 dS m^{-1}	4.5 dS m^{-1}
V1—Boquinha	80.5 ± 14.41 Ab	42.3 ± 8.03 Ba	80.5 ± 0.90 Ab	42.3 ± 1.04 Ba
V2—Ceará	63.4 ± 5.90 Ac	38.5 ± 9.84 Aa	63.4 ± 0.97 Ac	38.5 ± 1.37 Aa
V3—Costela de Vaca	42.0 ± 13.61 Ac	26.8 ± 3.28 Aa	42.0 ± 2.17 Ac	26.8 ± 0.87 Aa
V4—Lisão	56.8 ± 6.65 Ac	28.3 ± 6.59 Ba	56.8 ± 2.09 Ac	28.3 ± 1.16 Ba
V5—Canário	83.7 ± 6.25 Ab	24.2 ± 3.07 Ba	83.7 ± 1.33 Ab	24.2 ± 0.54 Ba
V6—Pingo de Ouro	74.2 ± 15.67 Ab	33.6 ± 2.86 Ba	74.2 ± 1.50 Ab	33.6 ± 1.48 Ba
V7—Roxão	82.2 ± 9.74 Ab	23.9 ± 5.39 Ba	82.2 ± 1.20 Ab	23.9 ± 1.54 Ba
V8—Feijão Branco	91.4 ± 9.99 Ab	36.3 ± 3.36 Ba	91.4 ± 1.32 Ab	36.3 ± 0.22 Ba
V9—Canapu Branco	78.6 ± 14.86 Ab	36.3 ± 17.92 Ba	78.6 ± 1.86 Ab	36.3 ± 1.29 Ba
V10—Canapu Miúdo	96.8 ± 10.20 Ab	37.7 ± 8.78 Ba	96.8 ± 1.32 Ab	37.7 ± 1.17 Ba
V11—Ovo de Peru	30.8 ± 2.27 Ac	23.0 ± 0.95 Aa	30.8 ± 2.12 Ac	23.0 ± 0.34 Aa
V12—Baeta	113.8 ± 8.74 Aa	45.7 ± 3.88 Ba	113.8 ± 2.14 Aa	45.7 ± 0.55 Ba
V13—Coruja	121.0 ± 14.88 Aa	50.0 ± 11.61 Ba	121.0 ± 1.15 Aa	50.0 ± 1.67 Ba
V14—Paulistinha	126.6 ± 9.92 Aa	35.5 ± 3.99 Ba	126.6 ± 1.82 Aa	35.5 ± 0.88 Ba
V15—Sempre Verde	43.0 ± 12.79 Ac	28.0 ± 5.36 Aa	43.0 ± 2.46 Ac	28.0 ± 0.88 Aa

Equal uppercase letters in the rows and lowercase letters in the column do not differ by Student's *t*-test and Scott–Knott test at 5% probability level, respectively.

3. Discussion

The soil salinity limit for cowpea culture is 4.8 dS m^{-1} [4]. Irrigation water with 4.5 dS m^{-1} increased by 1.4 to 2.15 times the electrical conductivity of the soil saturation extract considering the soil salinity limit. Therefore, every cowpea variety was under possible salt stress. This behavior was similar to that observed in irrigated areas with lower leaching fractions where irrigation-water salinity influenced the soil salinity at the end of the cycle [7,14].

Soil from pots with the highest SDM levels presented the highest ECse values (Canário and Roxão). This behavior relates to the diversity of salinity tolerance mechanisms [15], e.g., water consumption restrictions, selective salt absorption, and salt exclusion at the roots. Soil from pots with the lowest SDM levels exhibited the lowest ECse values (Costela de Vaca, Lisão, Canapu Miúdo, and Paulistinha) because of vacuolar ion compartmentation and non-selective ion absorption in the most susceptible varieties [16,17].

Irrigation water with 4.5 dS m^{-1} reduced the photosynthetic rate (A_N) in the Boquinha, Roxão, Feijão Branco, Canapu Miúdo, Coruja, and Paulistinha varieties. The A_N reductions were related to stomatal factors, i.e., the reduction of stomatal conductance limited the influx of CO_2 and consequently the internal concentration of CO_2 (Ci) in the substomatal cavity, reducing water absorption and transpiration [6,17,18]. The reduction of stomatal conductance occurred with increased soil salt concentrations and led to a decrease in osmotic and water potentials, resulting in toxicity of specific ions—Na^+ and Cl^- [5,8,9]. Plants experience issues absorbing water from the soil under salt stress and tend to reduce water loss by closing stomates and reducing transpiration [6,18].

The reduced A_N in the Boquinha, Feijão Branco, Canapu Miúdo, Coruja, and Paulistinha varieties was also related to non-stomatal factors. There were reductions in the *A/Ci*, indicating decreased ribulose 1,5-bisphosphate carboxylase/oxygenase (Rubisco) enzymic activity under stress conditions, e.g., the lack of adenosine triphosphate (ATP) and nicotinamide adenine dinucleotide phosphate (NADPH) from the electron transport chain of photosystem II [6,19].

The Ceará, Canário, Pingo de Ouro, Canapu Branco, Baeta, and Sempre Verde varieties increased A_N under salt stress. The increased production of photo-assimilates under salt stress improves the energetic input and allows the plant to use mechanisms to tolerate energy expenditure, such as vacuolar ion compartmentation, the exclusion of specific ions, and attempting to attain ionic homeostasis [16,17].

The increased A_N in the Ceará, Canário, Pingo de Ouro, Canapu Branco, Baeta, and Sempre Verde varieties coincided with increases in *gs*, CO_2 influx, transpiration, and the consequent water absorption. Increased water loss because of transpiration results in reduced water potential at the roots, overcoming the osmotic stress and aiding in water absorption [9–11]. The A_N of the Pingo de Ouro, Canapu Branco, Baeta, and Sempre Verde varieties coincided with increased A_N/Ci and *WUEi*, improved water usage, and increased Rubisco activity. However, the increased A_N in the Canapu Branco variety also coincided with a decrease in gs and E; in this case, the higher salt concentration, improved water usage (*WUEi*), and increased Rubisco activity (*A/Ci*) resulted in water consumption restrictions.

The A_N of the Costela de Vaca, Lisão, and Ovo de Peru varieties was not affected by salinity, but the *WUEi* increased because of lower transpiration. Soil salinity impairs water absorption by cowpea bean plants; therefore, lower transpiration is a strategy to reduce water losses [6,18].

Chlorophyll fluorescence did not differ among varieties, indicating similar photochemical activities. Increases of *Fv* and *ETR* in cowpea bean varieties under salt stress compared with control demonstrated an increased ability to transfer energy from the excited electrons of chlorophyll molecules to assemble NADPH and ATP and had reduced ferredoxin (Fdr). This increased energy transfer was vital for preventing the decrease in photosynthesis or improving photosynthesis under salt stress once the quantum efficiency of photosystem II (*Y(II)*) in cowpea bean varieties decreased under salt stress, indicating a decrease in the fraction of energy absorption by chlorophyll in PSII [20]. However, some varieties susceptible to salt stress, e.g., Boquinha, Roxão, Feijão Branco, Canapu Miúdo, Coruja, and Paulistinha, exhibited decreased A_N and A/Ci, despite this mechanism for increasing the photochemical energy transfer.

Cowpea bean plants increased their photoprotective capability under salt stress because of a higher quantum yield of regulated photochemical quenching ($Y_{(NPQ)}$) through thermal energy dissipation by the xanthophyll cycle [21]. This photoprotective mechanism is efficient in cowpea bean plants once the Fv/Fm values are greater than 0.75, indicating a lack of degradation of the photosynthetic apparatus [17]. Compared with the control, the Fo values were lower under salt stress, corroborating the absence of damage in the PSII reaction centers [20].

Despite improvements in A_N, *WUE*, and A/Ci, the growth, biomass accumulation, and production of cowpea beans decreased under salt stress, especially the primary branch length in the Costela de Vaca, Pingo de Ouro, Roxão, Canapu Branco, Canapu Miúdo, and Baeta varieties, characterized by their prostrate and semi-prostrate size and indeterminate growth habit. The reduced growth resulted from lower energetic stability because of decreased $Y_{(II)}$ under salt stress. The authors in [22] reported that reduced biomass in cowpea bean plants under salt stress relates to an energy bypass because of the metabolic cost incurred during acclimation.

In plants, salt stress results in morphologic and anatomic modifications with strategies for adapting to adverse conditions, e.g., fewer leaves and shorter branches, reflecting decreased transpiration to improve water absorption [14,23].

Roxão was the only variety without SDM modifications caused by salinity and was unique in the susceptible group with an investment in biomass at the expense of grain production. The authors in [7,8,24,25] and other authors observed reduced growth and biomass accumulation in cowpea beans under salt stress. These reductions are part of the acclimation process of cowpea bean plants that have the potential to tolerate salt stress as they seek to secure production and perpetuate the species. In this sense, varieties such as Ovo de Peru, Pingo de Ouro, Sempre Verde, Costela de Vaca, and Ceará presented reduced growth and biomass accumulation but exhibited the highest production compared with the control.

Grain production was reduced ($p < 0.05$) in cowpea bean varieties under salt stress, except for the Ceará, Costela de Vaca, Pingo de Ouro, Ovo de Peru, and Sempre Verde varieties. Reduced water absorption associated with specific ion toxicity and physiological effects of salinity resulted in reduced growth and production [6,26]. The stress from salt accumulation in plants resulted in fewer reproductive branches and higher abortion rates [14,27]. All varieties produced fewer pods under salt stress; however, the Ceará, Pingo de Ouro, and Sempre Verde had more seeds per pod, compensating for the final grain production.

Cluster analysis revealed heterogenicity among plants under saline water irrigation with the Ceará, Costela de Vaca, Lisão, Canário, Canapu Branco, Canapu Miúdo, Paulistinha, and Sempre Verde varieties presenting similar SDM and production compared with the control, indicating tolerance to high irrigation-water salinity. The tolerance of Lisão, Canário, Canapu Branco, Canapu Miúdo, and Paulistinha occurred due to SDM. The Ceará, Costela de Vaca, and Semper Verde varieties are more tolerant to saline stress for SDM and grain production. We recommend the Ceará, Costa de Vaca, Pingo de Ouro, Ovo de Peru, and Semper Verde varieties for grain production under saline stress conditions. Those results differed from studies with conventional cowpea bean varieties under salt stress, e.g., [28] (EPACE 10); [29] (MNC04-762F-9, MNC04-762F-3, MNC04-762F-21, MNC04-769F-62, and MNC04-765F-153); [30] (IPA-206 and BRS Guariba); [27] (BRS Pajeu); [24] (BRS Imponente, MNC04-795F-168, and MNC04-795F-161); [31] (CE 790 and CE 104); [32] (BRS Pajeú); and [33] (BRS Itaim), who reported that these conventional varieties were susceptible to high irrigation-water salinity (5.0, 4.8, 5.0, 4.5, 6.4, 5.0, 6.0, 4.5, and 6.0 dS m^{-1}, respectively). These findings support the hypothesis that traditional varieties are more tolerant of salt stress than conventional varieties; however, further field studies are necessary. The Boquinha, Pingo de Ouro, Roxão, Feijão Branco, Ovo de Peru, Baeta, and Coruja varieties presented significantly different values compared to the control, indicating high susceptibility to irrigation-water salinity (4.5 dS m^{-1}).

In summary, the reduced photosynthetic rates in cowpea bean varieties are mainly caused by reductions in stomatal conductance resulting from salt stress. Salt stress increases the energy transferability of photosystem II in cowpea bean varieties, increasing the CO_2 assimilation rate and the instantaneous carboxylation efficiency in varieties more tolerant to salt stress. Salt stress decreases 26% of the production of tolerant varieties to salt stress and 54% of susceptible varieties. The Ceará, Costela de Vaca, Pingo de Ouro, Ovo de Peru, and Sempre Verde varieties exhibited the best physiological and production performance under salt stress; therefore, these varieties are tolerant to salt stress. The Lisão, Canário, Canapu Branco, Canapu Miúdo, Paulistinha, Boquinha, Roxão, Feijão Branco, Baeta, and Coruja present the worst physiological and production performances under salt stress; therefore, those varieties are susceptible to salt stress.

4. Material and Methods

4.1. Location, Experimental Design, and Plant Material

The experiment was conducted in a greenhouse at the Federal Rural University of the Semi-Arid Region—UFERSA, East campus, Mossoró/RN, Brazil, from May to August 2019. The municipality is located at the geographical coordinates of 5°12′ S and 37°19′ W, with an average altitude of 18 m. According to Köppen's classification, the climate of the region is BSwh′, and maximum and minimum temperatures of 44.2 and 20.4 °C and maximum and minimum relative humidity (RH) of 86 and 22%, respectively, were recorded during the experimental period. The average temperature and average daily relative humidity throughout the experiment were 33.8 °C and 49% RH, respectively.

The experimental design used was randomized blocks, with treatments arranged in a 15 × 2 factorial scheme, consisting of the combination of fifteen cowpea varieties (V1 (Boquinha), V2 (Ceará), V3 (Costela de Vaca), V4 (Lisão), V5 (Canário), V6 (Pingo de Ouro), V7 (Roxão), V8 (Feijão Branco), V9 (Canapu Branco), V10 (Canapu Miúdo), V11 (Ovo de Peru), V12 (Baeta), V13 (Coruja), V14 (Paulistinha), and V15 (Sempre Verde)) with two levels of salinity of irrigation water (0.5 dS m^{-1} and 4.5 dS m^{-1}), with five replicates.

The seeds used were acquired from collections from Traditional Seed Guardians belonging to rural communities located in municipalities of the western region of the Rio Grande do Norte state. The seeds came from the 2018 season and were stored in PET bottles, which were sealed to avoid any change in the degree of moisture and stored in dry, well-ventilated warehouses without the use of preservatives. The varieties used in this study were chosen based on a preliminary study conducted on the germination and initial growth stages of cowpea [2].

4.2. Experiment Setup and Fertilization Management

Sowing was performed using 9 seeds, with the first thinning performed at 4 days after germination, leaving 3 plants per pot, and the second thinning 15 days later, leaving only one plant per pot.

Each experimental unit consisted of a plastic pot with a capacity of 12.0 L, with 1.0 L filled with crushed stone at the bottom, 1.0 L free at the top, and 10.0 L filled with soil classified as *Latossolo Vermelho Amarelo distrófico* (Oxisol), sandy loam texture [34], whose physical and chemical characteristics are presented in Table 8.

Table 8. Chemical and physical analysis of the soil used in the experiment.

pH	OM (%)	P	K^+	Na^+	Ca^{2+}	Mg^{2+}	Al^{3+}	H + Al	SB	t	CEC	V	ESP
		—(mg dm^{-3})—						—(cmol$_c$ dm^{-3})—				—%—	
5.30	1.67	2.1	54.2	21.6	2.70	0.90	0.05	1.82	3.83	3.88	5.65	68	2.0

Density (kg dm^{-3})	Sand	Silt	Clay
		—(g kg^{-1})—	
1.60	820	30	150

Soil acidity was corrected with calcium hydroxide (Ca(OH)$_2$), with 54% calcium. The soil was corrected to increase base saturation to 90%. After 15 days, the soil was fertilized according to the recommendations of [35] for pots in protected cultivation, applying 300 mg of P$_2$O$_5{}^-$, 150 mg of K$_2$O, and 100 mg of N per dm^3 of soil through fertigation, using urea (45% of N), potassium chloride (KCl = 60% of K$_2$O), and monoammonium phosphate (MAP = 12% of N and 50% of P$_2$O$_5{}^-$). Fertilization with micronutrients was performed by foliar application in pre-flowering and 15 days after flowering, with the foliar fertilizer Liqui-Plex Fruit® in the proportion of the 3 mL L^{-1} of the solution, following the manufacturer's recommendation (Table 9).

Table 9. Chemical characterization of Liqui-Plex Fruit® foliar fertilizer.

				Parameters					
N	Ca	S	B	Cu	Mn	Mo	Zn	OC
				—g L^{-1}—					%
73.50	14.70	78.63	14.17	0.74	73.50	1.47	73.50	2.45

N—nitrogen; Ca—calcium; S—sulfur; B—boron; Cu—copper; Mn—manganese; Mo—molybdenum; Zn—zinc; OC—organic carbon.

4.3. Saline Waters and Irrigation and Drainage Management

In the preparation of irrigation waters, local-supply water (ECw = 0.50 dS m^{-1}) was used for the lowest level of salinity. For the highest level of salinity (ECw = 4.50 dS m^{-1}), local-supply water was mixed with reject brine from brackish water desalination (ECw = 9.50 dS m^{-1}). The desalination reject brine was obtained at the Jurema Settlement, located beside the RN-013 highway, km 4 (Table 10). Local supply water and brine tailings were stored in water tanks with a volume of 2000 L. We monitored the electrical conductivity during mixing with a portable conductivity meter.

Table 10. Physical–chemical characterization of the water sources used in the experiment.

Water Sources					Parameters					
	pH	EC	K^+	Na^+	Mg^{2+}	Ca^{2+}	Cl^-	$CO_3{}^{2-}$	$HCO_3{}^-$	SAR
		dS m^{-1}			—mmol$_c$ L^{-1}—					
1	7.57	0.50	0.31	3.74	1.20	0.83	2.40	0.60	3.20	2.62
2	7.10	9.50	0.83	54.13	24.20	37.80	116.00	0.00	3.40	9.70

Water source 1—local-supply water; water source 2—reject brine; pH (H$_2$O)—hydrogen potential in water; EC—electrical conductivity; K^+—potassium; Na^+—sodium; Mg^{2+}—magnesium; Ca^{2+}—calcium; Cl^-—chlorine; $CO_3{}^{2-}$—carbonate; $HCO_3{}^-$—bicarbonate; SAR—sodium adsorption ratio, (mmol$_c$ L^{-1})$^{-0.5}$.

Irrigation management was based on the drainage lysimeter method to leave the soil with moisture close to the maximum retention capacity, and irrigations were performed once a day, applying a leaching fraction (LF) of 15% every seven days along with the applied depth. The volume applied (Va) per container was obtained by the difference between the previous depth applied (La) minus the mean drainage (D), divided by the number of containers (n), as indicated in Equation (1):

$$Va = \frac{La - D}{n(1 - LF)} \tag{1}$$

The irrigation system comprised a self-venting Metalcorte/Eberle circulation motor pump, driven by a single-phase motor, 210 V voltage, 60 Hz frequency, installed in a reservoir with a capacity of 50 L and 16 mm diameter hoses with pressure-compensating drippers with a flow rate of 1.3 L h^{-1}.

The electrical conductivity of the saturation extract (ECse) was estimated according to the methodology suggested by [4] for medium-textured soils. For this, at 80 days after sowing, an additional leaching fraction (15%) was applied, the drained volume was collected, and the electrical conductivity of the drainage water (ECd) was measured using a benchtop conductivity meter, with the data expressed in dS m^{-1} adjusted to the temperature of 25 °C. The data were applied in Equation (2):

$$ECse = \frac{ECd}{2} \tag{2}$$

4.4. Analysis of Gas Exchange and Chlorophyll a Fluorescence

Physiological analyses were performed during the flowering stage of the plants, at 58 days after sowing. Gas exchange was analyzed in the period from 6 to 9 a.m., with evaluations on fully expanded leaves located in the upper third of each plant, using a portable infrared gas analyzer (IRGA), LCPro^{+} Portable Photosynthesis System$^{®}$ (ADC BioScientific Limited, Hertfordshire, UK) with temperature control at 25 °C, irradiation of 1200 µmol photons m^{-2} s^{-1}, and airflow of 200 mL min^{-1}. The quantified variables were CO_2 assimilation rate (A_N) (µmol (CO_2) m^{-2} s^{-1}), transpiration (E) (mmol (H_2O) m^{-2} s^{-1}), stomatal conductance (gs) (mol (H_2O) m^{-2} s^{-1}), and internal CO_2 concentration (Ci) (mol m^{-2} s^{-1}). These data were then used to estimate the instantaneous water use efficiency (WUEi) (A_N/E) [(µmol (CO_2) m^{-2} s^{-1}) (mmol (H_2O) m^{-2} s^{-1})$^{-1}$] and instantaneous carboxylation efficiency (CEi) (A_N/Ci) [(µmol (CO_2) m^{-2} s^{-1}) (mol (CO_2) m^{-2} s^{-1})$^{-1}$] [6].

Immediately after gas exchange measurements, chlorophyll *a* fluorescence was evaluated using the OS5p pulse-modulated fluorometer from Opti science; the Fv/Fm protocol was used for evaluations under dark conditions. Under these conditions, the following fluorescence induction variables were estimated: initial fluorescence (Fo) (µmol (photons) m^{-2} s^{-1}), maximum fluorescence (Fm) (µmol (photons) m^{-2} s^{-1}), variable fluorescence (Fv = Fm-Fo) (µmol (photons) m^{-2} s^{-1}), and the maximum quantum efficiency of PSII (Fv/Fm) [6].

The pulse-modulated fluorometer was also used to perform evaluations under light conditions, through the yield protocol. Readings were taken by applying the actinic light source with a multi-flash saturating pulse, coupled to a photosynthetically active radiation determination clip (PAR-Clip) to estimate the following variables: initial fluorescence before the saturation pulse (F′), maximum fluorescence after adaptation to saturating light (Fm′), electron transport rate (ETR) (µmol (photons) m^{-2} s^{-1}), and quantum efficiency of photosystem II (Y(II)). With these data, the following parameters were determined: minimum fluorescence of the illuminated plant tissue (Fo′) [36], photochemical quenching coefficient by the lake model (qL) [37], quantum yield of regulated photochemical quenching (Y(NPQ)) [37], and the quantum yield of non-regulated photochemical quenching (Y(NO)) [37].

4.5. Growth Analysis and Biomass Accumulation

At 58 DAP, the following parameters were determined: main branch length (MBL), using a measuring tape and measured from the plant collar to the last leaf insertion; stem diameter (SD), measured at 1.0 cm from the plant collar using a digital caliper; and the number of leaves (NL). After harvesting the pods of all varieties 80 days after sowing, the aerial part of the plants was collected and dried in an oven with forced air circulation, at a temperature of 65 °C, until reaching constant weight, to quantify the values of shoot dry mass (SDM).

4.6. Production Quantification

The pods were harvested as each traditional variety reached the phenological stage R9 (maturity stage), when the fruits were dry with the color and brightness that are characteristic of the genotype. The pods were transported to the laboratory, where the number of pods per plant (NPP), the number of seeds per pod (NSPo), the number of seeds per plant (NSPl), and production per plant (PP) (g) were counted.

4.7. Statistical Analysis

The data were subjected to analysis of variance and F-test. In cases of significant effect, the Scott–Knott test ($p < 0.05$) was performed for the variety factor, and Student's *t*-test ($p < 0.05$) was performed for the salinity factor, using SISVAR® statistical analysis software [38]. The data of shoot dry mass and grain production per plant were used to classify salinity tolerance; for this, the data were subjected to standardization, leaving mean zero ($\overline{X} = 0$) and variance one ($S^2 = 1$). Subsequently, cluster analysis was performed by hierarchical method, Ward's minimum variance, using the Euclidean distance as a measure of dissimilarity. PAST 3 free software was used for univariate and multivariate statistical analyses.

Author Contributions: Conceptualization, S.S.C.P., M.F.N. and F.V.d.S.S.; methodology, F.V.d.S.S.; software, S.S.C.P. and F.V.d.S.S.; validation, F.V.d.S.S.; formal analysis, S.S.C.P., A.T.L., F.J.Q.S., B.F.U., L.d.A.S. and R.C.L.M.; investigation, S.S.C.P., A.T.L., F.J.Q.S., B.F.U., L.d.A.S., R.C.L.M., M.F.N., P.D.F. and F.V.d.S.S.; resources, M.F.N., N.d.S.D., P.D.F. and F.V.d.S.S.; data curation, S.S.C.P. and F.V.d.S.S.; writing—original draft preparation, S.S.C.P., M.F.N. and F.V.d.S.S.; writing—review and editing, A.S.d.M., C.F.d.L., P.D.F. and N.d.S.D.; visualization, M.F.N. and F.V.d.S.S.; supervision, M.F.N. and F.V.d.S.S.; project administration, S.S.C.P. and A.T.L.; funding acquisition, M.F.N., N.d.S.D. and F.V.d.S.S. All authors have read and agreed to the published version of the manuscript.

Funding: Coordenação de Aperfeicoamento de Pessoal de Nível Superior: 001.

Institutional Review Board Statement: Not applicable.

Informed Consent Statement: Not applicable.

Data Availability Statement: All other data are presented in the paper.

Conflicts of Interest: The authors declare no conflict of interest.

References

1. Melo, A.S.; Silva, A.R.F.; Dutra, A.F.; Dutra, W.F.; Brito, M.E.B.; Sá, F.V.S. Photosynthetic efficiency and production of cowpea cultivars under deficit irrigation. *Rev. Ambient. Água* **2018**, *13*, e2133. [CrossRef]
2. Praxedes, S.S.C.; Sá, F.V.S.; Ferreira Neto, M.; Loiola, A.T.; Reges, L.B.L.; Jales, G.D.; Melo, A.S. Tolerance of seedlings traditional varieties of cowpea (*Vigna unguiculata* (L.) Walp.) to salt stress. *Semin. Cienc. Agrar* **2020**, *41*, 1963–1974. [CrossRef]
3. Porto Filho, F.Q.; Medeiros, J.F.; Sousa Neto, E.R.; Gheyi, H.R.; Matos, J.A. Feasibility of irrigation of musk melon with salinity water in different phenological stages. *Ciênc. Rural* **2006**, *36*, 453–459. [CrossRef]
4. Ayers, R.S.; Westcot, D.W. *Water Quality for Agriculture*; Food and Agriculture Organization of the United Nations: Rome, Italy, 1985.
5. Leite, J.V.Q.; Fernandes, P.D.; Oliveira, W.J.; Souza, E.R.; Santos, D.P.; Santos, C.S. Effect of salt stress and ionic composition of irrigation wateron morphophysiological variables in cowpea. *Rev. Bras. Agric. Irrig.* **2017**, *11*, 1825–1833.
6. Sá, F.V.S.; Ferreira Neto, M.; Lima, Y.B.; Paiva, E.P.; Prata, R.C.; Lacerda, C.F.; Brito, M.E.B. Growth, gas exchange and photochemical efficiency of the cowpea bean under salt stress and phosphorus fertilization. *Comun. Sci.* **2018**, *9*, 668–679. [CrossRef]
7. Lima, Y.B.; Sá, F.V.S.; Ferreira Neto, M.; Paiva, E.P.; Gheyi, H.R. Accumulation of salts in the soil and growth of cowpea under salinity and phosphorus fertilization. *Rev. Ciênc. Agron.* **2017**, *48*, 765–773. [CrossRef]

8. Sá, F.V.S.; Ferreira Neto, M.; Lima, Y.B.; Paiva, E.P.; Silva, A.C.; Dias, N.S.; Souza, F.M.; Melo, A.S.; Moreira, R.C.L.; Silva, L.A. Phytomass accumulation and mineral composition of cowpea (*Vigna unguiculata*) under salt stress and phosphate fertilization. *Aust. J. Crop Sci.* **2019**, *13*, 1149–1154. [CrossRef]

9. Munns, R.; Tester, M. Mechanisms of salinity tolerance. *Annu. Rev. Plant Biol.* **2008**, *59*, 651–681. [CrossRef] [PubMed]

10. Annunziata, M.G.; Ciarmiello, L.F.; Woodrow, P.; Maximova, E.; Fuggi, A.; Carillo, P. Durum wheat roots adapt to salinity remodeling the cellular content of nitrogen metabolites and sucrose. *Front. Plant Sci.* **2017**, *7*, 2035. [CrossRef]

11. Dell'Aversana, E.; Hessini, K.; Ferchichi, S.; Fusco, G.M.; Woodrow, P.; Ciarmiello, L.F.; Abdelly, C.; Carillo, P. Salinity Duration Differently Modulates Physiological Parameters and Metabolites Profile in Roots of Two Contrasting Barley Genotypes. *Plants* **2021**, *10*, 307. [CrossRef]

12. Sá, F.V.S.; Paiva, E.P.; Torres, S.B.; Brito, M.E.B.; Nogueira, N.W.; Frade, L.J.G.; Freitas, R.M.O. Seed germination and vigor of different cowpea cultivars under salt stress. *Comun. Sci.* **2016**, *7*, 450–455. [CrossRef]

13. Sá, F.V.S.; Nascimento, R.; Pereira, M.O.; Borges, V.E.; Guimaraes, R.F.B.; Ramos, J.G.; Mendes, J.S.; Penha, J.L. Vigor and tolerance of cowpea (*Vigna unguiculata*) genotypes under salt stress. *Biosci. J.* **2017**, *33*, 1488–1494. [CrossRef]

14. Carvalho, J.F.; Silva, E.F.F.; Silva, G.F.; Rolim, M.M.; Pedrosa, E.M.R. Production components of *Vigna unguiculata* (L. Walp) irrigated with brackish water under different leaching fractions. *Rev. Caatinga* **2016**, *29*, 966–975. [CrossRef]

15. Munns, R. Comparative physiology of salt and water stress. *Plant Cell Environ.* **2002**, *25*, 239–250. [CrossRef]

16. Gupta, B.; Huang, B. Mechanism of salinity tolerance in plants: Physiological, biochemical, and molecular characterization. *Int. J. Genom.* **2014**, *2014*, 701596. [CrossRef]

17. Sá, F.V.S.; Brito, M.E.B.; Silva, L.A.; Moreira, R.C.L.; Fernandes, P.D.; Figueiredo, L.C. Physiology of perception of saline stress in 'Common Sunki' mandarinhybrids under saline hydroponic solution. *Comun. Sci.* **2015**, *6*, 463–470. [CrossRef]

18. Andrade, J.R.; Maia Júnior, S.O.; Silva, R.F.B.; Barbosa, J.W.S.; Nascimento, R.; Alencar, A.E.V. Gas exchanges in cowpea genotypes irrigated with saline water. *Rev. Bras. Agric. Irrig.* **2018**, *12*, 2653–2660.

19. Galmés, J.; Aranjuelo, I.; Medrano, H.; Flexas, J. Variation in Rubisco content and activity under variable climatic factors. *Photosynth. Res.* **2013**, *117*, 73–90. [CrossRef]

20. Santos, S.T.; Oliveira, F.A.; Oliveira, G.B.S.; Sá, F.V.S.; Costa, J.P.B.M.; Fernandes, P.D. Photochemical efficiency of basil cultivars fertigated with salinized nutrient solutions. *Rev. Bras. Eng. Agríc. Ambient.* **2020**, *24*, 320–325. [CrossRef]

21. Stirbet, A.; Govindjee. On the relation between the Kautsky effect (chlorophyll a fluorescence induction) and photosystem II: Basics and applications of the OJIP fluorescence transient. *J. Photochem. Photobiol. B Biol.* **2011**, *104*, 236–257. [CrossRef]

22. Lacerda, C.F.; Sousa, G.G.; Silva, F.L.B.; Guimarães, F.V.A.; Silva, G.L.; Cavalcante, L.F. Soil salinization and maize and cowpea yield in the crop rotation system using saline waters. *Eng. Agríc.* **2011**, *31*, 663–675. [CrossRef]

23. Oliveira, F.A.; Medeiros, J.F.; Oliveira, M.K.T.; Souza, A.A.T.; Ferreira, J.A.; Souza, M.S. Interaction between water salinity and biostimulant in the cowpea plants. *Rev. Bras. Eng. Agríc. Ambi.* **2013**, *17*, 465–471. [CrossRef]

24. Aquino, J.P.A.; Bezerra, A.A.C.; Alcântara Neto, F.; Lima, C.J.G.S.; Sousa, R.R. Morphophysiological responses of cowpea genotypes to irrigation water salinity. *Rev. Caatinga* **2017**, *30*, 1001–1008. [CrossRef]

25. Pereira, E.D.; Marinho, A.B.; Ramos, E.G.; Fernandes, C.N.D.; Borges, F.R.M.; Adriano, J.N.J. Saline stress effect on cowpea beans growth under biofertilizer correction. *Biosci. J.* **2019**, *35*, 1328–1338. [CrossRef]

26. Oliveira, F.A.; Oliveira, M.K.T.; Lima, L.A.; Alves, R.C.; Régis, L.R.L.; Santos, S.T. Salt stress and plant bioregulators in cowpea crop. *Irriga* **2017**, *22*, 314–329. [CrossRef]

27. Furtado, G.F.; Sousa Júnior, J.R.; Xavier, D.A.; Andrade, E.M.G.; Sousa, J.R.M. Photosynthetic pigments and yeld of cowpea *Vigna unguiculada* L. Walp under salinity and nitrogen fertilization. *Rev. Verde Agroec. Desenvolv. Sustent.* **2014**, *9*, 291–299.

28. Silva, F.L.B.; Lacerda, C.F.; Neves, A.L.R.; Sousa, G.G.; Sousa, C.H.C.; Ferreira, F.J. Irrigation with saline water plus bovine biofertilizer in the gas exchanges and productivity of cowpea. *Irriga* **2013**, *18*, 304–317. [CrossRef]

29. Brito, K.Q.D.; Nascimento, R.; Santos, J.E.A.; Silva, I.A.C.; Dantas Junior, G.J. Componentes de produção de genótipos de feijão-caupi irrigados com água salina. *Rev. Verde Agroec. Desenvolv. Sustent.* **2015**, *10*, 01–05. [CrossRef]

30. Sousa, L.V.; Ribeiro, R.M.P.; Santos, M.G.; Oliveira, F.S.; Ferreira, H.; Gerônimo, F.R.R.; Araújo, A.G.R.; Silva, T.I.; Barros Júnior, A.P. Physiological responses of cowpea (*Vigna unguiculata*) under irrigation with saline water and biostimulant treatment. *J. Agric. Sci.* **2018**, *10*, 24–33. [CrossRef]

31. Prazeres, S.S.; Lacerda, C.F.; Barbosa, F.E.L.; Amorim, A.V.; Araújo, I.C.S.; Cavalcante, L.F. Crescimento e trocas gasosas de plantas de feijão-caupi sob irrigação salina e doses de potássio. *Rev. Agro@mbiente On-Line* **2015**, *9*, 111–118. [CrossRef]

32. Sousa, J.R.M.; Andrade, E.M.G.; Furtado, G.F.; Soares, L.A.A.; Silva, S.S.; Sousa Júnior, J.R. Vegetative growth of cowpea under nitrogen irrigated with saline water. *Agropec. Cient. Semi-Árido* **2013**, *9*, 94–98.

33. Sousa, G.G.; Viana, T.V.A.; Lacerda, C.F.; Azevedo, B.M.; Silva, G.L.; Costa, F.R.B. Salt stress in cowpea plants in soil with organic fertilizers. *Rev. Agro@mbiente On-Line* **2014**, *8*, 359–367. [CrossRef]

34. Santos, H.G.; Jacomine, P.K.T.; Anjos, L.H.C.; Oliveira, V.A.; Lumbreras, J.F.; Coelho, M.R.; Almeida, J.A.; Araujo Filho, J.C.; Oliveira, J.B.; Cunha, T.J.F. *Sistema Brasileiro de Classificação de Solos*, 5th ed.; Embrapa: Brasília, Brazil, 2018.

35. Novais, R.F.; Neves, J.C.L.; Barros, N.F. Ensaio em Ambiente Controlado. In *Métodos de Pesquisa em Fertilidade do Solo*; Oliveira, A.J., Garrido, W.E., Araújo, J.D., Lourenço, S., Eds.; Embrapa-SEA: Brasília, Brazil, 1991; pp. 189–254.

36. Oxborough, K.; Baker, N.R. An instrument capable of imaging chlorophyll a fluorescence from leaves at very low irradiance and at cellular and subcellular levels of organization. *Plant Cell Environ.* **1997**, *20*, 1473–1483. [CrossRef]

37. Kramer, D.M.; Johnson, G.; Kiirats, O.; Edwads, G.E. New fluorecence parameters for the determination of QA redos state and excitation energy fluxes. *Photosynth. Res.* **2004**, *79*, 209–218. [CrossRef] [PubMed]
38. Ferreira, D.F. Sisvar: A Guide for its Bootstrap procedures in multiple comparisons. *Ciênc. Agrotec.* **2014**, *38*, 109–112. [CrossRef]

MDPI

St. Alban-Anlage 66

4052 Basel

Switzerland

www.mdpi.com

Plants Editorial Office

E-mail: plants@mdpi.com

www.mdpi.com/journal/plants